*Introductory*
*Fourier*
*Transform*
*Spectroscopy*

# *Introductory Fourier Transform Spectroscopy*

**ROBERT JOHN BELL**

*Department of Physics, University of Missouri-Rolla*
*Rolla, Missouri*

**1972**

**ACADEMIC PRESS New York and London**

ACADEMIC PRESS, INC.
111 Fifth Avenue, New York, New York 10003

*United Kingdom Edition published by*
ACADEMIC PRESS, INC. (LONDON) LTD.
24/28 Oval Road, London NW1

LIBRARY OF CONGRESS CATALOG CARD NUMBER: 70-182606

PRINTED IN THE UNITED STATES OF AMERICA

*For my Parents*

MASON AND AMORET BELL

*For my Wife*

CAROLYN

*For my Children*

SARAH AND ROBERT

# Contents

CHAPTER THREE

## *Fourier Analysis and Interferometry*



CHAPTER FOUR

## *Sample Calculations of Spectra from Interferograms*



CHAPTER FIVE

## *Apodization—Mathematical Filtering*



CHAPTER SIX

## *Resolution*

CHAPTER THIRTEEN

## *Procedures for Choosing Experimental Parameters*



CHAPTER FOURTEEN

## *Sample Interferograms and Spectra*



CHAPTER FIFTEEN

## *Lamellar Grating Interferometers*

CHAPTER SIXTEEN

## *Computation Techniques*



CHAPTER SEVENTEEN

## *The Cooley–Tukey Algorithm*

*Ralph W. Alexander and Robert J. Bell*



CHAPTER EIGHTEEN

## *Minicomputers and Real-Time Fourier Analysis*

*Ralph W. Alexander*

CHAPTER NINETEEN

## Commercial Instruments



APPENDIX A

## Optical Alignment of a Michelson Interferometer



APPENDIX B

## Computer Programs

*Ralph W. Alexander and Harold V. Romero*



APPENDIX C

## Mirror Tilt, the Cat's-Eye Retroreflector

# *Preface*

THIS BOOK introduces the subject of Fourier transform spectroscopy from a level that requires a knowledge of *only* introductory optics and mathematics and proceeds to the development of optical theory and equations to the extent required by the advanced student or researcher. Material is included for the physicist, chemist, astronomer, and others who are interested in spectroscopy.

The subject is approached through optical principles, not abstract mathematics. Information theory, Fourier analysis, and mathematical theorems, a very cursory knowledge of which is sufficient, are presented only to complete derivations or to give alternate views of an individual subject.

In some chapters the subject matter is approached in two ways. The first approach is through simple optics and physical intuition. The second, through a knowledge of Fourier analysis and the concepts of convolution and autocorrelation. This dual treatment bridges the gap between the introductory material in the book and the advanced material in the journals.

There is no longer any question about the wide applicability of Fourier transform spectroscopy. For the visible wavenumber range, the requirements of Fourier transform spectroscopy have been met by several laboratories, and at this writing, at least one company markets equipment which spans the spectral region from the visible to the millimeter wavelengths. This span has applications in many fields of study, and because of this, the readers' interests most certainly will be diverse. Thus, a genuine understanding of the techniques can be obtained only through the most basic assumptions.

To aid in the reader's quest for reference material, almost every chapter has a set of pivotal references appended. Bibliographies and indexes, one indexed by author and the other by subject, appear at the end of the book.

If the reader wishes an in-depth treatment of Fourier analysis, Bracewell's book, "The Fourier Transform and Its Application," is recommended. For studies of interferometers, Steel's book, "Interferometry," should be consulted. For Fourier analysis in optics, Mertz's book, "Transformations in

Optics," is an excellent reference. To obtain information about the Cooley–Tukey algorithm, the *IEEE Transactions on Audio and Electroacoustics*, **15** (2), 76, 1967) should be consulted. For the latest developments, the papers that were presented at the Aspen International Conference on Fourier Spectroscopy, 1970 (G. A. Vanasse, A. T. Stair, Jr., and D. J. Baker, eds., AFCRL-71-0019, 5 Jan. 1971, Spec. Rep. No. 114, L. G. Hanscom Field, Bedford, Massachusetts, 1971) should be read. For the physics and chemistry of the far infrared, the book "Far-Infrared Spectroscopy" by K. D. Möller and W. G. Rothschild is recommended.

The author hopes that this book will help students reach a level at which they can undertake advanced research programs in spectroscopy. He has tried to cover most of the major subjects that are needed for such development. However, such subjects as internal modulation, chirping, and other specialized problems not normally encountered have been excluded. The references given at the end of the book should help the reader with such subjects.

# Acknowledgments

I AM indebted to Dr. Ralph W. Alexander for his many contributions to and proof reading of the manuscript. Dr. Harold V. Romero has also given aid. There are many who have helped me in the preparation of the manuscript: Dr. Robert McFarland, Dr. Harold Q. Fuller, Dr. John D. Strong, Dr. Ely E. Bell, Dr. Ludwig Genzel, Dr. Ernest V. Loewenstein, Dr. R. Sanderson, Mr. Ken Mustain, Dr. W. Anderson, Mr. G. Kovenor, Mr. A. Purdes, Mr. J. Blea, Dr. T. J. McMahon, Dr. William F. Parks, Mr. John Koenig, Dr. John Bastin, Mr. C. E. Bryson III, Mr. C. Irslinger, Dr. Doran Baker, and many others. I wish also to thank Mrs. Carol Roman, Mrs. Lois Vastano, Miss Mary Meachum, and Miss Lois Watson for typing the manuscript. The staff of Academic Press has made writing this book somewhat tolerable through patience and professional expertise. The University of Missouri–Rolla and the Max-Planck-Institut für Festkörperforschung provided the proper, invigorating environments. Sarah and Rob bore well their father-imposed, involuntary sacrifices. My wife, Carolyn, did the most for me in this effort, and the results are on the following pages.

CHAPTER ONE

# *Fourier Transform Spectroscopy*

## INTRODUCTION

The subject of Fourier transform spectroscopy (FTS) is best introduced by discussing the merits and applications of Fourier transform spectrometers in terms of (1) general advantages over conventional instruments, (2) specific advantages and disadvantages, (3) the resolving power of two-beam interferometers, (4) quality factors, (5) spectral ranges, and (6) utilization in science and industry.

Some mention also needs to be made of the basic similarity of lamellar gratings and Michelson interferometers, the methods of judging interferometers, and the extension of spectral studies into the infrared and far-infrared regions through the the development of interferometry or Fourier transform spectroscopy.

A short article by P. Connes [1] that describes Fourier transform spectroscopy is recommended as supplementary reading. Dr. Pierre Connes and his wife, Dr. Janine Connes, are pioneers in the fields of infrared astronomy and interferometry.

## GENERAL ADVANTAGES OF FOURIER TRANSFORM SPECTROMETERS

Basically, the advantages of Fourier transform spectrometers arise from two major concepts known as the Fellgett and Jacquinot advantages. These are discussed in more detail in the following chapter, but they are mentioned here because their physical aspects are easily understood if the reader has a cursory knowledge of spectroscopy.

An interferometer receives information from the entire range of a given spectrum during each time element of a scan, whereas a conventional grating spectrometer receives information from only the very narrow region which lies within the exit slit of the instrument. Thus, the interferometer receives information about the entire spectral range during an *entire* scan, while the grating instrument receives information only in a narrow band at a given time. This is a statement of the Fellgett or multiplex advantage.

The interferometer can have a large circular source at the input or entrance aperture of the instrument with no strong limitation on the resolution. Also, it can be operated with small *f*/numbers or with large solid angles at the source and detector. However, the resolution of a conventional grating-type spectrometer depends linearly on the instrument's slit width, and the detected power depends on the square of the area of equal slits. A grating-type spectrometer requires long and narrow slits which never can have the same area for the same resolving power as the interferometer. Also, for high resolution, a spectrometer requires large radii for the collimation mirrors, and this condition in turn necessitates large *f*/numbers or small solid angles. Quantitatively, the ability of interferometers to collect large amounts of energy at high resolution was expressed by Jacquinot as a throughput or *étendue* advantage of interferometers over spectrometers.

## SPECIFIC ADVANTAGES AND DISADVANTAGES OF INTERFEROMETERS

Several additional advantages follow from the Fellgett and Jacquinot advantages and can be listed as follows:

1. Very large resolving power.
2. High wavenumber accuracy.
3. Vastly reduced stray or unwanted flux problems.
4. Fast scanning time, which increases the probability of successfully completing an experiment.
5. Large wavenumber range per scan.
6. Possibility of making weak-signal measurements at millimeter wavelengths.
7. Use of small images in sample compartments without requiring special measures.
8. Measurement in amplitude spectroscopy of complex reflection or transmission coefficients.

9. Low cost of basic optical equipment.
10. Smaller size and lower weight of interferometers than spectrometers.

The few disadvantages of interferometers are that they do require access to computer facilities, and computer costs do have to be considered as factors in their operation. They are also somewhat deficient because absolute magnitudes of flux are sometimes in error by a few per cent, and the interferograms (recorded signals versus interferometer arm displacement) sometimes cannot be visually interpreted, which makes it difficult for an operator to judge quickly whether or not an experiment is satisfactory.

The large resolving power of an interferometer is a result of the Fellgett and Jacquinot advantages and depends linearly on the relative arm displacement of the instrument's movable mirror. Relative mirror displacements of the order of 2 m can be attained with some interferometers. This displacement ability has made it possible to observe weak lines with resolving powers of the order of $10^5$ or higher.

The high wavenumber accuracy and the problem of reduced stray light both result from the interference phenomena inherent in the instrument. Accurate movement of an interferometer's movable mirror carriage produces a precise change in the interference pattern which can be capitalized on with excellent wavenumber accuracy in the computed spectrum. Because the unwanted waves which reach the instrument's detector have a definite wavelength, they produce distinct interference patterns which, when transformed into spectra, are identifiable. It is not uncommon to obtain transmittance measurements which are reliable to as low as 0.3%.

Fast scan times (sometimes less than 1 sec), a large wavenumber range (sometimes as large as a decade from the minimum to maximum wave number), and a measurement capability in millimeter wavelengths even when the source is very weak are all gained through the Fellgett and Jacquinot advantages. The ability to use small images at the sample is also derived from the multiplex and *étendue* advantages.

Complex reflection or transmission coefficients can be measured directly in amplitude spectroscopy by placing the sample in one arm of the interferometer. The amplitude and phase angles of the complex reflection or transmission coefficients can be obtained without any special data manipulation, such as is required in a Kramers–Kronig analysis. Thus, complex indices of refraction can be noted experimentally. Also, it is possible to make accurate flux calculations even when the transmitted (reflected) flux is as low as 0.01%. Effects of different boundaries in a sample can also be separated.

Partly because interferometers are relatively simple instruments, their space requirement and weight are small. For example, the U.S. Nimbus III satellite carried a Michelson interferometer which weighed, with its power supply, 14.5 kg and required only 1 or 2 $ft^3$ of space. This instrument operated for several months in orbit around the earth and took 1% accurate radiometric data between 400 and 2000 $cm^{-1}$. In the Viking Project, one of the basic instruments in the probe will be an interferometer which will scan the surface of Mars. The optical systems of interferometers are frequently less expensive than those of spectrometers; however, the cost of high-speed data handling systems that are needed for the operation of interferometers can nullify these savings.

The disadvantages of interferometers are few, and these are rapidly being minimized. For instance, when transmittance or reflectance measurements are made, the results can be in error up to 5% of the absolute value. Fluctuations in the interferogram, or recorded signal versus the instrument's arm displacement, can produce this amount or error. If the error is random, repetition of the experiment can reduce the deviation. With some of the commercial instruments, which have scan times of less than 1 sec and computer controls, hundreds of repeated experiments can be performed in minutes and computer-averaged in the laboratory.

Often, a spectrum can be so complicated that the experimenter cannot immediately learn much from the interferogram. This problem can be solved with one of two methods. The low-cost method is through the experimenter's experience and in his knowing a few of the interferogram's signatures or tell-tale features which presage the particular spectral features sought. He can then apply Fourier analysis to a few simple cases, such as a Gaussian spectrum, and can make sliderule estimates of the experimental progress. The more expensive but more satisfying technique is to use real-time analyses or rapid computer calculations. If the experimenter has a minicomputer in his laboratory with a memory of 4000 or more words, he can compute the spectrum over the entire wavenumber range while the data are being recorded. In fact, as the interferometer's mirror carriage moves to larger displacements, he can observe the increase in resolving power, or if the experiment is especially long, he can watch the desired spectral features develop. If it is short, he can obtain the results of the entire scan almost instantly. In addition to the real-time analyses, he can wait until the end of the experiment and make on-line computer calculations of the spectrum in a very short time of seconds or minutes.

Large computers are used by most experimenters to transform the interferograms into spectra if the data points exceed 1024. If on-line computer time-sharing is available with suitable graphic and tabular data return, there is no problem, but if the data have to be recorded on paper tapes and converted to cards, there is usually a delay of one or two days. The computation time can be shortened considerably if the Cooley–Tukey algorithm is used. This algorithm makes it possible to make rapid computations of Fourier transforms without approximations. For example, about 9 min are required to compute the spectrum from an interferogram of one million points (see Chapter 16). Nevertheless, many computerized plotters are slow and have large backlogs which introduce delays; this can be circumvented if minicomputers with plotters are used in the laboratory.

Although the costs of computation time vary considerably, a rule-of-thumb cost reference can be established on the basis that an IBM 360 model 50 computer can handle 256 data points for a sample and background scan and produce the ratio for less than $1 in about 20 sec, providing that the programmer uses the Cooley–Tukey algorithm. Minicomputers, with the necessary plotters and oscilloscopes, can be purchased for less than $25,000–$30,000, but they are versatile instruments and can be used to operate the interferometer, compare data, and average spectra. When they are not being used to transform interferograms, other research applications can be found for them. With judicious use, minicomputers can be used economically even in small research facilities.

## TWO-BEAM INTERFEROMETERS, THE ULTIMATE IN SPECTROMETERS

Gebbie [2] suggests a direct comparison of spectral sorting techniques of various instruments (Table 1-1). The prism and grating spectrometers and the Fabry–Perot and two-beam interferometers are compared in terms of their ultimate resolving power.

The prism has the lowest resolution because its resolving power is limited by the size of the transparent prisms. The resolving power is the product of the length of the prism base $b$ and the dispersion of the prism material $dn/d\lambda$. The best resolving powers that can be obtained with a prism-type instrument are of the order of only a few thousand.

The resolving power of the grating instrument depends on its slit width $s$, the focal length $F$ of the collimating mirror in the monochromator section, and the angle of rotation $\theta$ of the grating away from the zero order. For good gratings, in the near-infrared, resolving powers of the order of $10^5$ can be attained. This resolving power has a diffraction limited maximum of $Nm$, which is the product of the total number of grooves in the grating $N$ and the grating order number $m$.

The Fabry–Perot interferometer used to be superior to the grating instrument in resolving power. However, large arm displacements used with

TABLE 1-1

Comparison of Spectral Sorting Techniques of Different Instruments in Order of Increasing Resolving Power[a]

| Instrument | Number of resolved lines or beams | Resolving power |
|---|---|---|
| Prism | $\infty$ | $b\, dn/d\lambda$ |
| Grating | $N \equiv$ Number of grooves ($\leq 10^5$–$10^6$) | $(mS/2F) \cot \theta \leq Nm$ |
| Fabry–Perot interferometer | $m =$ Order ($\leq 10^4$–$10^6$) (Finesse) | $m\pi\, R^{1/2}/(1 - R)$ |
| Two-beam interferometers (Michelson, lamellar, etc.) | $2 \equiv$ Number of beams or arms | $L\sigma$ |

[a] $b \equiv$ length of base; $dn/d\lambda \equiv$ change of index of refraction with respect to wavelength; $s \equiv$ slit width; $F \equiv$ focal length of collimator; $\theta \equiv$ angle of rotation of grating; $m \equiv$ order; $R \equiv$ single surface reflectance; $L \equiv$ maximum path difference; $\sigma \equiv$ wave number.

Michelson interferometers by J. Pinard (of Centre National de la Recherche) allow the latter to surpass the Fabry–Perot instruments.

The two-beam interferometers have resolving powers that are limited only by the arm displacement of the instrument. If a measurement is made for a frequency of $10^5$ cm$^{-1}$, a displacement of 10 cm will give a resolving power of $10^6$. Such measurements have been made, and the researchers at the Air Force Cambridge Laboratories and at the Centre National de la Recherche Scientifique at Campus D'Orsay in France are working on instruments that have carriage travels of 2 m. Thus, resolving powers of the order of $10^7$–$10^8$ are possible. Of course, the Michelson interferometer, in its spectroscopic mode of operation, surpasses all other instruments in its ultimate resolving power and in its unhampered wavenumber sorting ability over large frequency ranges.

It is interesting to note that the prism-type instrument sorts a source beam into an infinite number of beams. The Michelson interferometer produces only two beams—one for each arm. Considering the sorting ability of the four instruments in order, the prism, the grating, the Fabry–Perot interferometer, and Michelson interferometer (Table 1-1), the number of beams decreases monotonically, but the energy throughput increases, thereby implying that the fewer the number of beams, the greater the energy. The function of all these instruments ultimately depends on interference, and the phenomenon of interference requires at least two beams; therefore, two-beam interferometers have ultimate throughput or efficiency. Also, it is worth noting that all optical systems can be Fourier-analyzed [3, 4]. The prism-type instrument presents the simplest signal, the spectrum, and has a complicated Fourier analysis. At the other extreme, the two-beam interferometer yields a complicated signal, the interferogram, and has a simple sine or cosine Fourier transform. The grating-type instrument and the Fabry–Perot interferometer are intermediate. There is great difficulty or long delay in obtaining the spectrum from a two-beam instrument, but it has the greatest possible efficiency, which negates the necessity of finding fundamental improvements in instrumentation. Design modifications are in order, but it would be futile to try to improve the function of the two-beam instrument.

The Michelson interferometer is often favored over the lamellar grating interferometer. This has its roots in the construction and detailed operation of the two instruments. The lamellar grating interferometer is, however, a very fine and usable instrument in the far-infrared region and splits wave-fronts, whereas the Michelson interferometer divides amplitudes. Nonetheless, both instruments produce the same physical phenomenon, interference, and, hence, obey the same basic equations.

## QUALITY FACTORS

The comparison of the different types of spectroscopic instruments is important because there are so many different species of interferometer. There are so many that some, as Steel [5] points out, have been only designed and never built. Connes [6] proposed that interferometers be compared by using a quality factor $Q$ which will be large for the better instruments. As a point of reference, Connes in 1969 set his best spectral data at a quality factor of $5 \times 10^7$.

$Q$ can be defined with the aid of a graph as shown in Fig. 1-1. The curves of both spectra in Fig. 1-1 are plots of power $B(\sigma)$ versus the wave number $\sigma$. The curve on the left is the power spectrum of a bandwidth $\Delta\sigma$ that has an absorption region in it of full-width at one-half peak intensity, $\delta\sigma$. It is assumed that $\delta\sigma$ is the apparent full-width at one-half peak absorption of the *narrowest* absorption line that can be detected, i.e., $\delta\sigma$ is the resolution.

The number $M$ of spectral elements in the bandwidth is represented by $M = \Delta\sigma/\delta\sigma$. $M$ is also called the "multiplex factor." The curve on the left of Fig. 1-1 portrays the ideal case in which there is no noise, because of an assumed, ideally efficient, electronic system, detector, and source. The right-hand curve is for an identical case except that it represents a real or an actual spectrum. That is, the data exhibit random noise with a root mean square value indicated by $B_{\mathrm{rms}}$. The average signal level is designated by $B_{\mathrm{av}}$. The

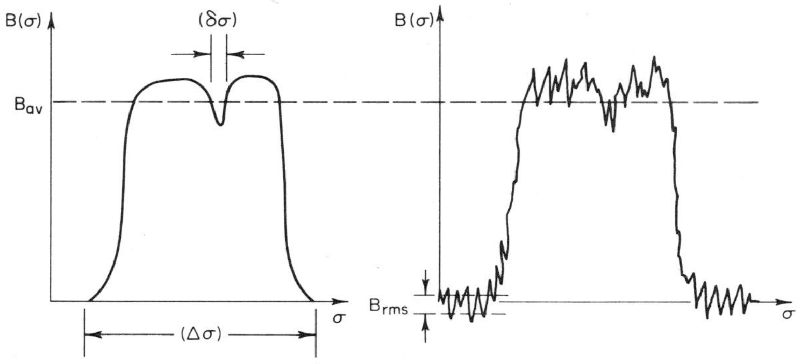


FIG. 1-1. Plot of spectra showing a small absorption line in midband. The curve on the left is for a noise-free spectrum, and the curve on the right is an actual spectrum with noise. The parameters are defined in the text and are used to calculate the quality factor, $Q$.

signal-to-noise ratio of the actual spectrum that is computed from the interferogram is then $B_{av}/B_{rms}$. If $Q$ is defined as

$$Q \equiv MB_{av}/B_{rms}, \tag{1-1}$$

then

$$Q = (\Delta\sigma)B_{av}/(\delta\sigma)B_{rms}. \tag{1-2}$$

Physically, the numerator of Eq. (1-2) is proportional to the area under the whole curve or is proportional to the energy in the entire spectrum of the bandwidth $\Delta\sigma$. For the denominator, it is assumed that the smallest detectable absorption line has a depth of about the magnitude of the rms noise level $B_{rms}$. Assuming a triangular line shape, the spectral area removed by the absorption would be about $(\delta\sigma)B_{rms}$. Thus, $Q$ can be physically represented by

$$Q \simeq \frac{\text{total energy in the spectrum}}{\text{energy in the smallest spectral line}} \tag{1-3}$$

The quality factor of an instrument takes into account both the resolving power and the signal-to-noise ratio and can be expressed in a number that is accessible to experimentalists, who, unfortunately, do not make wide use of it. However, Connes always accompanies his data with quality factors to assess the value of his experiments. A more detailed treatment of signal-to-noise $S/N$ factors can be found in a paper by Sakai [7]. To the experimentalist, the signal-to-noise factor in the interferogram is roughly maintained in the spectrum, and the overall performance can be estimated with $Q$. As mentioned previously, the value of a superior $Q$ is about $5 \times 10^7$; for a spectrum of modest resolution, $Q \simeq 1 \times 10^3$. If necessary, quality factors can be applied to grating or prism spectrometer data as well.

## SPECTRAL RANGES

The entire 10–10,000-$cm^{-1}$ spectral region can now be scanned with commercial instruments, Extensive wavelength ranges can be covered with any one instrument. Also, it is a simple matter to extend the coverage of commercial equipment to 5 $cm^{-1}$. Manufacturers can supply interferometers with or without complete controls or computers.

Far-infrared spectra (5–800 $cm^{-1}$) can be obtained with instruments currently being manufactured. By simply interchanging the source and beam splitter of these instruments, spectra down to 5 $cm^{-1}$ can be measured. Chemists use the far-infrared spectra to study weak chemical bonds, charge transfer bonds, and absorbed species. They also use them for studies of molecules with heavy atoms or long chains because many of the infrared absorption bands of these molecules are in the far-infrared region. Many solid-state measurements, which require the calibration of crystal lattice vibrations, use far-infrared spectra.

Near-infrared spectra (400–10,000 $cm^{-1}$) are commonly within the measurement capability of commercial equipment. By using deposits of different films, such as germanium or iron oxide, on transparent materials, such as fluorite or quartz, it is possible to make beam splitters so that the interferometers can be used in the measurement of spectra up to the red visible region. In the near-infrared, photoconductive detectors, such as lead sulfide or lead selenide, can be used in either a cooled or uncooled state. These detectors are an order of magnitude more sensitive than thermal detectors which are used in the far-infrared.

Visible spectra (15,000–33,000 $cm^{-1}$) may soon be obtained with the Digilab and Idealab interferometers. Other companies should follow soon. Although the multiplex or Fellgett advantage is cancelled by increased photon noise when background-limited detectors such as photomultipliers are used, the multiplex spectrometer is more tolerant of photomultiplier dark current. It is more tolerant of background noise by a factor equal to $M$ (the number of resolution elements in the spectrum) than a dispersive spectrometer with the same resolution. A large angular acceptance and aperture give the interferometer a Jacquinot advantage of approximately $M^{1/2}$ (Chapter 2).

The visible range that interferometers measure is comprised of displacements which are precisely controlled by a laser. The mirror tilt can also be controlled with three inexpensive He–Ne lasers which have simple feedback circuits connected to stacked piezoelectric ceramics or the cat's eye retroreflector (Appendix C). All the instruments have large, simultaneous wavelength coverage, no order-sorting problems, tolerance to source fluctuations, a circular field of view, small size, and low weight. The reader should consult Chapter 19 on commercial instruments for further details.

## APPLICATIONS OF FOURIER TRANSFORM SPECTROSCOPY

The applications of Fourier transform spectroscopy or interferometry in science and industry are extensive. The usefulness and simplicity of the technique lend it to almost every field into which the spectroscopist has ventured and into some where he has not gone before. It has fared better in most instances than conventional techniques.

Examples of the various applications of the technique are given in many parts of this book. Additional examples can be found in the books and articles referred to in each of the chapters and in the bibliographies.

Data on wavelength molecular absorption can be found in the work of Gebbie, Dowling, Chamberlain, Sanderson, and Connes. Gebbie, by placing gases in long tubes, has used path lengths of tens of meters to study the spectral absorption of atmospheric constituents. Dowling has performed similar experiments in shorter gas cells, but has used double-beam Fourier transform interferometry for his studies. Dowling has examined the pure rotational spectra of CO, $CO_2$, and NO as well as the spectra of other gases. Connes has made extraordinarily high-resolution measurements both in the laboratory and with telescopes of the molecular spectra of stars and planets. His laboratory data have enabled him to see quadrupole and spin–ground-state splitting in rotational spectra. He has obtained vibrational and rotational spectra from planets with such precision that the data are as good as if taken entirely from laboratory samples. Chamberlain, Gibbs, Chantry and Gebbie at the National Physical Laboratory in England and Bell and Sanderson at Ohio State University have used amplitude spectroscopy (Chapter 8) to obtain the complex indices of refraction of gases, even in regions of large absorption.

The individuals cited above have been singled out primarily for their pioneering efforts and worthy accomplishments. Other people, such as those who are or have been employed with the Air Force Cambridge Research Laboratory (AFCRL), the National Aeronautics and Space Administration (NASA), and the National Physical Laboratory (NPL) in England have also been instrumental in making several significant contributions to molecular physics and chemistry with Fourier transform interferometry. Their spectral measurements were or are being made at high and low temperatures, and under laboratory and field conditions.

Environmental problems, such as water, atmospheric, gasous, particulate, and thermal pollution, as well as natural disturbances, are being examined by the AFCRL and NASA groups. Their work is very comprehensive and is definitely not restricted to the singular effects of "nuclear disturbances." Because they are obtaining a great deal of information about the ecological system of the earth, their studies will definitely affect the direction of future environmental studies.

Liquid-phase studies are progressing slowly, but there is no fundamental reason for this. Liquid cells which are similar to the ones used for grating and prism instruments can be used satisfactorily with interferometers. The taking of measurements, with samples in or out of the instruments, can proceed in the same manner as in conventional spectroscopy. It is noteworthy that amplitude Fourier spectroscopy (Chapter 8) has something unique to contribute to liquid studies. Chamberlain *et al.* have studied liquids which they placed on a mirror in one arm of an asymmetric interferometer. From the interferograms, which were taken in reflection and transmission amplitude spectroscopy, they were able to distinguish first from second surface effects, bulk absorption, and multiple internal reflection phenomena. Amplitude spectroscopy has a tremendous potential in the surface-state studies of liquids and solids.

In solid-state studies, the applications of Fourier transform spectroscopy are countless. It has been used for superconducter energy-gap measurements, infrared detector and source development projects, magnetic resonance, infrared filter development, phonon or *reststrahlen* measurements, plasmon–phonon interaction studies, local mode problems, and metallic surface property studies. Measurements have been made under a variety of conditions in transmission and reflection at high, room, and low temperatures and in and out of electric and magnetic fields. The interferometer provides the experimentalist with easily obtained and usable, small radiation beams at the sample. Sometimes, samples of 1 $mm^2$ area can be used at 10 $cm^{-1}$. Because of the reduction or even elimination of the stray or unwanted light, which plagues grating and prism spectroscopy, very low-transmittance measurements can be made with Fourier transform interferometry. Again, amplitude spectroscopy stands out because the complex reflection coefficient which the technique yields is easily used to obtain the complex indices of refraction of a sample without having to apply Kramers–Kronig calculations or making assumptions concerning a physical model.

Many workers have made a variety of astronomical measurements, but those of Pierre Connes and his wife Janine Connes, L. Mertz, and R. Hanel are truly outstanding. These scientists have observed the rotational spectra of gases on Venus, the spectra of $CO_2$ on Mars, the ice–water phase of the "frost" particles in the rings of Saturn, and satellite altitude spectra of the earth's atmosphere.

Deep-space measurements of stellar spectra open many new avenues of research to astronomers. The Fellgett and Jacquinot advantages are fully utilized in astronomical investigations, and they make it possible to couple spectrometers to telescopes. Pierre and Janine Connes are working, at present, with a 60-in. telescope. Several observatories, such as the McDonnell Observatory of the University of Texas, have coupled Fourier transform spectrometers to telescopes, and new data are being rapidly acquired. Interferometers have been placed in satellites by R. Hanel *et al.* of the Goddard Space Flight Center (NASA) and in rockets by D. McNutt and P. Feldman of the Naval Research Laboratories. So far, much of the data obtained by satellite-borne instruments were obtained through telescopes aimed either at the earth or at the sun; however, it will be only a short time before other good extra-solar or extraterrestial data will be gathered. For instance, the Viking probe to the planet Mars will carry a near-infrared, Michelson Fourier-transform spectrometer.

For some low-temperature studies, a worthy and fruitful effort has been made by G. Augason of Ames Research Center (NASA) to make cryogenically cooled Michelson interferometers. When the entire instrument is cooled, background radiation levels are practically eliminated, and very weak emission or absorption problems can be studied at low temperatures. It may be some time before this technique, with its very special cryogenic applications, advances to a prominent position. For the cosmic background problem in astronomy, cooled interferometers are important.

Some people used to think that Fourier transform spectroscopy could not be applied to rapidly changing phenomena, but techniques have been developed to study plasma effects in steady flow tubes, relaxation phenomena of a wide variety, and exhaust plumes of jet engines. J. Strong of the University of Massachusetts is studying gases in a tube which is 100 ft long. His instrument is a spherical, lamellar grating interferometer (Chapter 15). R. Murphy and H. Sakai at AFCRL have developed a technique for obtaining excellent time and spectral resolution in relaxation processes. They excite gases and then observe the relaxation process at each step as the interferometer arm is advanced to its maximum displacement. At each step, the signal for each desired time increment is stored. At the end of the scan, which may include as many steps as the experiment requires for high spectral resolution, the temporal and spectral features are separated on a computer. The results are that one can observe competing, sequential, and cascading processes as a function of time over a relatively large wavenumber range with excellent spectral resolution.

The field of rapid-scan Fourier transform spectroscopy (Appendix D) has made possible extremely fast and accurate spectral measurements. It is possible to record complete scans in less that $\frac{1}{8}$ sec. The laboratory experimentalist can make "on-line" measurements, and the astronomer can also make very quick spectral recordings. The measurements can be repeated to improve signal-to-noise averages and then computer culled and averaged.

Miscrosampling, which is an important selling point of Fourier transform interferometry, has several major advantages:

1. Small quantities of a pure compound in a KBr disk can be studied. The ability to obtain a high signal-to-noise ratio with conventional commercial beam condensers make it possible to characterize small quantities of sample, in that the size of the sample is reduced in proportion to the area of the sample disk.

2. Very small holes can be used to mask the beam for small samples. The fact that the signal-to-noise ratio for routine samples is very high means that the radiation which emerges through a very small hole can be detected. Samples have been studied for which the hole diameter was only 150 $\mu$m. When samples are scanned on grating or prism instruments, if the chopper is behind the sample with the same beam condenser, the only spectrum that is measured is an emission spectrum from the whole sample region, which is heated by the condensed beam. The absorption spectrum from the infrared radiation which passes through the hole in the mask is not measured. In rapid-scan Fourier transform spectroscopy (Appendix D), one is able to measure the very low energy which passes through the sample, and problems caused by the sample being heated in the beam are essentially eliminated. By this means, the amount of pure compound that is necessary for the measurement of an infrared spectrum is substantially reduced, and liquid microsample quantities can be measured at the submicrogram level.

3. Another use which workers in research laboratories envisage is the examination of compounds that can be held under pressure between diamond windows. Because the diamonds would be limited in size, very little flux would reach the detector. However, by using interferometers under these circumstances, even high-resolution spectra with a very low noise level could be measured in an hour.

4. Attenuated total reflection (ATR) techniques have been used in Fourier transform spectroscopy for several years to measure the spectra of small amounts of material.

Reflection measurements are widespread. Many laboratories have examined spectra of many compounds by using specular reflectance techniques on the paints of painted metals, corroded surfaces, and optically opaque substances. Even fibers and cords, for which it would be thought almost impossible to measure spectra, have been characterized by winding them round a flat sampling post. Samples, such as paper, that have strong scattering tendencies can be used as substrates for reflector studies, and the spectra of oils and pigments on small areas of paper have been characterized.

Another use which has been considered is the study of catalysts and heterogeneous reactions in which the change in the molecular structure of an absorbed species can be investigated through its infrared spectrum.

Scattering and strong absorption by samples can be studied by transmission measurements. In addition to the ability to measure the reflection spectra of strongly scattering species, the great light-gathering power of a Fourier transform spectrometer enables transmission spectra of strongly scattering materials, such as paper, or strongly absorbing compounds, such as water, to be measured. Thus, by transmission as well as reflection, the examination of solutes in aqueous solutions becomes possible.

Weak emission spectra can be measured by making minor alterations in the source housing of many of the commercial instruments. The *f*/number of the input optics should be matched; but change is not too difficult. Usually, only an extra, inexpensive mirror is required. Interferometers have been connected to telescopes for measuring spectra from remote sources such as smokestacks, stars, and planets. In the field of chemistry, several workers are using Fourier transform spectroscopy to measure infrared fluorescence and chemiluminescence.

Kinetic studies are often desirable. Until now, if the concentration of any species was to be monitored by using infrared absorption bands, the spectrometer was not scanned, but the intensity of one band was monitored continuously. Because the measurement time for each interferogram can be less than 1 sec, an entire spectrum can be stored in the memory of a computer, and concentrations of several species can be simultaneously measured instead of measuring just one band of one component.

Process and quality control is one of the largest commercial applications of interferometry. The ability to measure a complete spectrum at high speed and to examine it with a computer for extraneous bands means that multicomponent mixtures can be monitored and a production line altered on very short notice. Also, because in routine spectra, additive and impurity bands, even at low concentrations, can be measured quickly, quality control applications are also possible. This latter ability should have a bearing on the chemical, plastics, coatings, petroleum, and pharmaceutical industries.

## CONCLUSIONS

In this chapter, we have attempted to introduce Fourier transform spectroscopy or interferometry, with emphasis on its merits. These merits are going to make Fourier transform interferometry the most widely applied technique to spectroscopy. The versatility of the technique is tremendous. Because of its general applicability, the interferometer stands beside Newton's prism, Fraunhofer's grating, and Towne's maser as hallmark developments in optics.

## *References*


*1.* P. Connes, "Lasers and Light." Freeman, San Francisco, California, 1969.

*2.* H. A. Gebbie, Fourier Transform versus Grating Spectroscopy, *Appl. Opt.* **8**, 501 (1969).

*3.* J. Stone, "Radiation Optics." McGraw-Hill, New York, 1963.

*4.* J. W. Goodman, "Introduction to Fourier Optics." McGraw-Hill, New York, 1968.

*5.* W. A. Steel, "Interferometry." Cambridge Univ. Press, London and New York, 1967.

*6.* P. Connes, High Resolution and High Information Fourier Spectroscopy, *Aspen Int. Conf. on Fourier Spectrosc., 1970*, p. 121. (G. A. Vanasse, A. T. Stair, Jr., and D. J. Baker, eds.). AFCRL [Air Force Cambridge Research Laboratories, Hanscomb Field, Cambridge, Massachusetts] -71-0019, 5 Jan. 1971, Spec. Rep. No. 114.

*7.* H. Sakai, Consideration of the Signal-to-Noise Ratio in Fourier Spectroscopy, *Aspen Int. Conf. on Fourier Spectrosc., 1970*, p. 19. (G. A. Vanasse, A. T. Stair, Jr., and D. J. Baker, eds.). AFCRL-71-0019, 5 Jan. 1971, Spec. Rep. No. 114.


*Note added in Proof*: The book by K. D. Möller and W. G. Rothschild ("Far-Infrared Spectroscopy." Wiley-Interscience, New York, 1971.) is an especially good reference on the physics and chemistry of far-infrared problems that can be studied by Fourier transform spectroscopy.

## CHAPTER TWO

# *Historical Sketch and Crucial Ideas*

## INTRODUCTION

The purpose of this chapter is to bring to the reader's attention significant papers and books on the subject of interferometry. The major contributions of specific persons or groups are emphasized and the reviews of historical developments that have been collected by the more recent writers of comprehensive studies are blended into this chapter. For the convenience of the reader, the titles of these reviews and other references are included in the text.

There are several histories of Fourier transform spectroscopy, but only three are mentioned. These three augment the references given in this chapter and are as follows: Dr. Ernest V. Loewenstein of AFCRL in 1966 wrote a paper on The History and Current Status of Fourier Transform Spectroscopy, which was published in Volume 5 of *Appl. Opt.* [1]. Dr. Lawrence Mertz of the Smithsonian Astronomical Observatory gave an invited paper on Fourier Spectroscopy Past, Present, and Future at the *1970 International Conference on Fourier Spectroscopy at Aspen, Colorado* [2]. Dr. Pierre Connes of CNRS, Campus, D'Orsay, France, has reviewed many of the early papers and books and has made several major contributions of his own; therefore, his presentation, How Light is Analyzed *in* "Lasers and Light," edited by A. Schawlow and published by W. H. Freeman and Co., San Francisco in 1969 [3] gives perspective to the subject.

## MICHELSON AND HIS INTERFEROMETER

The science of Fourier transform interferometry was initiated in 1880 when Dr. Albert A. Michelson invented the interferometer. The instrument and its uses were well received by the scientific community of the day, and Michelson received the Nobel Prize for his useful concepts. Michelson used the interferometer to measure, with greater precision than anyone before him, the velocity of light in a vacuum. He proved, in the famous Michelson–Morley experiment, that the earth does not move through an ether, which many scientists thought had to exist for the propagation of light in outer space. The Michelson–Morley experiment was not important to Dr. Albert Einstein's creative concepts, but it did help make Einstein's fellow scientists more comfortable with the subject of relativity. Because Michelson lacked computers and electronic detectors, he could not exploit the field of Fourier transform spectroscopy. He had to work with visibility curves, and, as Lord Rayleigh pointed out, unique spectral distributions cannot be obtained from the visibility curve alone, except in the case of symmetric resonances. Michelson perceived the basic techniques of Fourier transform spectroscopy and could have developed the field 90 years ago† if the rest of the prerequisite technology had not been so far behind. He measured intensity levels with his eye and made crude estimates of the spectrum of the Zeeman effect [*Astrophys. J.* **7**, 131 (1898)] [4]. To compute Fourier transforms, he invented a mechanical sine and cosine analog computer which could handle about 80 data points. For some reason, he did not extensively use the analog computer, which he termed a "harmonic analyzer," for Fourier transform spectroscopy. However, using the techniques available to him, Michelson resolved several spectra which were doublets. A readily available source of information on the subject is a note in *Appl. Opt.* Vol. 6, p. 692 (1967), which contains excerpts from Michelson's book "Light Waves and Their Uses." Univ. of Chicago Press, 1902 [6]. To verify Michelson's rightful claim to having realized and illuminated the basic concepts of Fourier transform spectroscopy, the reader can refer to A. A. Michelson, *Phil. Mag. Ser. 5* **31**, 256 (1891) [7] and A. A. Michelson, On the Application of Interference Methods to Spectroscopic Measurements, *Phil. Mag.* **34**, 280 (1892) [8]. If the reader wishes to see what Michelson initiated in terms of interferometer varieties, he should consult

† Because Michelson worked with visibility curves and used his eye as the detector, the first true interferogram was not made until 1911, when it was recorded with a microradiometer by Rubens and Wood, who obtained crude spectral data by trial and error. They guessed spectra, computed the interferogram, and compared the results with experiments. Their work is described in H. Rubens and R. W. Wood, *Phil. Mag.* **21**, 249 (1911) [5].

Dr. William H. Steel's book "Interferometry," Cambridge Univ. Press, New York, 1967 [9]. Steel's book contains many details about a number of different types of interferometers that are designed for a variety of experiments.

## INTERFEROMETERS

Figure 2-1 is an optical diagram of the basic Michelson interferometer. There are a multitude of variations, but the different instruments are basically similar. In some variations, a part is omitted, such as the compensator, or a piece is added. In others, a considerable amount of study is required to determine whether or not an instrument is indeed an interferometer.

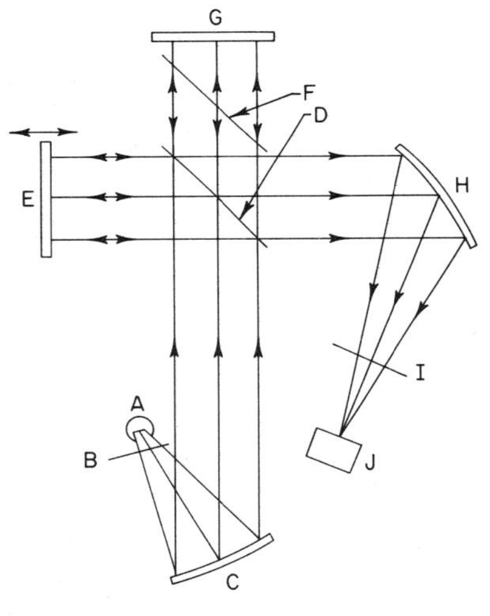


FIG. 2-1. Basic Michelson interferometer. A: Source. B: Chopper. C: Collimator. D: Beamsplitter. E: Movable mirror. F: Compensator. G: Fixed mirror. H: Focusing mirror. I: Spectral filters. J: Detector.

To understand the mechanics of the interferometer, one can start with the source, which is usually mated with a desired spectral region—mercury lamps for the far-infrared, glowers for the near-infrared, and a variety of lamps in the visible. A chopper is added to give an ac infrared flux which can be used with phase-sensitive detection and reference systems to eliminate background radiation and electronic fluctuations and drift. After the radiation beam is collimated by a mirror or lens, it is amplitude-divided at the beamsplitter. The thickness and coatings of the beamsplitter must be considered carefully (Chapter 9). One part of the radiation beam goes through a compensator plate to a fixed mirror, which reflects the beam back through the compensator plate. The beam then is reflected by a beamsplitter toward a focusing mirror. The other part of the radiation beam is reflected from the beamsplitter, goes to and returns from a movable mirror, is transmitted through the beamsplitter, and goes to the focusing mirror. The compensator plate is sometimes introduced to keep the optical paths in the two arms at approximately equal length including in the *beamsplitter material*. The focusing mirror merely focuses the recombined radiation beams on the detector. The spectral filters mainly eliminate the high-wavenumber flux, which is unwanted.

The various degrees of interference for different wavelengths are produced by an optical path difference in the two arms. The path difference is twice the carriage or arm displacement away from the balanced position. The balanced optical path position of the movable mirror is called the "origin," "white light position," or "grand maximum." The carriage can be moved to either side of the origin.

## FUNDAMENTALS OF FOURIER TRANSFORM SPECTROSCOPY

Michelson realized that each wavelength in the interferometer produces its own characteristic interference pattern as the movable mirror is displaced. A monochromatic source yields a cosine variation in the flux of the combined beams at the detector. The period of the cosine function is uniquely determined by the wavelength and optical path difference for the radiation beams in the two arms. Each wavenumber band has its own characteristic cosine flux pattern with a particular magnitude. The recording of the detected signal versus optical path difference is the *interferogram*. For a source of many frequencies, the interferogram is the sum of the fluxes of each wavelength pattern. Fourier analysis, as outlined in Chapter 3, enables one to convert the interferogram into a spectrum, i.e., signal versus frequency. That is, Fourier analysis of the interferogram picks out the pattern for each frequency and determines the magnitude of the flux at that frequency, the Fourier coefficient.

## JACQUINOT ADVANTAGE

The Jacquinot or throughput advantage (*étendue*) states that in a lossless optical system, the brightness of an object equals the brightness of the image; therefore, the flux throughput and brightness can be considered at any point in a lossless optical system. The validity of this concept can be illustrated with the following example:

First, consider a *point* source of radiant flux $dF$ (with MKS units of watts) which is collected by an increment of area $d\mathscr{A}$ of an optical component (a mirror) at a distance $r$ from the source. Assume that the normal to the mirror $n$ is at an angle $\varphi$ with respect to the central incident ray (Fig. 2-2).

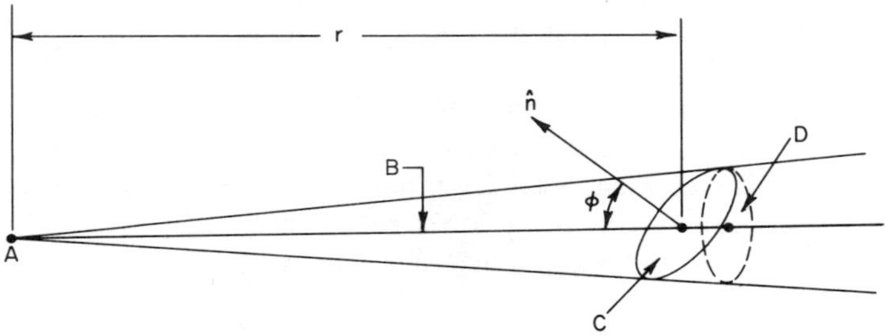


FIG. 2-2. Optical diagram of a point source and a collecting optical element. A: Point source. B: Optical axis. C: Optical element area $\equiv d\mathscr{A}$. D: Projected area $= \cos\varphi \, d\mathscr{A} \equiv dA$.

The flux striking the optical component is proportional to the projected area of the optical element and inversely proportional to the square of the distance from the source. Now, extend the source from a point to an area $ds$ which is inclined at an angle $\theta$ between the source normal and the optical axis. The flux is emitted by the finite source which is inclined at an angle $\theta$. The source is said to be a Lambert-law radiator when the flux is proportional to $\cos\theta \, ds$. Thus, the flux from a finite, inclined source striking an inclined optical element at distance $r$ is

$$dF = B(\cos\theta \, ds)(\cos\varphi \, d\mathscr{A})/r^2. \tag{2-1}$$

The proportionality constant is the brightness $B$ of the source, having MKS units of W $m^{-2}$ $ster^{-1}$. Equation (2-1) applies to any general pair of optical elements.

By reassociating terms in Eq. (2-1), the solid angle subtended by the *source* can be defined as

$$d\Omega = (\cos\theta \, ds)/r^2 \tag{2-2}$$

and the projected area of the collimator is given by

$$dA = (\cos\varphi \, d\mathscr{A}). \tag{2-3}$$

The radiant flux is now given by

$$dF = B \, dA \, d\Omega. \tag{2-4}$$

If there is a series of optical elements with no losses between elements, the same basic arguments which lead to Eq. (2-1) result in the following set of equations for the étendue $E$:

$$E \equiv (dF/B)_S = dA\, d\Omega = dA'\, d\Omega' = dA''\, d\Omega'' = \cdots = (dF/B)_D. \qquad (2\text{-}5)$$

A good reference for Eq. (2-5) is F. A. Jenkins and H. E. White, "Fundamentals of Optics," 3rd ed. McGraw-Hill, New York (1957), p.111 [10]. In the reasoning used to obtain Eq. (2-5), the first optical element becomes the source for the second optical element, and so on. In Eq. (2-5), the subscript S refers to the source and the subscript D refers to the detector. If no flux is lost in the optical system, $B_S = B_D$.

For the interferometer, Jacquinot focused attention on the fact that $dA\, d\Omega$ is a constant for the instrument from the source to the detector. He pointed out that it is possible to use large values of $dA\, d\Omega$ in an interferometer and have high resolution which is virtually independent of $dA$ and $d\Omega$. The product $dA\, d\Omega$ was called the "étendue" or "throughput."

If the power through the instrument is proportional to the étendue for sources of the same brightness $B$, then the étendue of grating spectrometers and interferometers can be compared. First, consider the étendue of the Michelson interferometer, $E_M$, as it is derived in Chapter 11, Field of View, of this book. The solid angle of the collimator mirror subtended by the source is

$$\Omega_M = \pi h^2/4F^2, \qquad (2\text{-}6)$$

where $h$ is the diameter of the circular source and $F$ is the focal length of the collimator mirror. The resolving power of the Michelson interferometer $R_M$ is given by

$$R_M \equiv \sigma/\delta\sigma = 8F^2/h^2; \qquad (2\text{-}7)$$

therefore, the solid angle–resolving power product is given by

$$R_M\, \Omega_M = 2\pi. \qquad (2\text{-}8)$$

The étendue of a Michelson interferometer, after combining Eqs. (2-5) and (2-8), is given by

$$E_M = A_M\, \Omega_M = 2\pi A_M/R_M \qquad (2\text{-}9)$$

where $A_M$ is the area of the collimator mirror and $R_M$ is the resolving power of the interferometer. Note that the incremental parameters of Eq. (2-5) are replaced by their total values in obtaining Eq. (2-9).

For the grating spectrometer, the power through the instrument is limited by the area of the entrance slit. So the effective source area becomes the slit area, and the slit subtends the solid angle from the collimating mirror. With the slit in the focal plane of the collimator, the grating solid angle $\Omega_G$ becomes

$$\Omega_G = Wl'/F^2. \tag{2-10}$$

The slit width is $W$, the slit length is $l'$, and $F$ is the focal length of the collimator. The grating equation for a grating spectrometer system is

$$1/\lambda = \sigma = 1/(2md \cos \alpha \sin \theta) \tag{2-11}$$

where $m$ is the order, $d$ is the grating constant, $\alpha$ is the half-angle between incident and diffracted rays from the grating—entirely an instrument and not a grating parameter—and $\theta$ is the angle of rotation of the grating. By differentiation of Eq. (2-11), the resolving power can be found in terms of the increment of angle $\delta\theta$ through which the grating must be rotated to move a bandwidth $\delta\sigma$ across the equal exit slit. This small angle $\delta\theta$ is given by $W/2F$. So the resolving power $R_G$ of a grating spectrometer is given by

$$R_G = (2F/W) \tan \theta. \tag{2-12}$$

Substituting $W$ from Eq. (2-12) in Eq. (2-10) and using the definition of the étendue in Eq. (2-5) gives

$$E_G = A_G \Omega_G = (l'/F_G)(A_G/R_G)(2 \tan \theta). \tag{2-13}$$

For grating instruments, the maximum efficiency is achieved when $\theta$ is near the blaze angle. With blaze angles of the order of 30° one has 2 tan $\theta$ of the order of unity. The grating spectrometer étendue becomes approximately

$$E_G \simeq (l'/F)(A_G/R_G). \tag{2-14}$$

Assuming the same area and focal length for the collimators and the same resolving power, the ratio of the interferometer and grating étendues becomes

$$E_M/E_G \simeq 2\pi(F/l'). \tag{2-15}$$

In the very best grating spectrometers, $F/l'$ is never less than 30; therefore, $E_M/E_G$ is of the order of 190 for the best grating instruments. That is, about 200 times more power can be put through the interferometer than through the best grating spectrometers. Also, the optical system can be mush smaller for the interferometer than for the grating spectrometer.

Jacquinot realized and wrote about the throughput advantage and other ideas in the following papers: P. Jacquinot and C. J. Dufour, *J. Rech. C.N.R.S.* **6**, 91 (1948) [11] and P. Jacquinot, *Rep. Prog. Phys.* **23**, 267 (1960) [12]. A clear statement of the étendue can be found in a book by A. Girard and P. Jacquinot on "Advanced Optical Techniques" (A.C.S. Van Hell, ed.), North Holland Publ., 1967, p. 71, 109 [13].

Jacquinot not only brought the throughput advantage to the attention of the scientific community, but he trained many fine students, such as Drs. Pierre and Janine Connes. Jacquinot deserves much credit for the revitalization and injection of major concepts into the field of Fourier transform spectroscopy. Later, he was made Director General of the Centre National de Recherche Scientifique.

## FELLGETT ADVANTAGE

To study the multiplex or Fellgett advantage in quantitative terms, only a few basic ideas need be considered in addition to the physical description of the advantage that was given in Chapter 1. Fellgett, who was the first person to transform interferograms numerically, put forward his ideas shortly after Jacquinot stated the throughput advantage. The most comprehensive treatment of Fellgett's advantage appears in the dissertation which he submitted in 1951 to the faculty of Cambridge University in England. A more accessible article was published by P. Fellgett in *J. Phys. Radium*, Vol. 19, p. 187 (1958) [14].

The multiplex principle can be explained as follows: Suppose one is interested in measuring a broad spectrum between the wave numbers $\sigma_1$ and $\sigma_2$ with a resolution $\delta\sigma$. The number of spectral elements $M$ in the broad band is then

$$M = (\sigma_2 - \sigma_1)/\delta\sigma \equiv (\Delta\sigma)/\delta\sigma. \tag{2-16}$$

If a grating or prism instrument is being used, each small band of width $\delta\sigma$ can be observed for a time $T/M$, where $T$ is the total time required for a scan from $\sigma_1$ to $\sigma_2$. One can then say that the integrated signal received in a small band $\delta\sigma$ is proportional to $T/M$. If the noise is random and independent of the signal level, the signal noise should be proportional to $(T/M)^{1/2}$. Thus, for a grating instrument, the signal-to-noise ratio $S/N$ would be

$$(S/N)_G \propto (T/M)^{1/2}. \tag{2-17}$$

The situation for the interferometer is different because it detects in the broad band $\sigma_2 - \sigma_1$ *all* small bands of resolution $\sigma\delta$ all the time. So the integrated signal in a small band $\delta\sigma$ is proportional to $T$. If the noise is random and independent of the signal level, the signal noise is proportional to $T^{1/2}$. Thus, for an interferometer, the signal-to-noise ratio would be

$$(S/N)_I \propto T^{1/2} \tag{2-18}$$

with the same proportionality constant as Eq. (2-17). Comparing Eqs. (2-17) and (2-18), the ratio of the $(S/N)_I$ for interferometers to the $(S/N)_G$ for gratings is

$$\frac{(S/N)_I}{(S/N)_G} = M^{1/2} \tag{2-19}$$

Since $M$ is the number of spectral elements of width $\delta\sigma$ in the broad band $\sigma_2 - \sigma_1 = \Delta\sigma$, Eq. (2-19) predicts that the interferometer has a much higher signal-to-noise ratio than the grating or prism instruments. If $\sigma_2 - \sigma_1$ is of the order of the mean wave number $\sigma$, then by considering Eqs. (2-7) and (2-16), $M$ is of the order of the resolving power $R$. Thus, if high resolving power is sought, say $10^4$–$10^6$, the interferometer will have a better signal-to-noise ratio than the grating spectrometer by a factor of about $R^{1/2}$, or $10^2$–$10^3$ for the hypothetical cases.

In deriving Eq. (2-19), it was assumed that the noise was independent of signal and, therefore, only depended on the measuring time. For the infrared region, the noise is usually detector noise, which is independent of signal. Hence, Eq. (2-19) is valid in infrared studies.

In the visible wavelength region, the detectors are much better than in the infrared, and individual photons can be detected. The noise is governed by the statistical fluctuations in the number of photons emitted by the source during a measurement period. Because the fluctuation in the number of photons is proportional to the total number of photons (random emission), the source illumination noise level is proportional to the square root of the source intensity.

For a grating instrument in which the photon noise dominates, if $I(\delta\sigma)$ is the intensity in the small band $\delta\sigma$, the signal recorded for the small band is proportional to $(T/M)I(\delta\sigma)$ and the noise is proportional to $[(T/M)I(\delta\sigma)]^{1/2}$. Thus, for grating instruments, the signal-to-noise ratio in the small band is

$$(S/N)_G \propto [(T/M)I(\delta\sigma)]^{1/2}. \tag{2-20}$$

A signal-to-noise ratio which is proportional to the square root of the source intensity is well known to spectroscopists who work in the visible wavelength range.

For interferometers with the noise proportional to the square root of the source intensity, the signal-to-noise ratio is entirely different than indicated by Eqs. (2-18) and (2-19). One still has the spectral signal *in a small band* proportional to $TI(\delta\sigma)$. At a particular position of the movable mirror in the interferometer, one can record a composite signal which has contributions from *all* wave numbers in the band from $\sigma_1$ to $\sigma_2$. This signal in the interferogram is proportional to $TMI(\delta\sigma)$ if all $M$ small bands of width $\delta\sigma$ have the same intensity $I(\delta\sigma)$. The composite signal noise is then proportional to

$[TMI(\delta\sigma)]^{1/2}$. In Fourier-transforming the composite recorded signal into a spectrum, the noise in the spectrum remains about the same as it was in the composite recorded signal. The computed noise $[TMI(\delta\sigma)]^{1/2}$ is the same for each small band but the signal in *a* band is $[TI(\delta\sigma)]$; so, for the interferometer, the signal-to-noise ratio *in the small band* is proportional to

$$(S/N)_{\text{I}} \propto [(T/M)I(\delta\sigma)]^{1/2}. \tag{2-21}$$

It is obvious that when the noise is proportional to the square root of the source intensity, there is no Fellgett advantage because Eqs. (2-20) and (2-21) are the same. Thus, in the visible wavelength range, where photon noise dominates, the Fellgett advantage is lost. However, the Jacquinot advantage is not affected; therefore, interferometers still have a large throughput advantage in the visible wavelength range.

## STRONG'S GROUP

In the early 1950's, experimental efforts to make Fourier transform spectroscopy a reality were undertaken at Johns Hopkins University in Baltimore, Maryland, where Dr. John D. Strong had a group of outstanding investigators consisting of Drs. H. A. Gebbie, E. V. Loewenstein, and G. Vanasse. These four men† showed that the Fellgett and Jacquinot advantages could be actually achieved. They built instruments and obtained sample data which showed that excellent resolution could be obtained. Drs. Loewenstein and Vanasse worked with Dr. Strong for a few years and then joined the Air Force Cambridge Research Laboratories, where they established Fourier transform spectroscopy. Dr. Gebbie returned to the National Physics Laboratories at Teddington, England, and worked with Drs. J. E. Chamberlain, G. W. Chantry, and J. E. Gibbs. These four men developed Fourier transform spectroscopy with Michelson interferometers and interested such companies as Grubbs–Parsons in manufacturing some of their instruments. Dr. Gebbie is now with the National Bureau of Standards in Boulder, Colorado, where he is setting up a large, millimeter-wave telescope at Pike's Peak.

In addition to popularizing experimental Fourier transform spectroscopy, Dr. Strong's group developed the lamellar grating interferometer and published several articles concerning it. One article by G. A. Vanasse and J. D. Strong appeared as Appendix F, Application of Fourier Transforms in

† Dr. L. Wigglesworth was with the group for a while, and in an internal report, he applied Lord Rayleigh's concepts of apodization to Fourier transform spectroscopy. Apodization is described in Chapter 5.

Optics: Interferometric Spectroscopy, in Strong's book [15]. "Classical Concepts of Optics" (W. H. Freeman and Company, San Francisco, 1958) [15]. The article presents the basic concepts that are needed to obtain and transform interferograms, and it utilizes as an example instrument the lamellar grating interferometer. The importance of the article lies in the fact that it was widely circulated among people of varying interests and diverse backgrounds. Because the lamellar grating interferometer is an important instrument, it will be described in detail in Chapter 15.

Simultaneously with Strong's group, Prof. L. Genzel at Freiburg University in W. Germany was developing lamellar grating interferometers. Genzel was interested in *both* the grating and Fourier transform spectroscopy uses of the lamellar grating [L. Genzel and R. Weber, *Z. angew. Phys.* **10**, 127 (1958); **10**, 195 (1958).].

Lamellar grating interferometers are very useful for low-wavenumber research. Their range extends from near 100 $cm^{-1}$ down to 2 or 3 $cm^{-1}$; they are very efficient instruments. There are two types of lamellar interferometers: the plane (1950's) and the spherical (1970) grating instruments. Both were developed by Dr. Strong and his associates.

## OTHER PIONEERING FOURIER TRANSFORM SPECTROSCOPISTS

In the late 1950's and early 1960's, several users and developers of Fourier transform spectroscopy came into prominence. Drs. Pierre and Janine Connes applied the techniques to astronomy. They, with J. P. Maillard, measured excellent spectra from the planets and published a great deal of the data [16] in "Atlas des Spectres dans le Proche Infrarouge de Venus, Mars, Jupiter, et Saturne." Editions du Centre National de la Recherche Scientifique, Vol. 15, Paris (1969), [16]. Their many contributions include major equipment developments in the near-infrared to solve problems associated with the telescope, the earth's atmosphere, and extraterrestrial sources. They have contributed many improvements to data handling and computational techniques, including practical real-time analysis equipment for $3 \times 10^4$ data points and quick computer techniques for the large machines so that $5 \times 10^6$ samples can be transformed in 9 min.

Other references to their work are: J. Connes, *Rev. Opt. Thero. Instrum.* **40**, 45, 116, 171, 231 (1961) [17]; P. Connes, Lasers and Light *in* "How Light is Analyzed" (A. Schawlow, ed.) Freeman, San Francisco (1969), p. 35 [3]; and D. M. Hunten, Fourier Spectroscopy of Planets, *Science* **162**, 313 (1968) [18].

In England, Drs. J. E. Chamberlain, J. E. Gibbs, and H. A. Gebbie concentrated on developing the Michelson interferometer equipment by conducting experiments in molecular spectroscopy and making atmospheric studies. Their experiments on ozone in the laboratory and in the upper atmosphere are excellent. One notable contribution was their work on asymmetric Fourier transform spectroscopy,† which is used to measure refractive indices of gases. Some of the early asymmetric results were reported in their article on The Determination of Refractive Index Spectra by Fourier Spectroscopy, *Infrared Phys.* **9**, 185 (1969) [19]. The technique enabled one to obtain the real part of the index of refraction of materials directly from the analysis of the interferogram. The possibility of obtaining, with high accuracy, both the real and imaginary parts of the index of refraction of gases, liquids, and solids became a reality. Physical or chemical models are not needed nor are Kramers–Kronig type of calculations used in asymmetric Fourier transform spectroscopy.

Asymmetric Fourier transform spectroscopy has been highly developed by an Ohio State University group under the direction of Drs. E. E. Bell and R. Sanderson. They have studied a wide variety of compounds and optical elements with their equipment. For a comprehensive review of the method, the reader is referred to E. E. Bell's paper on the Measurement of the Far Infrared Optical Properties of Solids with a Michelson Interferometer Used in the Asymmetric Mode: Part 1, Mathematical Formulation, *Infrared Phys.* **6**, 57 (1966) [20]. The subject is discussed in Chapter 8 of this book, where an explanation is given on how the indices of refraction are calculated and how very low reflection and transmission coefficients are measured.

In industrial and governmental laboratories in the United States, work on Fourier transform spectroscopy progressed rapidly during the 1960's. At Block Engineering Corporation, Dr. L. Mertz made several notable contributions to the development of what is now Digilab's current equipment. Digilab is a subsidiary of Block Engineering. Dr. Mertz, who is now at the Smithsonian Astrophysical Observatory in Cambridge, Massachusetts, also wrote a book entitled "Transformations in Optics," Wiley, New York (1965) [21], which has several chapters devoted to Fourier transform spectroscopy. At the Aerospace Corporation in Los Angeles, Drs. J. M. Dowling and C. M. Randall developed and demonstrated double-beam Fourier transform spectroscopy. Their lamellar grating instrument must be regarded as a developmental model interferometer and is too expensive for most laboratories. The pioneering efforts of Dowling and Randall should be studied because these

† Simultaneously Dr. Ely E. Bell of Ohio State University developed asymmetric interferometry for solids, liquids and gases. Bell reported his efforts in 1964 at a conference [E. E. Bell, *Japan J. Appl. Phys.* **4**, 412 (1965)], and J. E. Chamberlain, J. E. Gibbs, and H. A. Gebbie reported in 1963 in *Nature* **198**, 874 (1963).

two men have defined most of the principal problems of double-beam interferometry and have indicated how they can be solved. Their atmospheric studies on gases, such as NO, constitute worthy contributions to molecular spectroscopy. Their equipment and some experimental results are described in J. M. Dowling, Investigations in the Far Infrared with a Lamellar Grating Interferometer (March 1967), Air Force Report No. SSD-TR-67-30; Aerospace Report No. TR-1001 (9260-01)-7; J. M. Dowling, Space Physics Laboratory, Laboratories Division, Laboratory Operations, Aerospace Corp. [22], and J. M. Dowling, Signal and Noise in Two-Beam Interferometry, *Appl. Opt.* **6**, 1580 (1967) [23]. Two other reports which deal with molecular spectroscopy are J. M. Dowling, Direct Interpretation of Far Infrared Interferograms with Application to Diatomic and Linear Molecules, *J. Opt. Soc. Amer.* **54**, 663 (1964) [24] and J. M. Dowling and R. T. Hall, Far Infrared Interferometry: Upper Limit for Line Widths for the Pure Rotational Band of Carbon Monoxide. *J. Mol. Spectrosc.* **19**, 108 (1966) [25].

At the Air Force Cambridge Research Laboratories, Drs. E. V. Loewenstein, G. Vanasse, H. Sakai, R. Murphy, and A. T. Stair, Jr. made several notable contributions to theoretical and experimental Fourier transform spectroscopy. They studied *S/N* problems (see H. Sakai, Considerations on *S/N* in Fourier Spectroscopy, *1970 International Conference on Fourier Spectroscopy, Aspen, Colorado* [26]) and mathematical transforms in different mathematical spaces, *i.e.*, Hilbert space, and improved the theory of measurements in Fourier transform spectroscopy. They made a number of measurements on atmospheric gases in the laboratory and from airplanes, and studied molecular spectra, including relaxation processes (see R. Murphy and H. Sakai, Application of Fourier Spectroscopy Techniques to the Study of Relaxation Phenomena, *1970 International Conference on Fourier Spectroscopy, Aspen, Colorado* [27]) which are associated with environmental problems—natural and man-made. One development of this group which should be looked for in future publications is that of a Michelson interferometer with about a 2-m carriage movement.

Also in the 1960's. Dr. R. Hanel and his colleagues at the Goddard Space Flight Center (NASA), Greenbelt, Maryland, packaged a Michelson interferometer for the Nimbus III satellite. They obtained data for about three and one-half months from different parts of the earth (R. Hanel, B. Schlachman, F. Clark, C. Prokesh, J. Taylor, W. Wilson, and L. Chaney, The Nimbus III Michelson Interferometer, *1970 International Conference on Fourier Spectroscopy, Aspen, Colorado* [28]).

In 1964, when Dr. Paul L. Richards was associated with the Bell Telephone Laboratories at Murray Hill, New Jersey (he is now at the University of California–Berkeley), he published a comprehensive, experimental comparative study of interferometers and grating instruments. He built a Michelson interferometer, a lamellar grating interferometer, and a Randall-type far-infrared grating spectrometer. He proved that interferometers are superior to grating instruments. One of his major accomplishments was to demonstrate that the lamellar grating interferometer is an extremely efficient instrument at low wave numbers. P. L. Richards' paper is entitled High-Resolution Fourier Transform Spectroscopy in the Far-Infrared and was published in the *J. Opt. Soc. Amer.* **54**, 1474 (1964) [29].

The last major development of the 1960's was in the area of data handling and computational techniques. In 1965, the Cooley–Tukey algorithm was published by J. W. Cooley and J. W. Tukey, An Algorithm for the Machine Calculation of Complex Fourier Series, *Math Comput.* **19**, 297 (1965) [30]). The Cooley–Tukey algorithm, without any approximations, reduces the computation time of Fourier transforms by several orders of magnitude. A table indicating the efficiency of the algorithm is given at the end of Chapter 16 of this book. Dr. M. L. Forman called the algorithm to the attention of Fourier transform spectroscopists in an article entitled Fast Fourier Transform Technique and Its Application to Fourier Spectroscopy published in the *J. Opt. Soc. Amer.* **56**, 978 (1966) [31]. A summary of the fast Fourier transform (FFT) technique was published in the *IEEE Trans. on Audio and Electroacoustics* **AU-15(2)** (1967). In that journal on page 63, there is a particularly appropriate article concerning Fourier transform spectroscopy by E. O. Brigham and R. E. Morrow that is entitled The Fast Fourier Transform Spectrum [32]. Finally, in a talk on Computer Problems in Fourier Spectroscopy [33] that was delivered at the *1970 International Conference on Fourier Spectroscopy in Aspen, Colorado*, Dr. Janine Connes presented several mathematical techniques for further reducing computation times on machines. She noted the advantage of recording one-sided interferograms which are real (odd or even) functions and showed that further savings in time of a factor of four were possible. (Refer to Chapters 16–17 and Appendix B.)

While computational and mathematical techniques were being revolutionized in the 1960's, so were computers. Large computers were made faster and given larger memory capacities, and, just as importantly, minicomputers were developed. Dr. H. Yoshinaga and his colleagues created several small-computer techniques which they published in the following article: H. Yoshinaga, S. Fujita, S. Minami, Y. Suemoto, M. Inoue, K. Chiba, K. Nakano, S. Yoshida, and H. Sugimori, A Far-Infrared Interferometric Spectrometer with a Special Electronic Computer, *Appl. Opt.* **5**, 1159 (1966) [34]. Dr. Yoshinaga's group contributions strengthened instrumentation, electronic data handling, and spectral filters. In the late 1960's, the Data General Corporation, Varian Associates, Hewlett-Packard Corporation, and Minneapolis-Honeywell Corporation, among others, developed minicomputer systems which (including input and output equipment) can be purchased for under $25,000. These minicomputers have made real-time analysis possible. Dr. Pierre Connes at the Aspen conference mentioned that the real-time analysis system in the laboratory at the Centre National de Recherche Scientifique (CNRS) can handle up to $3.2 \times 10^4$ sample Fourier transformations.

In the 1960's, the Beckman Corporation, Grubbs-Parson, Digilab Inc., Coderg, and Idealab marketed Fourier transform spectrometers. All of these companies produce instruments which cover the range from 10 $cm^{-1}$ (or lower) to several hundred wave numbers. The Digilab equipment is the first commercial equipment capable of operating at the visible frequencies and has the first computer-controlled interferometers.

## CONCLUSIONS

The field of Fourier transform spectroscopy until now has been dominated by astronomers, physicists, and physical chemists; however, wide application in chemistry and chemical engineering is apparent, and such workers as Drs. F. Miller and W. G. Fateley of Carnegie-Mellon University, Dr. Walter Edgel of Purdue University, Dr. E. L. Lippencott of the University of Maryland, and Dr. M. J. D. Low of New York University are moving rapidly into Fourier transform spectroscopy as applied to chemistry and chemical physics. The history of the field indicates the great usefulness of this promising and highly developed tool. It should also be apparent that Fourier transform spectroscopy has supplanted grating and prism spectroscopy as the finest tool in near- and far-infrared spectroscopy.

There have been major conferences on Fourier transform spectroscopy such as those at CNRS, Bellevue, France in 1957; the *1965 Conference at Brooklyn Polytechnic Institute*; at the Mellon Institute on Far Infrared Transpose Spectroscopy; at Pittsburg in June 1965 (*Appl. Opt.* **4**, 1374 (1965) [35]); and the *1970 International Conference on Fourier Spectroscopy at Aspen, Colorado* sponsored by the Air Force Cambridge Research Center.

## *References*

*1.* E. V. Loewenstein, The History and Current Status of Fourier Transform Spectroscopy, *Appl. Opt.* **5**, 845 (1966).

*2.* L. Mertz, Fourier Spectroscopy, Past, Present, and Future, *Aspen Int. Conf. on Fourier Spectrosc., 1970*, p. 142. (G. A. Vanasse, A. T. Stair, Jr., and D. J. Baker, eds.). AFCRL-71-0019, 5 Jan. 1971, Spec. Rep. No. 114.

*3.* P. Connes, "Lasers and Light." Freeman, San Francisco, California, (1969).

*4.* A. A. Michelson, *Astrophys. J.* **7**, 131 (1898).

*5.* H. Rubens and R. W. Wood, *Phil. Mag.* **21**, 249 (1911).

*6.* A. A. Michelson, "Light Waves and Their Uses." Univ. Chicago Press, Chicago, Illinois, 1902.

*7.* A. A. Michelson, *Phil. Mag. Ser. 5*, **31**, 256 (1891).

*8.* A. A. Michelson, On the Application of Interference Methods to Spectroscopic Measurements, *Phil. Mag.* **34**, 280 (1892).

*9.* W. H. Steel, "Interferometry." Cambridge Univ. Press, New York, and London, 1967.

*10.* F. A. Jenkins and H. E. White, "Fundamentals of Optics," 3rd ed., p. 111. McGraw-Hill, New York, 1957.

*11.* P. Jacquinot and J. C. Dufour, *J. Rech. C.N.R.S.* **6**, 91 (1948).

*12.* P. Jacquinot, *Rep. Progr. Phys.* **23**, 267 (1960).

*13.* A. Firard and P. Jacquinot, "Advanced Optical Techniques" (A.C.S. Van Hell, ed.), p. 71–109. North Holland Publ., Amsterdam,.

*14.* P. Fellgett, *J. Phys. Radium* **19**, 187 (1958).

*15.* G. A. Vanasse and J. D. Strong, Appendix F, Application of Fourier Transforms in Optics: Interferometric Spectroscopy, in "Classical Concepts of Optics" (J. D. Strong, ed.). Freeman, San Francisco, California, 1958.

*16.* P. Connes, J. Connes, and J. P. Maillard, "Atlas des Spectres dans le Proche Infrarouge de Venus, Mars, Jupiter, et Saturne," Editions du Centre National de Recherche Scientifique, Paris, 1969.

*17.* J. Connes, *Rev. Opt. Theor. Instrum.* **40**, 45, 116, 171, 231 (1961).

*18.* D. M. Hunten, Fourier Spectroscopy of Planets, *Science* **162**, 313 (1968).

*19.* J. E. Chamberlain, J. E. Gibbs, and H. A. Gebbie, The Determination of Refractive Index Spectra by Fourier Spectroscopy, *Infrared Phys.* **9**, 185 (1969).

*20.* E. E. Bell, Measurement of the Far Infrared Optical Properties of Solids with a Michelson Interferometer Used in the Asymmetric Mode: Part 1, Mathematical Formulation *Infrared Phys.* **6**, 57 (1966).

*21.* L. Mertz, "Transforms in Optics." Wiley, New York, 1965.

*22.* J. M. Dowling Investigations in the Far Infrared with a Lamellar Grating Interferometer, Air Force Rep. No. SSD-TR-67-30; Aerospace Rep. No. TR-1001 (9260-01)-7 (March 1967).

*23.* J. M. Dowling, Signal and Noise in Two-Beam Interferometry, *Appl. Opt.* **6**, 1580 (1967).

*24.* J. M. Dowling, Direct Interpretation of Far Infrared Interferograms with Application to Diatomic and Linear Molecules, *J. Opt. Soc. Amer.* **54**, 663 (1964).

*25.* J. M. Dowling and R. T. Hall, Far Infrared Interferometry: Upper Limit for Time Widths for the Pure Rotational Band of Carbon Monoxide, *J. Mol. Spectrosc.* **19**, 108 (1966).

*26.* H. Sakai, Consideration of the Signal-To-Noise Ratio in Fourier Spectroscopy, *Aspen Int. Conf. on Fourier Spectrosc., 1970*, p. 19. (G. A. Vanasse, A. T. Stair, Jr., and D. J. Baker, eds.). AFCRL-71-0019, 5 Jan. 1971, Spec. Rep. No. 114.

*27.* R. Murphy and H. Sakai, Application of Fourier Spectroscopy Techniques to the Study of Relaxation Phenomena, *Int. Conf. Fourier Spectrosc. Aspen, Colorado* (March 1970).

*28.* R. A. Hanel, B. Schlachman, F. D. Clark, C. H. Prokesh, J. B. Taylor, W. M. Wilson, and L. Chaney, The Nimbus III Michelson Interferometer, *Aspen Int. Conf. on Fourier Spectrosc., 1970*, p. 231. (G. A. Vanasse, A. T. Stair, Jr., and D. J. Baker, eds.). AFCRL-71-0019, 5 Jan. 1971, Spec. Rep. No. 114.

*29.* P. L. Richards, High-Resolution Fourier Transform Spectroscopy in the Far-Infrared, *J. Opt. Soc. Amer.* **54**, 1474 (1964).

*30.* J. W. Cooley and J. W. Tukey, An Algorithm for the Machine Calculation of Complex Fourier Series, *Math Comput.* **19**, 297 (1965).

*31.* M. L. Forman, Fast Fourier Transform Technique and Its Application to Fourier Spectroscopy, *J. Opt. Soc. Amer.* **56**, 978 (1966).

*32.* E. O. Brigham and R. E. Morrow, The Fast Fourier Spectrum, *IEEE Trans. Audio Electroacoust.* **AU-15(2)**, 63 (1967).

*33.* J. Connes, Computing Problems in Fourier Spectroscopy, *Aspen Int. Conf. on Fourier Spectrosc., 1970*, p. 83. (G. A. Vanasse, A. T. Stair, Jr., and D. J. Baker, eds.). AFCRL-71-0019, 5 Jan. 1971, Spec. Rep. No. 114.

*34.* H. Yoshinaga, S. Fujita, S. Minami, Y. Suemoto, M. Inoue, K. Chiba, K. Nakano, S. Yoshida, and H. Sugimori, A Far Infrared Interferometric Spectrometer with a Special Electronic Computer, *Appl. Opt.* **5**, 1159 (1966).

*35.* A. Hadni, Far Infrared Transpose Spectroscopy, *Appl. Opt.* **4**, 1374 (1965).

CHAPTER THREE

# Fourier Analysis and Interferometry

## INTRODUCTION

This chapter begins with a very short review of the necessary definitions and restrictions of Fourier analysis. The mathematical review is followed by a detailed derivation of the integral equation which enables one to obtain the spectrum from the interferogram of a two-beam interferometer. The derivation highlights both the optical and mathematical assumptions. Since the derivation is long, a shorter and less rigorous treatment is presented later in the chapter. The final result, the basic integral equation of Fourier transform spectroscopy, enables one to transform an interferogram to a spectrum. For the experimentalist, a general explanation of the application of the integral is given. Finally, some comments about convolution and autocorrelation functions are made relative to Fourier transform spectroscopy (FTS).

Since Fourier analysis is essential to FTS, the statement of the Fourier transform integral theorem with the conditions of validity is made. There are many good references on Fourier analysis. One book which has often been used is *The Fourier Transform and Its Applications* by Bracewell [1]. A compact listing of theorems useful to Fourier transform spectroscopy is in *Radiation and Optics* by Stone [2].

## DERIVATION OF THE BASIC INTEGRAL FOR FOURIER TRANSFORM SPECTROSCOPY

The derivation of the basic integral equation for Fourier transform spectroscopy breaks into five parts (A–E). Parts A–D establish the definition of Fourier integrals and show that the electric field as a function of position is the Fourier transform of the electric field as a function of frequency. Therefore, the advanced reader can easily skip to Part E.

*Part A.* Fourier transform integrals:

Mathematically, a function $F(x)$ can be written

$$F(x) = \int_{-\infty}^{\infty} A(\xi)e^{-i2\pi\xi x}\,d\xi \equiv \mathscr{F}^{-1}\{A(\xi)\}; \tag{3-1}$$

if we know $F(x)$ and want $A(\xi)$, it is *uniquely* given by

$$A(\xi) = \int_{-\infty}^{\infty} F(x)e^{+i2\pi\xi x}\,dx \equiv \mathscr{F}\{F(x)\}, \tag{3-2}$$

where the second integral is called the Fourier transform and the first integral is called the inverse Fourier transform. The conditions of validity are:

(1) $F(x)$ is everywhere finite; discontinuities are finite in size and number.
(2) The derivatives of $F(x)$ and $A(\xi)$ must exist except at a finite number of points, and then, the left and right derivatives must exist.
(3) $\int_{-\infty}^{\infty} |F(x)|\,dx < \infty$, i.e., $F(x)$ is absolutely integrable.

This is a statement of the Fourier transform integral theorem.

*Part B* Establishment of the effect of the law of superposition on amplitudes (0 to $\infty$ limits for the integral):

An excellent reference to assist the reader in understanding Parts B–E of this derivation of the basic FTS integral is in Strong's book, "Concepts of Classical Optics" [3].

Consider a plane, monochromatic wave at a point $z$ and at time zero. The wave may be represented by

$$y(z, \sigma) = a(\sigma)\cos(kz) = a(\sigma)\cos(2\pi\sigma z), \tag{3-3}$$

where $k$ is the propagation constant and $\sigma$ is the wave number. If, at a point, one has many waves each of a different frequency and amplitude $a(\sigma)$, then, in a wave train, the resultant amplitude is

$$y(z) = (1/\bar{\sigma}) \int_0^\infty a(\sigma) \cos(2\pi\sigma z)\, d\sigma. \tag{3-4}$$

In Eq. (3-4), $\bar{\sigma}$ is some average wave number. Recall that an integral is a sum. Note that every function is real and visualize $a(\sigma)/\bar{\sigma}$ as an amplitude density.

*Part C.* Establishing that the amplitude limits can be $-\infty$ to $+\infty$, so that the Fourier transform integral theorem can be used:

Now, we want to replace $\cos(2\pi\sigma z)$ by $e^{i2\pi\sigma z}$ and extend the lower limit of the integral from 0 to $-\infty$, so we can make use of the Fourier integral theorem.

Consider the following separated integral:

$$\int_{-\infty}^\infty b(\sigma) e^{i2\pi\sigma z}\, d\sigma = \int_{-\infty}^0 b(\sigma) e^{i2\pi\sigma z}\, d\sigma + \int_0^\infty b(\sigma) e^{i2\pi\sigma z}\, d\sigma. \tag{3-5}$$

Note that, on reversing the limits of integration and using the conclusion of the footnote,† one has

$$\int_{-\infty}^0 b(\sigma) e^{i2\pi\sigma z}\, d\sigma = \int_0^\infty b^*(\sigma)(e^{i2\pi\sigma z})^*\, d\sigma. \tag{3-6}$$

† *Showing* $b^*(\sigma) = b(-\sigma)$: If $y(z) \equiv$ electric field at the point $z$, a *real* quantity, we assume that $y(z)$ can be expressed as

$$y(z) = (\text{real constant}) \int_{-\infty}^\infty b(\sigma) e^{i2\pi\sigma z}\, d\sigma,$$

with the restriction that $y(z)$ be *real*. By the Fourier transform integral theorem,

$$b(\sigma) = (\text{real constant})' \int_{-\infty}^\infty y(z) e^{-i2\pi\sigma z}\, dz.$$

Taking the complex conjugate of $b(\sigma)$ yields

$$b^*(\sigma) = (\text{real constant})' \int_{-\infty}^\infty y^*(z) e^{i2\pi\sigma z}\, dz.$$

Since $y(z)$ is *real*, one has

$$b^*(\sigma) = (\text{real constant})' \int_{-\infty}^\infty y(z) e^{i2\pi\sigma z}\, dz.$$

Also, changing signs in $b(\sigma)$ yields

$$b(-\sigma) = (\text{real constant})' \int_{-\infty}^\infty y(z) e^{i2\pi\sigma z}\, dz.$$

Thus, the integrals for $b^*(\sigma)$ and $b(-\sigma)$ are equal because $y(z)$ is *real*. With this property, $b(\sigma)$ is said to Hermitian. The real part of $b(\sigma)$ is an even function of $\sigma$, and the imaginary part is an odd function of $\sigma$.

Substituting Eq. (3-6) in Eq. (3-5) and noting that a complex product equals the complex conjugate of a product, one has

$$\int_{-\infty}^{\infty} b(\sigma)d^{i2\pi\sigma z}\, d\sigma = \int_{0}^{\infty} [b(\sigma)e^{i2\pi\sigma z} + \{b(\sigma)e^{i2\pi\sigma z}\}^{*}]d\sigma$$

$$= 2\int_{0}^{\infty} \mathrm{Re}[b(\sigma)e^{i2\pi\sigma z}]\, d\sigma, \tag{3-7}$$

where in Eq. (3-7) use has been made of

$$(A + iB) + (A + iB)^{*} = 2A = 2\,\mathrm{Re}(A + iB). \tag{3-8}$$

If $b(\sigma)$ is real (it is then also an even function), one has only the even part of the right integral in Eq. (3-7), i.e., the cosine or real part. Then, Eq. (3-7) can be rearranged as

$$\int_{0}^{\infty} b(\sigma)\cos(2\pi\sigma z)\, d\sigma = \tfrac{1}{2}\int_{-\infty}^{\infty} b(\sigma)e^{i2\pi\sigma z}\, d\sigma. \tag{3-9}$$

*Part D.* Combining Parts A–C to show that one can write the amplitude ($z$-dependent) in terms of the electric field ($\sigma$-dependent) with integral limits of $-\infty$ and $+\infty$:

Let $b(\sigma) = a(\sigma)/\bar{\sigma}$, where $b(\sigma)$ was introduced in Part C and $a(\sigma)/\bar{\sigma}$ was introduced in Part B. Using Eq. (3-9), the resultant amplitude in Eq. (3-4) may be expressed as an integral from $-\infty$ to $+\infty$ with the cosine function replaced by $e^{i2\pi\sigma z}$. The result is

$$y(z) = \tfrac{1}{2}\int_{-\infty}^{\infty} b(\sigma)e^{i2\pi\sigma z}\, d\sigma. \tag{3-10}$$

Equation (3-10), or (3-11) which follows, results from the field being real, and the equations are often just assumed. Note that $y(z)$ can be used to represent a wave composed of many frequencies of different amplitudes. Define $\tfrac{1}{2}b(\sigma) \equiv \mathscr{E}(\sigma) \equiv$ electric field amplitude of wave number $\sigma$. Recall that $b(\sigma)$ is an arbitrary real and symmetric function; so, we may set $2\pi\sigma = \xi$ in Eqs. (3-1) and (3-2) for the Fourier transform integral theorem. Thus, the electric field as a function of position is the Fourier transform of the electric field given as a function of frequency.

$$\boxed{y(z) = \int_{-\infty}^{\infty} \mathscr{E}(\sigma)e^{i2\pi\sigma z}\, d\sigma \equiv \mathscr{F}\{\mathscr{E}(\sigma)\}.} \tag{3-11}$$

Note that one can set $\mathscr{E}(\sigma) = A(\xi)$, and a Fourier transform can be used, yielding

$$\mathscr{E}(\sigma) = \int_{-\infty}^{\infty} y(z)e^{-i2\pi\sigma z}\,dz. \tag{3-12}$$

Now, one is ready to apply Fourier analysis to interferometers.

*Part E.* Combining waves from the two arms of the interferometer, finding the resultant electric field, computing the flux density, and manipulating the equations to find the flux versus wave number, i.e., the spectrum:

The amplitudes of two coherent waves which at time zero have the same amplitude $\mathscr{E}(\sigma)$ at wave number $\sigma$, but which are separated by a phase difference $k\delta = 2\pi\sigma\delta$, can be written as [using Eq. (3-11)]

$$y_1(z) = \int_{-\infty}^{\infty} \mathscr{E}(\sigma)e^{i2\pi\sigma z}\,d\sigma \tag{3-13}$$

and

$$y_2(z) = \int_{-\infty}^{\infty} \mathscr{E}(\sigma)e^{i2\pi\sigma(z-\delta)}\,d\sigma. \tag{3-14}$$

In Eq. (3-14), $\delta$ is the *optical path difference between the two waves.* Equations (3-13) and (3-14) represent the waves from each arm of the interferometer after they have been recombined. Using the law of superposition, one has

$$y(z) = y_1(z) + y_2(z) = \int_{-\infty}^{\infty} [\mathscr{E}(\sigma)(1 + e^{-i2\pi\sigma\delta})e^{i2\pi\sigma z}]\,d\sigma. \tag{3-15}$$

Define the resultant field $\mathscr{E}_R(\delta, \sigma)$ as

$$y(z) \equiv \int_{-\infty}^{\infty} \mathscr{E}_R(\delta, \sigma)e^{i2\pi\sigma z}\,d\sigma. \tag{3-16}$$

So, by the uniqueness of Fourier transforms, we have, on comparing Eqs. (3-15) and (3-16),

$$\mathscr{E}_R(\delta, \sigma) = \mathscr{E}(\sigma)(1 + e^{-i2\pi\sigma\delta}). \tag{3-17}$$

But the intensity $B(\sigma, \delta)$ (irradiance or flux density) is

$$B_R(\sigma, \delta) \equiv \tfrac{1}{2}cK_0\,\mathscr{E}_R{}^*(\delta, \sigma)\mathscr{E}_R(\delta, \sigma) \tag{3-18}$$

as stated by Strong [3, p. 47]. In Eq. (3-18), $K_0$ is the electric permittivity of free space in MKS units ($8.85 \times 10^{-12}$ MKS) and $c$ is the velocity of light in vacuum.

Substituting $\mathscr{E}_R(\delta, \sigma)$ from Eq. (3-17) into the expression for $B(\sigma, \delta)$, multiplying the terms, and using the definition of the cosine, one has

$$B(\sigma, \delta) = cK_0 \mathscr{E}^2(\sigma)[1 + \cos(2\pi\sigma\delta)]. \tag{3-19}$$

There is no phase relation between the fluxes of different wave numbers at a given path difference $\delta$ from a white light source. So, we add the fluxes of different frequencies and define $I_R(\delta)$ as

$$I_R(\delta) \equiv (1/\bar{\sigma}) \int_0^\infty B(\sigma, \delta)\, d\sigma. \tag{3-20}$$

Substituting $B(\sigma, \delta)$ from Eq. (13-19) in (3-20), $I_R(\delta)$ becomes

$$I_R(\delta) = (cK_0/\bar{\sigma}) \left\{ \int_0^\infty \mathscr{E}^2(\sigma)\, d\sigma + \int_0^\infty \mathscr{E}^2(\sigma) \cos(2\pi\sigma\delta)\, d\sigma \right\}. \tag{3-21}$$

Note from Eq. (3-21) that the value of $I_R(\delta)$ for $\delta = 0$ is given by†

$$I_R(0)/2 = (cK_0/\bar{\sigma}) \int_0^\infty \mathscr{E}^2(\sigma)\, d\sigma. \tag{3-22}$$

Substituting Eq. (3-22) in Eq. (3-21), one has

$$[I_R(\delta) - \tfrac{1}{2}I_R(0)] = (2/\pi)^{1/2} \int_0^\infty [(cK_0/\bar{\sigma})\mathscr{E}^2(\sigma)(\pi/2)^{1/2}] \cos(2\pi\sigma\delta)\, d\sigma. \tag{3-23}$$

Writing the Fourier transform of this integral using Eqs. (3-1) and (3-2), we have

$$[(cK_0/\bar{\sigma})(\pi/2)^{1/2}\mathscr{E}^2(\sigma)] = (2/\pi)^{1/2} \int_0^\infty [I_R(\delta) - \tfrac{1}{2}I_R(0)] \cos(2\pi\sigma\delta)\, d\delta. \tag{3-24}$$

Using the definition expressed in Eq. (3-18), the final result is obtained:

$$\boxed{B(\sigma) = (\text{constant}) \int_0^\infty \{[I_R(\delta) - \tfrac{1}{2}I_R(0)] \cos(2\pi\sigma\delta)\}\, d\delta.} \tag{3-25}$$

Equation (3-25), at a given wave number $\sigma$, states that if the flux versus optical path $I_R(\delta)$ is known as a function of $\delta$, the Fourier cosine transform of $[I_R(\delta) - \tfrac{1}{2}I_R(0)]$ yields $B(\sigma)$, the flux density at the wave number $\sigma$. In order to obtain the spectrum, it is only necessary to repeat the calculation of the

† In Eq. (3-21), one may note, for $\delta$ approaching infinity, that the $\cos(2\pi\sigma\text{-}\delta)$ term oscillates very rapidly, and, therefore, averages to zero. Thus, one would have

$$I_R(\infty) = (cK_0/\bar{\sigma}) \int_0^\infty \mathscr{E}^2(\sigma)\, d\sigma = I_R(0)/2.$$

Fourier transform using Eq. (3-25) for each wave number in the range of interest. Equation (3-25) is the basic equation of all of Fourier transform spectroscopy! Only two assumptions were needed in its derivation, namely (1) the law of superposition [in Eq. (3-15)], and (2) the statement that the electric field is a real quantity [in Eq. (3-6)]. The rest of the derivation of Eq. (3-25) relied on the Fourier transform integral theorem and mathematical manipulations.

The complex Fourier transform of $[I_R(\delta) = \frac{1}{2}I_R(0)]$ will be derived in Chapter 8 on asymmetric interferometers and amplitude spectroscopy. It will be shown in Chapter 8 that the general case of unequal reflectivities of the mirrors in the two arms requires, not the cosine Fourier transform, but the complex Fourier transform for the computation of the spectrum. The results of the complex transform of the interferogram is not just $B(\sigma)$, but is $[r(\sigma)B(r)]$, where $r(\sigma)$ is the *complex* reflection coefficient coefficient of one mirror relative to the other. For single-beam, conventional spectroscopy with the sample not in one arm but in the combined beams, the factor $r$ would cancel in taking ratios for sample-in to sample-out spectra. Thus, for conventional spectroscopy, "*symmetric interferometry*," either the Fourier cosine or complex transforms may be used.

As discussed in Chapter 8, for the general case, one has

$$B(\sigma) = \int_{-\infty}^{\infty} [I_R(\delta) - \tfrac{1}{2}I_R(0)]e^{-i2\pi\sigma\delta}\, d\delta, \tag{3-26}$$

where the proportionality constants have been dropped. If the interferometer arms are optically unequal in the same way for all wave numbers, one has $r^*(\sigma) = r_0(\sigma)e^{-i\varphi(\sigma)}$ equal to some complex constant at each wave number in Eq. (8-15). Thus, the parameters $r_0(\sigma)$ and $\varphi_r(\sigma)$ were eliminated in Eq. (3-26) for single-beam spectroscopy with the sample in the combined beams. For more details, the reader is referred to Chapter 8.

## SHORT DERIVATION OF THE BASIC INTEGRAL FOR FOURIER TRANSFORM SPECTROSCOPY

As a shorter, alternate derivation of the spectral equation, consider the following procedure. In this alternate derivation, proper units will be assumed to be provided by omitted normalizing factors. Let the incident amplitude on the beamsplitter be

$$\mathscr{E}(z, \sigma)\, d\sigma = \mathscr{E}_0(\sigma)e^{i(\omega t - 2\pi z\sigma)}\, d\sigma. \tag{3-27}$$

In a Michelson interferometer after amplitude splitting, there are two beams which have traveled distances $z_1$ and $z_2$ as they recombine. Each beam has undergone one reflection from and one transmission through the beamsplitter. If $r$ is the reflection coefficient and $t$ is the transmission coefficient of the beamsplitter, the recombined amplitude is

$$\mathscr{E}_R(z_1, z_2, \sigma)\, d\sigma = rt\mathscr{E}_0(\sigma)[e^{i(\omega t - 2\pi\sigma z_1)} + e^{i(\omega t - 2\pi\sigma z_2)}]\, d\sigma. \quad (3\text{-}28)$$

The flux for the fixed spectral range $d\sigma$ after recombination of the beams can be defined as (within multiplying constants)

$$\begin{aligned} I(z_1, z_2, \sigma)\, d\sigma &= \mathscr{E}_R(z_1, z_2, \sigma)\mathscr{E}_R{}^*(z_1, z_2, \sigma)\, d\sigma \\ &= 2\mathscr{E}_0{}^2(\sigma)|rt|^2\{1 + \cos[2\pi(z_1 - z_2)\sigma]\}\, d\sigma, \end{aligned} \quad (3\text{-}29)$$

So, the total flux at any path difference $\delta = (z_1 - z_2)$ is

$$\begin{aligned} I_R(\delta) &= \int_0^\infty I(\delta, \sigma)\, d\sigma \\ &= 2|rt|^2 \int_0^\infty \mathscr{E}_0{}^2(\sigma)\, d\sigma + \int_0^\infty \mathscr{E}_0{}^2(\sigma) \cos(2\pi\sigma\delta)\, d\sigma]. \end{aligned} \quad (3\text{-}30)$$

Observing the value of $I_R(\delta)$ for $\delta = 0$, from Eq. (3-30), the interferogram is found to be

$$[I_R(\delta) - \tfrac{1}{2}I_R(0)] = 2|rt|^2 \int_0^\infty \mathscr{E}_0{}^2(\sigma) \cos(2\pi\delta\sigma)\, d\sigma, \quad (3\text{-}31)$$

and the spectrum from the Fourier cosine transform is

$$B(\sigma) \propto \mathscr{E}_0{}^2(\sigma) = (1/\pi|rt|^2) \int_0^\infty [I_R(\delta) - \tfrac{1}{2}I_R(0)] \cos(2\pi\sigma\delta)\, d\delta. \quad (3\text{-}32)$$

This is the same equation for the spectrum as found in Eq. (3-25). This second derivation is much shorter that the first and avoids the detailed exposure to the basic assumptions. The longer derivation is very useful for later derivations such as for the spectrum of an asymmetric interferometer.

## COMPUTING SPECTRA

The general method of obtaining a spectrum using an interferometer is as follows:

1. Measure $I_R(\delta)$ by recording the signal versus the interferometer arm displacement noting that $\delta = 2$ (arm displacement). (The arm is usually moved in discrete steps—see Chapter 12 on phase error and sampling problems.)

2. At zero path difference, experimentally determine $I_R(0)$. (See the discussion that follows.)
3. Substitute $[I_R(\delta) - \frac{1}{2}I_R(0)]$ in Eq. (3-25) and evaluate the integral for a chosen $\sigma$. The integration is performed on a computer. (See the discussion that follows.)
4. Perform the integration of Eq. (3-25) for each chosen $\sigma$.
5. As a result of steps 1–4, one has, within a multiplicitive constant, computed $B(\sigma)$ versus $\sigma$, i.e., the spectrum.

It is usual to call $[I_R(\delta) - \frac{1}{2}I_R(0)]$ the *interferogram*. That is, the oscillations of the signal about the value $[\frac{1}{2}I_R(0)]$ are referred to as the interferogram. However, $I_R(\delta)$ is sometimes called the interferogram.

## COHERENCE IN THE INTERFEROMETER

At zero arm displacement, the waves for all frequencies will interact coherently. At infinite arm displacement, the waves from the two arms after recombination are separated by more than a wavetrain-length and, therefore, will interact incoherently. So, $I_R(0)$ is the flux associated with coherent interference. Designate the flux for infinite arm displacement as $I_R(\infty)$, noting that $I_R(\infty)$ is associated with incoherent interference.

It is well known that the flux, after coherent interference of $s$ equal waves of electric field $E$, is proportional to $s^2E^2$. The flux after incoherent interference is proportional to $sE^2$. Therefore, from the above considerations, one has

$$I_R(\infty)/I_R(0) = 1/s. \tag{3-33}$$

At any given wave number $\sigma$, there are two waves, one from each arm, so that $s = 2$ for the two-beam interferometer. Thus, one may write the interferogram as

$$[I_R(\delta) - \tfrac{1}{2}I_R(0)] = [I_R(\delta) - I_R(\infty)]. \tag{3-34}$$

This equation shows that the interferogram is the fluctuation of the signal about the incoherent signal level found at large arm displacements.

If the recorded signal versus arm displacement is in error by an additive constant $\Delta I$, for all $\delta$, one has

$$[I_R(\delta) - I_R(\infty)] = \{[I_R(\delta) + \Delta I] - [I_R(\infty) + \Delta I]\}. \tag{3-35}$$

If the interferometer is not perfectly aligned, there will be a constant flux $\Delta I$ reaching the detector that is independent of the arm position. Equation

(3-35) indicates that the true interferogram may than be redefined as the magnitude of the signal fluctuations $I_R(\delta) + \Delta I$ about the large arm displacement asymptote $I_R(\infty) + \Delta I$. Thus, Eq. (3-35) is of great importance to the experimentalist and affects steps 2 and 3 listed before.

## APPLICABILITY OF THE BASIC INTEGRAL EQUATION OF FOURIER TRANSFORM SPECTROSCOPY

For experimental purposes, the mathematics and physics involved in deriving Eq. (3-25) do not depend on how the phase $2\pi\sigma\delta$ was obtained. As long as there are two beams interfering and there is a path difference $\delta$, Eq. (3-25) applies. Therefore, Eq. (3-25) works equally well for the lamellar grating interferometers (Chapter 15), which divide a wavefront, and for the Michelson interferometer, which splits amplitudes. Both instruments introduce optical path differences between two beams in a controlled way.

Equation (3-25) applies to essentially the "single-beam" operation in spectroscopy. Single-beam spectroscopy refers to determining sample transmittances (reflectances) by measuring the signal with the sample in the beam and then again with the sample out of the beam. The transmittance is the ratio of the sample-in signal to sample-out signal. For Fourier transform spectroscopy, the sample is usually in the beam either before or after the beamsplitter. There is double-beam Fourier transform spectroscopy, as mentioned in Chapters 1, 2, and 15, but the technique is, as yet, rather difficult. Asymmetric spectroscopy (Chapter 8) dictates that the sample be in the optical beam in one of the arms of the interferometer. Equation (3-25) does not apply to asymmetric interferometry.

## PROVING THAT THE INTERFEROGRAM IS THE AUTOCORRELATION FUNCTION OF THE ELECTRIC FIELD

We are ready to provide more insight into the fundamental integral equation of Fourier transform spectroscopy. In order to do this, we must introduce the concepts of convolution and autocorrelation. In this chapter, we want to make a connection between autocorrelation functions, interferogram, and spectra. In later chapters, we shall expand the concepts in more detail to provide links between the optical and mathematical approaches to FTS.

First, consider the concept of the convolution, which is independent of Fourier analysis. The convolution of two functions $f(z)$ and $g(z)$, written $f(z) \star g(z)$, is defined as

$$f(z) \star g(z) \equiv \int_{-\infty}^{\infty} f^*(u)g(z+u)\,du, \tag{3-36}$$

where the * on $f(u)$ implies complex conjugation. If $f(z)$ and $g(z)$ are the same function, one may define the self-convolution of a function $f(x)$ as the autocorrelation function, written $f(z) \star f(z)$, with the equation

$$f(z) \star f(z) = \int_{-\infty}^{\infty} f^*(u)f(z+u)\,du. \tag{3-37}$$

What we want to show first is that the inverse Fourier transform of the autocorrelation function of the electric field amplitude as a function of position is the spectrum, i.e., flux density as a function of frequency or wave number or $B(\sigma)$. This statement is quite general and is called the Wiener–Khinchine theorem [1]. Define the autocorrelation function ACF as

$$\text{ACF} \equiv E(\delta) \star E(\delta) \equiv \int_{-\infty}^{\infty} E^*(z)E(z+\delta)\,dz, \tag{3-38}$$

where $E(z)$ is the electric field (amplitude) as a function of position $z$. Taking the inverse Fourier transform of the ACF, $\mathscr{F}^{-1}\{\text{ACF}\}$, we have

$$\mathscr{F}^{-1}\{\text{ACF}\} = \int_{-\infty}^{\infty}\left[\int_{-\infty}^{\infty} E^*(z)E(z+\delta)\,dz\right]e^{-i2\pi\sigma\delta}\,d\delta. \tag{3-39}$$

Rearranging terms, one has

$$\mathscr{F}^{-1}\{\text{ACF}\} = \int_{-\infty}^{\infty} E^*(z)\left[\int_{-\infty}^{\infty} E(z+\delta)e^{-i2\pi\sigma\delta}\,d\delta\right]dz. \tag{3-40}$$

The result of integrating over $\delta$, when the shift rule (Stone [2], Chapter 11, Theorem 2a) is applied, is the function $e^{i2\pi\sigma z}\,\mathscr{F}^{-1}\{\mathscr{E}(\delta)\}$. But from Eq. (3-11) we have $\mathscr{F}^{-1}\{E(\delta)\} = E(\sigma)$, which is true in general. So, the $\mathscr{F}^{-1}\{\text{ACF}\}$ simplifies to

$$\mathscr{F}^{-1}\{\text{ACF}\} = E(\sigma)\left[\int_{-\infty}^{\infty} E(z)e^{-i2\pi\sigma z}\,dz\right]^* = E(\sigma)E^*(\sigma). \tag{3-41}$$

Reapplying Eq. (3-11) and recalling that the flux density $B(\sigma)$ is given by $E(\sigma)E^*(\sigma)$, we have

$$\mathscr{F}^{-1}\{\text{ACF}\} = B(\sigma). \tag{3-42}$$

Thus, we have shown that the inverse Fourier transform of the autocorrelation function is the spectrum. Recall that Eq. (3-26) is equivalent to saying that the inverse Fourier transform of an interferogram is also the spectrum. Therefore, the interferogram is an autocorrelation function; from Eq. (3-38), we have that it is an autocorrelation function of the electric field (amplitude) as a function of position.

## CONCLUSIONS

In this chapter, we have defined Fourier transform integrals, have shown that the inverse Fourier transform of the interferogram is the spectrum (by two different methods), and have proved that the interferogram is an autocorrelation function. At all stages of the development, all terms and operations were explicitly defined. Also, we have outlined how the basic integral of FTS, Eq. (3-25), can be applied to obtain spectra experimentally. The procedure for obtaining the spectra, as modified by closer considerations, will be expanded upon in subsequent chapters. Nonetheless, Eq. (3-25) will stand as the basic equation, with minor conceptual changes.

### *References*

*1.* R. Bracewell, "The Fourier Transform and Its Applications." McGraw-Hill, New York, 1965.

*2.* J. M. Stone, "Radiation and Optics," Chapter 11. McGraw-Hill, New York, 1963.

*3.* J. D. Strong, "Concepts of Classical Optics." Freeman, San Francisco, California, 1958. Appendix F coauthored with G. A. Vanasse.

CHAPTER FOUR

# *Sample Calculations of Spectra from Interferograms*

## INTRODUCTION

In this chapter, the basic integral of Fourier transform spectroscopy will be applied to analytic interferograms to obtain the spectra. Sometimes, the interferogram curves can be accurately fit with empirical, analytic functions. Taking analytically known interferograms, two examples of the use of Eq. (3-25) are given. One example is rather academic in that the spectrum turns out to be an exponentially decaying function with increasing frequency. The other example is applicable to the spectrum of an emission doublet such as the sodium D lines. The sodium doublet serves as the first example of interpreting spectral features directly from the interferogram without using a computer. For both examples, which are developed in detail, plots of typical interferograms and their spectra are shown.

## ACADEMIC EXAMPLE OF THE USE OF EQ. (3-25)

Assume that a Michelson interferometer yielded the interferogram $[I_R(\delta) - \frac{1}{2}I_R(0)]$ from a mercury lamp in the range $10\ \text{cm}^{-1} < \sigma < 100\ \text{cm}^{-1}$. Suppose the interferogram was $\{1/[1 + (\delta/\delta_0)^2]\}$, where $\delta$ was the optical path difference of the beams in the two arms. What is the spectrum?

Recall the basic Fourier transform spectroscopy integral, Eq. (3-25), namely

$$B(\sigma) = (\text{constant}) \int_0^\infty [I_R(\delta) - \tfrac{1}{2} I_R(0)] \cos(2\pi\sigma\delta)\, d\delta. \tag{4-1}$$

If one substitutes the interferogram in Eq. (4-1), he has

$$B(\sigma) = (\text{constant})\delta_0 \int_0^\infty \left\{ \left[ \frac{1}{1 + (\delta/\delta_0)^2} \right] \cos\left( \delta_0 2\pi\sigma \frac{\delta}{\delta_0} \right) d\left( \frac{\delta}{\delta_0} \right) \right. . \tag{4-2}$$

Upon integrating Eq. (4-2), the spectrum is obtained:

$$B(\sigma) = (\text{constant})' \, e^{-2\pi\sigma\delta_0}. \tag{4-3}$$

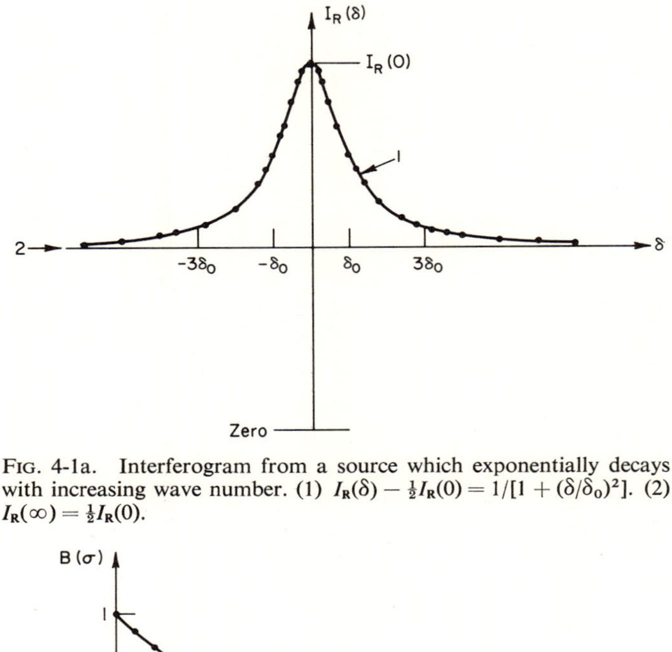


FIG. 4-1a. Interferogram from a source which exponentially decays with increasing wave number. (1) $I_R(\delta) - \frac{1}{2}I_R(0) = 1/[1 + (\delta/\delta_0)^2]$. (2) $I_R(\infty) = \frac{1}{2}I_R(0)$.

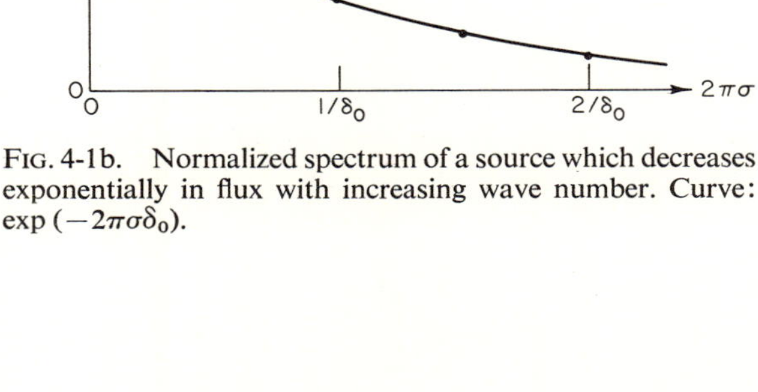


FIG. 4-1b. Normalized spectrum of a source which decreases exponentially in flux with increasing wave number. Curve: $\exp(-2\pi\sigma\delta_0)$.

The interferogram is shown in Fig. 4-1a and the spectrum is exhibited in Fig. 4-1b. The interferogram is a Lorenztian profile about the origin $\delta = 0$ and the spectrum is an exponentially decaying function with increasing wave number. Such a spectrum has not been observed, but it is simple enough to serve as an example calculation.

## PRACTICAL EXAMPLE OF THE USE OF EQ. (3-25) THE DOUBLET PROBLEM

Suppose an interferogram from a Michelson interferometer is given by

$$[I_R(\delta) - \tfrac{1}{2}I_R(0)] = [\exp - (\delta/a)^2] \cos(2\pi b\delta) \cos(2\pi g\delta), \qquad (4\text{-}4)$$

where $b \gg g$, $a^2g \gg 1$. The constants $a$, $b$, and $g$ are real, positive numbers.

We want to find the spectrum within a multiplying constant. If this example is to be an exercise for the reader, there are three general comments to be made. First, a very broad outline of the procedure is stated below in four steps. This outline is followed by a presentation of the major equations with a description of the conclusions. The four steps are:

1. Consider no approximations until the spectrum is plotted that produced the interferogram.
2. Successive use of the identity $(2\cos A \cos B) = [\cos(A + B) + \cos(A - B)]$ will eventually lead to a sum of four integrals replacing the original single integral in Eq. (3-25).
3. Consider that all integrals are of the form

$$\int_0^\infty [\exp(-q^2x^2)] \cos(px)\, dx = (\pi^{1/2}/2q) \exp(-p^2/4q^2) \qquad (4\text{-}5)$$

4. Using the approximations, one can obtain the amplitudes, half-widths, and separation of the peaks in terms of $a$, $b$, and $g$. One should leave the functions in exponential form, and note that all of the exponential arguments are negative.

The first two steps result in the following four terms in the integral, and the spectrum becomes

$$B(\sigma) = (\text{constant}) \int_0^\infty [\exp - (\delta/a)^2] \{\cos[(b + g + \sigma)2\pi\delta]$$
$$+ \cos[(b + g + \sigma)2\pi\delta] + \cos[(b + g - \sigma)2\pi\delta]$$
$$+ \cos[(b - g - \sigma)2\pi\delta]\} d\delta, \qquad (4\text{-}6)$$

after the trigonometric identity has been used twice. Using the suggested integral in step 3, one has

$$B(\sigma) = (\text{constant})\{\exp[-(\pi a)^2(b+g+\sigma)^2] + \exp[-(\pi a)^2(b-g+\sigma)^2] + \exp[-(\pi a)^2(b+g-\sigma)^2] + \exp[-(\pi a)^2(b-g-\sigma)^2]\}. \tag{4-7}$$

The exponentials in Eq. (4-7) are Gaussian functions $G(x)$, which are of the general form

$$G(x) \equiv \{\exp -\tfrac{1}{2}[(x-\mu)/\sigma']^2\}/\{\sigma'(2\pi)^{1/2}\} \tag{4-8}$$

where $\mu$ is the mean value of $x$ and $\sigma'$ is the standard deviation of $x$, or $2\sigma'$ is the rms half-width of the resonance, i.e., $2\sigma'$ is slightly less than the full-width at one-half the peak height. By comparing Eqs. (4-8) and (4-7), keeping in mind that all the constants are real and positive, and noting from the initial conditions that $a^2b \gg a^2g \gg 1$, one has that only the last two terms in Eq. (4-7) are appreciably greater than zero for $\sigma$ greater than zero. We ignore the negative $\sigma$ solutions as physically unreal. Thus, there are two nearly *equal* Gaussian resonances of peak amplitude equal to approximately unity. For $a^2b \gg 1$ and positive $\sigma$, the first two terms in Eq. (4-7) never approach unit magnitude and may be ignored. The two resonances are centered at $\sigma = b \pm g$, are separated by $2g$, and have standard deviations of $\sigma' = \sqrt{2}/2\pi a$. Converting the standard deviation into the half-width using Eq. (4-8) results in the half-width given by $[2\sigma'(2 \ln 2)^{1/2}]$.

The interferogram and the spectrum are plotted in Figs. 4-2a and 4-2b. In Fig. 4-2a for the interferogram, the various constituent functions of the interferogram function are plotted separately. The periodicity of the low-

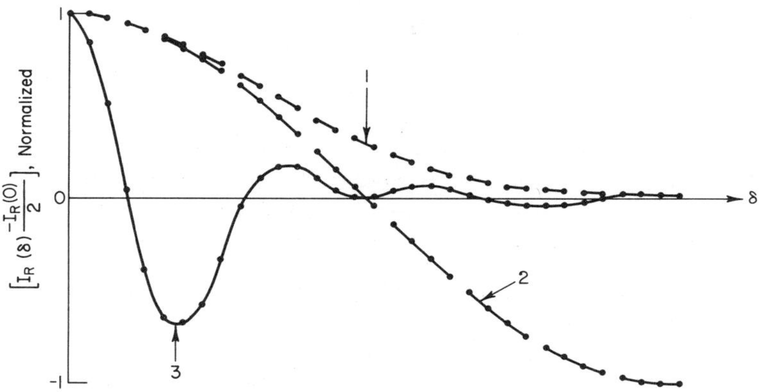


FIG. 4-2a. Interferogram produced by two equal Gaussian spectral resonances. The low-frequency cosine portion [curve 2: $\cos(2\pi g\delta)$] and the damping term [curve 1: $\exp -(\delta/a)^2$] are shown separately. Curve 3: $[\exp -(\delta/a)^2] \cos(2\pi g\delta) \cos(2\pi b\delta)$.

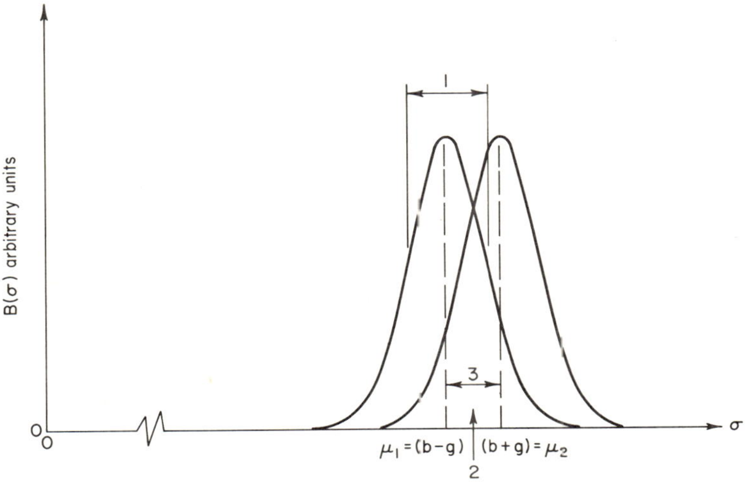


FIG. 4-2b. Two equal Gaussian spectral resonances with (1) standard deviation wave number $2\sigma' = \sqrt{2}/\pi a$; (2) mean wavenumber $\bar{\sigma} = 1/\lambda_{\text{mean}} = b$; and (3) peak separation wavenumber $\Delta\sigma = 2g = \Delta\lambda/\lambda^2$.

frequency $\cos(2\pi g\delta)$ envelope yields $g$ and, hence, the doublet separation frequency. The high-frequency envelope is determined by $b$ and gives the mid-range frequency between the resonances. The exponential decay function yields $a$, which enables one to compute the half-widths. Thus, if one had recorded the interferogram from an emission doublet over sufficiently large optical path differences, the main features of the Gaussian profiles could be calculated from only *inspection*. This is the first example of "reading" an interferogram, and other cases are presented in Chapter 15. For the sodium doublet, the theory outlined in this second example has been applied in a teaching laboratory to the interferogram to obtain the spectrum directly. A simple instructional laboratory Michelson interferometer was used.

## CONCLUSIONS

Examples of Fourier optics were presented in this chapter. The first example was essentially an academic exercise illustrating the use of Eq. (3-25). It was assumed that the recording of signal versus optical path displacement had been empirically fit to obtain an analytic function of the interferogram. Usually, no analytic function of the interferogram is known, and a numerical integration of Eq. (3-25) is performed on the computer.

The second example exhibits what can be done with an emission doublet and, most importantly, illustrates how an interferogram can be quickly interpreted without a computer. If the experimentalist is repeatedly studying spectra which are similar, he would like to have a cheap, easy, and quick way of judging the quality of the main spectral features directly from the interferogram. From example two, he can see the general procedure. The general analytic form of the interferogram is obtained. This may be done by fitting the interferogram *or* by knowing the analytic function of the spectrum and taking its Fourier transform to obtain the interferogram by using Eq. (3-24). Thus, by using either Eq. (3-24) or (3-25), both the interferogram and spectrum can be known analytically. Intercepts, envelopes, relative magnitudes, etc. in the interferogram can then be used to predict special spectral features such as resonance frequencies, half-widths, frequency separation, etc.

## *References*

*1.* J. D. Strong, "Concepts of Classical Optics," p. 419. Freeman, San Francisco, 1958.
*2.* J. M. Stone, "Radiation and Optics." McGraw-Hill, New York, 1963.

CHAPTER FIVE

# Apodization—Mathematical Filtering

## INTRODUCTION

In this chapter, the subject of mathematically treating the interferogram in order to modify the spectrum will be discussed. Since the basic Fourier transform integrals, Eq. (3-25) or Eq. (3-26), have *infinite* limits for the optical path difference and the experimental limits are *finite*, modifications to the computational theory must be made to use Eq. (3-25) or Eq. (3-26). The corrective procedure for modifying the basic Fourier transform integral is called "apodization." The interferogram is usually multiplied by a function, the apodizing function, which removes false sidelobes introduced into transformed spectra because of the *finite* optical path displacements. The false sidelobes look like feet in the spectrum; so, the word apodization is appropriate (apod or apodal means having no feet). The main apodization function used for illustration is the triangular function, but other functions which are useful are mentioned. The required properties of the apodization function are listed. Apodization degrades the resolution that can be realized in a spectrum, and this degradation is discussed.

Another topic of interest concerning mathematical filtering is the convolution of ideal spectra with instrumental line shapes. It will be shown that such a convolution yields the instrumental spectra. That is, the effect of the instrument is to broaden the ideal spectrum and give the actual experimental spectrum.

Though apodization and mathematical filtering are treated in detail in his chapter, it should be noted that not everyone apodizes his data. Sometimes, the spectral features are so distinct that sidelobes are unimportant, and, as will be seen, apodization broadens lines, which makes it more difficult to observe sharp spectral features. If real-time analysis is being performed, it is not always easy to program the apodizing function and operations into the small computers. Nonetheless, mathematic data processing is extremely useful, instructive, and easy under almost all but the above-mentioned conditions.

There are three good references concerning the subject of apodization. They are: *Infra Red Instrumentation and Techniques* by Martin [1]; "Computer Problems in Fourier Spectroscopy" by Connes [2]; and "Interferometry" by Steel [3].

## INTERFEROGRAM PRODUCED BY A MONOCHROMATIC SOURCE

To begin the discussion of apodization, one should first consider a monochromatic source and ask, "What kind of interferogram does it produce?" The question may then be inverted and one may ask, "What spectrum from a monochromatic source is produced by the Fourier transform of an interferogram which has a finite maximum optical displacement $L$?" Taking the questions in order, recall that the spectrum in Eq. (3-26) may be written as

$$B(\sigma) = \int_{-\infty}^{\infty} [I_R(\delta) - \tfrac{1}{2}I_R(0)]e^{-i2\pi\sigma\delta}\, d\sigma, \tag{5-1}$$

with its Fourier transform as

$$[I_R(\delta) - \tfrac{1}{2}I_R(0)] = \int_{-\infty}^{\infty} B(\sigma)e^{i2\pi\sigma\delta}\, d\sigma, \tag{5-2}$$

where $\sigma$ is the wave number and $\delta$ is the optical path difference. Equations (5-1) and (5-2) were discussed in Chapter 3 and are derived in Chapter 8 on asymmetric interferometers and amplitude spectroscopy. It is necessary to change from the cosine Fourier transform to the complex Fourier transform in order to describe the most general optical situation. Sometimes, the optical systems in the two arms of the interferometer are unbalanced. That is, the reflectivity of the fixed mirror might be different from the reflectivity of the movable mirror, etc. Then, from the discussion in Chapter 8, it can be seen that the complex Fourier transform is required.

What if the source is monochromatic of frequency $\sigma_1$, and $B(\sigma)$ is equal to $\frac{1}{2}[\delta'(\sigma - \sigma_1) + \delta'(\sigma + \sigma_1)]$, where $\delta'(\sigma - \sigma_1)$ is the Dirac delta function?† The

† If $f(x)$ is continuous, the Dirac delta function is defined by the integral

$$\int_{-\infty}^{\infty} \delta'(x - a)f(x)\, dx = f(a)$$

two delta functions are used to assure that energy is conserved in going from the interferogram to the spectrum. On substitution of $B(\sigma)$ in Eq. (5-2), one has

$$[I_R(\delta) - \tfrac{1}{2}I_R(0)] = \tfrac{1}{2}\int_{-\infty}^{\infty} [\delta'(\sigma - \sigma_1) + \delta'(\sigma + \sigma_1)]e^{+i2\pi\sigma\delta}\, d\delta. \tag{5-3}$$

Using the definition of the Dirac delta function as outlined in the footnote just cited, we have for the monochromatic source that

$$[I_R(\delta) - \tfrac{1}{2}I_R(0)] = 2\cos(2\pi\sigma_1\delta). \tag{5-4}$$

Figure 5-1a shows the $\cos(2\pi\sigma_1\delta)$ interferogram and Fig. 5-1b the Dirac delta-function spectrum. A monochromatic line produces a cosine interferogram and vice versa.

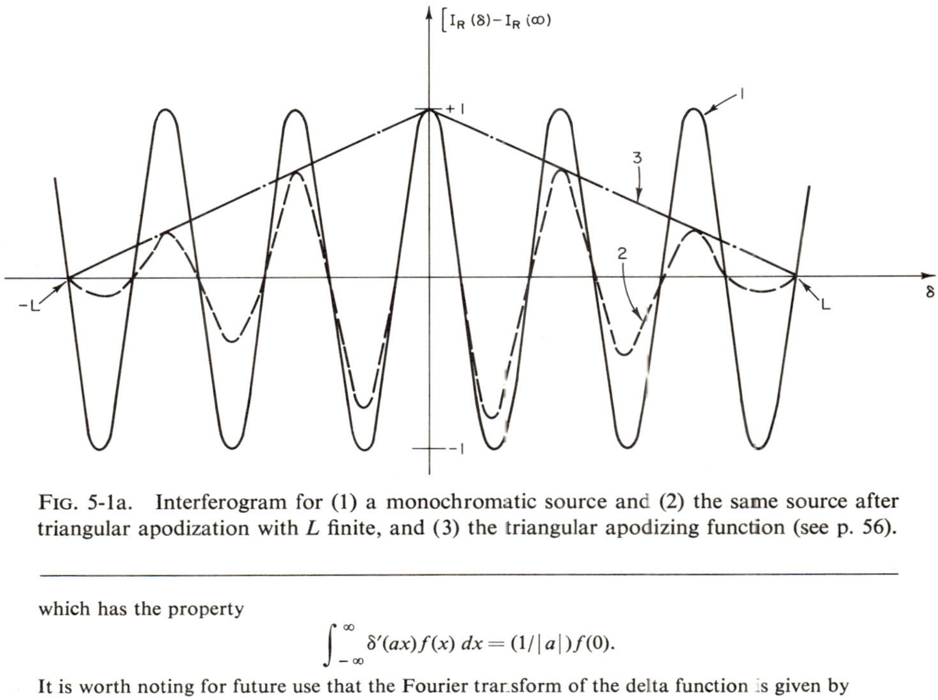


FIG. 5-1a. Interferogram for (1) a monochromatic source and (2) the same source after triangular apodization with $L$ finite, and (3) the triangular apodizing function (see p. 56).

---

which has the property

$$\int_{-\infty}^{\infty} \delta'(ax)f(x)\,dx = (1/|a|)f(0).$$

It is worth noting for future use that the Fourier transform of the delta function is given by

$$\mathscr{F}\{\delta'(x-a)\} \equiv \int_{-\infty}^{\infty} \delta'(x-a)e^{i2\pi\sigma x}\,dx = e^{i2\pi\sigma a}.$$

By the use of the above equation, one can show that

$$\tfrac{1}{2}\mathscr{F}\{\delta'(x-a) + \delta'(x+a)\} = \cos(2\pi\sigma a)$$

and that

$$(1/2i)\mathscr{F}\{\delta'(x-a) - \delta'(x+a)\} = \sin(2\pi\sigma a)$$

The inverse Fourier transform does not exist because the integral of $[\cos^2(2\pi\sigma x)]\,dx$ from $-\infty$ to $+\infty$ is not absolutely integrable.

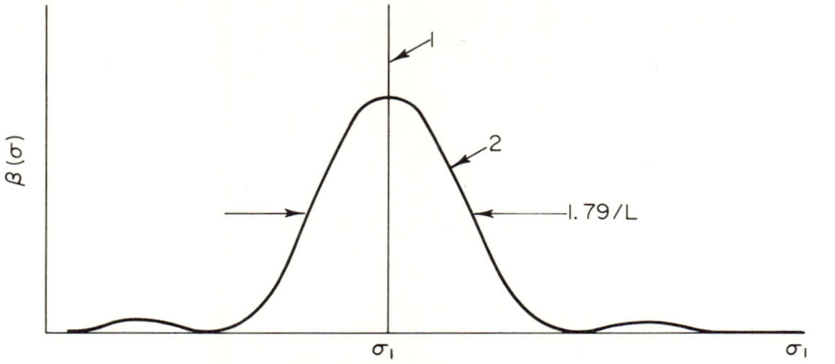


FIG. 5-1b. The spectra of (1) the monochromatic source if $L$ approaches infinity and of (2) the same source if $L$ is finite after triangular apodization (see pp. 56 and 57).

## COMPUTED SPECTRUM FROM INTERFEROGRAMS USING FINITE SCANS

Since the interferogram can *not* be scanned to infinity, but from $\delta = -L$ to $\delta = +L$, what effect is produced? In this case, put in the interferogram for a monochromatic source and ask, "What sort of spectrum do we obtain mathematically if we put the limits of the integral at $\pm L$ instead of at infinity?"

Substituting the interferogram function $2\cos(2\pi\sigma_1\delta)$ into Eq. (5-1), using Euler's formula,† noting that the integral of the odd function is zero, and using a trigonometric identity, one has

$$B(\sigma) = \int_{-L}^{L} \{\cos[2\pi(\sigma_1 + \sigma)\delta] + \cos[2\pi(\sigma_1 - \sigma)\delta]\}\, d\delta \tag{5-5}$$

Upon integrating, one finds that

$$B(\sigma) = 2L \left\{\frac{\sin[2\pi(\sigma_1 + \sigma)L]}{2\pi(\sigma_1 + \sigma)L} + \frac{\sin[2\pi(\sigma_1 - \sigma)L]}{2\pi(\sigma_1 - \sigma)L}\right\}, \tag{5-6}$$

In order to observe the pertinent features of the spectrum from a monochromatic source as produced by a shortened interferogram, we must consider some typical magnitudes of the parameters in Eq. (5-6). We want to show that the first term is much smaller than the second term.

Looking at the magnitudes, let us compute the largest value the first term can probably have. The smallest $\sigma$ would be zero. For far-infrared

† Euler's formula states that $e^{ix} = \cos x + i \sin x$.

spectroscopy for a 2-mm-wavelength source, one would have $\sigma_1 = 5\ \text{cm}^{-1}$. If one wishes modest resolution, an optical path of about 5 cm seems reasonable (see Chapter 6 on resolution). This combination of parameters gives $2\pi(\sigma_1 + \sigma)L \approx 150$. Likewise, in the near-infrared with modest resolution, one might have $\sigma_1 = 400\ \text{cm}^{-1}$ with $L = 0.5$ cm, so that $2\pi(\sigma_1 + \sigma)L \approx 125$. Considering that the numerator is never greater than unity, the first term is less than 0.01. The second term in Eq. (5-6) has a peak value of unity at $\sigma = \sigma_1$; so, we ignore the first term. Therefore, the spectrum of a monochromatic source produced by transforming an interferogram with a finite maximum optical path displacement is

$$B(\sigma) \approx 2L(\sin z)/z \equiv 2L \operatorname{sinc} z, \tag{5-7}$$

where

$$z \equiv 2\pi(\sigma_1 - \sigma)L. \tag{5-8}$$

Often one designates $2L \operatorname{sinc} z$ as the instrument line shape function (ILS).

Figure 5-2 shows a plot of the spectrum of a monochromatic source modified by a *finite* maximum path $L$, i.e., sinc $z$. The $\operatorname{sinc}^2 (z/2)$ function is plotted for later use.

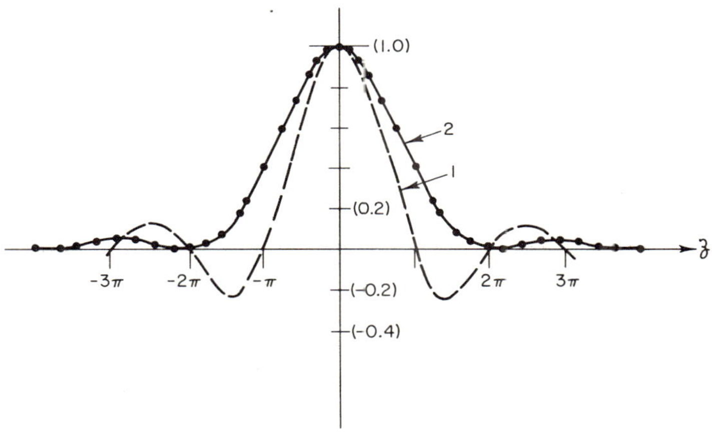


FIG. 5-2. Plots of (1) sinc $z \equiv [(\sin z)/z]$ versus $z$ and (2) $\operatorname{sinc}^2 (z/2)$ versus $z$.

The sinc $z$ function is an approximation we get to the monochromatic beam. The "sidelobes" or "feet" drop 22% below zero and that is rather large. We can tolerate the central peak of finite width as an approximation to an infinitely narrow bandwidth, but the "feet" would appear as false sources of energy at nearby wavelengths. To reduce this error, we reduce the size of the "feet" by the process called "apodization."

## APODIZATION AND RESOLUTION

Suppose one tries to correct the interferogram by multiplying it with the triangular function $\Lambda = (1 - |\delta|/L)$; then, we have

$$B(\sigma) = 2\int_{-L}^{L} [1 - (|\delta|/L)] \cos(2\pi\sigma_1\delta)e^{-i2\pi\sigma\delta}\, d\delta. \tag{5-9}$$

Using Euler's formula, noting that the integral of the odd function is zero, using a trigonometric identity, and dropping the $\cos[2\pi(\sigma_1 + \sigma)\delta]$ term because it yields a small integral, we have

$$\begin{aligned} B(\sigma) &= \int_{-L}^{L} [1 - (|\delta|/L)] \cos[2\pi(\sigma_1 - \sigma)\delta]\, d\delta \\ &= 2\{1 - \cos[2\pi(\sigma_1 - \sigma)L]\}/[2\pi(\sigma_1 - \sigma)]^2 L. \end{aligned} \tag{5-10}$$

Using a trigonometric identity, we have

$$B(\sigma) = L \operatorname{sinc}^2(z/2), \tag{5-11}$$

where $z$ is defined in Eq. (5-8).

This reduces the "feet" or sidelobes, as can be seen in Fig. 5-2. There are several results to note. The peaks of the "feet" are reduced by about a factor of four. Negative intensities are gone. The width is increased somewhat, but not seriously. So we can modify the interferogram of a monochromatic source by multiplying it with the apodizing function $[1 - (|\delta|/L)]$, and know that we will obtain a spectrum which is an acceptable approximation to the perfect monochromatic source. Multiplying constants cancel in transmittance and reflectance measurements by taking ratios. However, the experimentalist must be cautious, since $L$ appears in the coefficient. One must use the same scan distances in the background and sample-in measurements. We have been considering triangular apodization. There are also cosine, trapezoidal, Gaussian, etc., functions used for apodization. We use trapezoidal in our program described in Appendix B. Some comments concerning triangular apodization are in order (see Figs. 5-1a and 5-1b).

1. By maximizing the apodization at $\delta = 0$ and having it zero at $\delta = L$, we "weigh" the small $\delta$'s against the largest $\delta$'s; i.e., we favor the high signal-to-noise ratio data in the interferogram in most practical situations.

2. Not having a monochromatic source is no problem in the general theory of apodization. Each Dirac delta function $\delta'(\sigma - \sigma_1)$ picks out the frequency $\sigma_1$. Just letting $\sigma_1$ go to $\sigma$ automatically gives the spectrum $B(\sigma)$ for the entire range of wave numbers of interest.

3. We want the apodizing function to be an even function so we can scan for both positive and negative $\delta$'s.

It should be noted that the apodization process affects the resolution of the spectra. Considering Fig. 5-2 and setting the full half-width of the unapodized spectrum (sinc $z$) equal to unity, the width of the triangularly apodized spectrum increases to 1.48, i.e., the frequency spread is increased by about 48% by triangular apodization.

We may express the full-widths at one-half the peak intensity another way. In Fig. 5-2, the intensity drops to 0.5 at $z = 0.607\pi$; so, by using the definition of $z$, one sees that the full half-width is given by $1.21/L$. If the same procedure is followed for different apodizing functions, one may make a table of full half-widths as a function of the maximum optical path difference. Table 5-1 shows several examples, and also presents the percentage increases in the full half-widths of the apodized spectra compared to the unapodized spectra.

TABLE 5-1

Apodizing Functions and the Resultant Full-Widths at One-Half Peak Intensity as a Function of the Maximum Optical Path Difference $L$, together with the Percent Increase in Full Half-Width Compared to the Unapodized Spectrum.

| Apodizing function | Full half-width at one-half peak intensity | Percent increase in full half-width |
|---|---|---|
| Constant such as unity | $1.21/L$ | 0 |
| $\cos(\pi\delta/2L)$ | $1.58/L$ | 31 |
| $[1 - (\lvert\delta\rvert/L)]$ | $1.79/L$ | 48 |
| $[1 - (\delta/L)^2]^2$ | $1.91/L$ | 58 |

There is not a great deal of difference in the apodizing functions as far as resolution is concerned. In order to avoid phase errors for two-sided interferograms, one should have the peak of the apodizing function fall exactly at the origin as experimentally determined.

The apodizing function $A(\delta, L)$ can be introduced in a general fashion by the equations

$$B_{\mathrm{A}}(\sigma) = \int_{-L}^{L} A(\delta, L)[I_{\mathrm{R}}(\delta) - I_{\mathrm{R}}(\infty)]e^{-i2\pi\sigma\delta}\,d\delta. \tag{5-12}$$

In Eq. (5-12), $B_{\mathrm{A}}(\sigma)$ is the modified or apodized spectrum, which should be closer to the true spectrum than the unapodized spectrum $B_{\mathrm{U}}(\sigma)$ given by

$$B_{\mathrm{U}}(\sigma) = \int_{-L}^{L} [I_{\mathrm{R}}(\delta) - I_{\mathrm{R}}(\infty)]e^{-i2\pi\sigma\delta}\,d\delta. \tag{5-13}$$

Equation (5-13) differs from Eq. (3-26) because of the change of the infinite upper limit of the integral in Eq. (3-26) to the finite value $L$. Equation (5-13) would be used by the experimentalist who does *not* need apodized spectra.

From some of the discussions concerning the effects of finite optical path differences, we are also in a position to explore mathematical filtering from another aspect. As will be seen, instrumental effects for idealized situations can be expressed in terms of the convolution operation. One can know how an idealized spectrum should appear and, with a knowledge of the instrumental effects, predict what form the experimental spectrum will have.

## INSTRUMENT LINE SHAPE AND CONVOLUTIONS

We have seen from Eqs. (5-7) and (5-8) that the instrument and computer effects are such as to yield an experimental spectrum which is not monochromatic. Experimentally, the source was monochromatic of wave number $\sigma_1$, and after the instrument and computer operated on the wave, it had the spectral shape given by Eq. (5-7). Thus, the instrument–computer line shape function is $\text{ILS} \equiv 2L\,\text{sinc}(2\pi\sigma L)$.

It will now be shown that the spectrum actually measured and computed for a *finite* optical path difference, the instrumental spectrum, is the convolution of the idealized spectrum produced by *infinite* optical path differences with the instrument line shape function.

The instrumental spectrum is given by

$$B_{\mathrm{I}}(\sigma) \equiv \int_{-L}^{L} [I_{\mathrm{R}}(\delta) - I_{\mathrm{R}}(\infty)]e^{-i2\pi\sigma\delta}\, d\delta \tag{5-14}$$

or

$$B_{\mathrm{I}}(\sigma) = \int_{-\infty}^{\infty} [I_{\mathrm{R}}(\delta) - I_{\mathrm{R}}(\infty)]\,\text{rect}(\delta)e^{-i2\pi\sigma\delta}\, d\delta, \tag{5-15}$$

where the rectangular function is defined as

$$\text{rect}(\delta) = \begin{cases} 1, & |\delta| < L, \\ 0, & |\delta| > L. \end{cases} \tag{5-16}$$

The idealized spectrum for infinite optical path differences is

$$B(\sigma) = \int_{-\infty}^{\infty} [I_{\mathrm{R}}(\delta) - I_{\mathrm{R}}(\infty)]e^{-i2\pi\sigma\delta}\, d\delta. \tag{5-17}$$

If we take the Fourier transform of $B_I(\sigma)$ as expressed in Eq. (5-15), we have

$$\mathscr{F}\{B_I(\sigma)\} = [I_R(\delta) - I_R(\infty)]\,\text{rect}(\delta). \quad (5\text{-}18)$$

From Eq. (5-17), we have

$$\mathscr{F}\{B(\sigma)\} = [I_R(\delta) - I_R(\infty)]. \quad (5\text{-}19)$$

We may define a function $F(\sigma)$ such that its Fourier transform is the rectangular function; so, we write

$$\mathscr{F}\{F(\sigma)\} \equiv \text{rect}(\delta). \quad (5\text{-}20)$$

Substituting Eqs. (5-18) and (5-19) in Eq. (5-17), we have

$$\mathscr{F}\{B_I(\sigma)\} = \mathscr{F}\{B(\sigma)\}\mathscr{F}\{F(\sigma)\}. \quad (5\text{-}21)$$

Noting that the Fourier transform of the convolution of two functions is the product of the Fourier transforms of the functions, we have

$$\mathscr{F}\{B_I(\sigma)\} = \mathscr{F}\{B(\sigma) \star F(\sigma)\}. \quad (5\text{-}22)$$

So, we may extract the results that

$$B_{\mp}(\sigma) = B(\sigma) \star F(\sigma). \quad (5\text{-}23)$$

From the defining equations (5-16) and (5-20), one has the Fourier transform $F(\sigma)$ of the rectangular function given by

$$F(\sigma) = \int_{-L}^{L} e^{-i2\pi\sigma\delta}\, d\delta = 2L\,\text{sinc}(2\pi\sigma L). \quad (5\text{-}24)$$

By comparing Eq. (5-24) with the instrument line shape function, ILS given in Eq. (5-7), the instrumental spectrum $B_I(\sigma)$ becomes

$$B_I(\sigma) = B(\sigma) \star (\text{ILS}). \quad (5\text{-}25)$$

Thus, the instrumental spectrum is given by the convolution of the ideal spectrum with the instrumental line shape function.

## MATHEMATICAL FILTERING

Before the Cooley–Tukey algorithm (Chapters 16–18) was discovered, numerically computing Fourier transforms often involved a great deal of computer time. The computer costs were proportionately high, and sometimes the length of time the computer was needed was intolerable. A method of reducing the computer time was devised whereby the computed spectral range was limited with the use of rectangular functions. The method devised by Connes and Nozal [4] in Mathematical Filtering in Spectroscopy by the Fourier Transformation was faster for a narrow spectral range than other techniques available before the introduction of the Cooley–Tukey algorithm (Chapter 16). The technique is still usable when data points in the interferograms total into the millions. Then, one can use mathematical filtering to handle small spectral bands.

Suppose one has obtained an interferogram from a source emitting radiation over a wide spectral range and wishes to compute only the spectrum in a narrow range of wave numbers between $\sigma_1$ and $\sigma_2$. Mathematically, the spectrum $B_{12}(\sigma)$ over the narrow range can be extracted from the spectrum $B(\sigma)$ from over a very broad spectral range with the use of the rectangular function. Defining $\mathrm{rect}(\sigma_1, \sigma_2)$ as equal to unity over the range from $\sigma_1$ to $\sigma_2$ and zero elsewhere, we may state that

$$B_{12}(\sigma) = \mathrm{rect}(\sigma_1, \sigma_2)B(\sigma). \tag{5-26}$$

If we take the Fourier transform of Eq. (5-26), use the convolution theorem, note that the interferogram $I_{\mathrm{R}}(\delta) - I_{\mathrm{R}}(\infty)$ is the Fourier transform of the spectrum, and finally take the inverse Fourier transform $\mathscr{F}^{-1}$, we have

$$B_{12}(\sigma) = \mathscr{F}^{-1}\{[\mathscr{F}\{\mathrm{rect}(\sigma_1, \sigma_2)\}] \star [I_{\mathrm{R}}(\delta) - I_{\mathrm{R}}(\infty)]\} \tag{5-27}$$

where $\mathscr{F}$ implies taking the Fourier transform. By direct integration, we may compute the Fourier transform of the rectangular function. We obtain

$$\mathscr{F}\{\mathrm{rect}(\sigma_1, \sigma_2)\} = (\Delta\sigma)e^{i2\pi\sigma_{\mathrm{m}}} \,\mathrm{sinc}[\pi(\Delta\sigma)\delta], \tag{5-28}$$

where the small bandwidth desired is

$$\Delta\sigma \equiv \sigma_2 - \sigma_1 \tag{5-29}$$

and the mean frequency of the small band is

$$\sigma_{\mathrm{m}} = (\sigma_1 + \sigma_2)/2. \tag{5-30}$$

Substitution of Eq. (5-28) in Eq. (5-27) yields

$$B_{12}(\sigma) = (\Delta\sigma)\mathscr{F}^{-1}\{\{e^{i2\pi\sigma_{\mathrm{m}}\delta} \,\mathrm{sinc}[\pi(\Delta\sigma)\delta]\} \star [I_{\mathrm{R}}(\delta) - I_{\mathrm{R}}(\infty)]\} \tag{5-31}$$

Thus, Eq. (5-31) may be used to compute a spectrum on a limited, small band pass between $\sigma_1$ and $\sigma_2$. We must take into account the fact that the interferogram and the factor $[\exp(2\pi\sigma_{\mathrm{m}}\delta)]\,\mathrm{sinc}[\pi(\Delta\sigma)\delta]$ are to be convolved over finite ranges of $\delta$. If we performed the transform of the interferogram over a finite range of $\delta$ as we did in the first part of this chapter, we would have to apodize it to reduce sidelobe problems. Likewise, with the first term in the convolution, we would get oscillations in the inverse Fourier transform. The rectangular function would have peaks on it which are particularly large near the corners. The peaks are known as "Gibbs ears" or the "Gibbs phenomena." Thus, both functions are *separately* apodized.

We emphasize the separate apodization because for a given band $\Delta\sigma$, the sinc function might approach zero much faster than $[I_R(\delta) - I_R(\infty)]$. Partly, it is the rapid decrease in the sinc function that makes mathematical filtering techniques faster than an ordinary calculation of the Fourier transform for select wave numbers. In the ordinary calculation, one would choose the desired wave numbers and compute the inverse Fourier transforms of the *entire* interferogram. With the sinc function decreasing as determined by a given $\Delta\sigma$, the convolution does not need to proceed past an $L'$ for which $\text{sinc}[\pi(\Delta\sigma)L']$ is reduced to the same relative magnitude as the noise in the interferogram. If $\Delta\sigma$ is not too small and the sinc function not too broad, the number of terms in the convolution and in the inverse Fourier transform of Eq. (5-31) can be reduced.

Also, the number of terms in the computation may be reduced by the use of the sampling interval limitations as will be shown in Chapter 7. It will be shown in Eq. (7-21) that the experimental spacing $\Delta\sigma_{\text{exp}}$ in the sampled points of the interferogram must be less than the maximum wave number used in the spectrum. That is, we have

$$\Delta\delta_{\text{exp}} \leq 1/2\sigma_{\text{max}}. \tag{5-32}$$

In the small spectral band, we always have $\sigma_2 \leq \sigma_{\text{max}}$, so the steps of $(\Delta\delta)_{\text{comp}}$ in the interferogram used in the computation may be increased in size. If $\sigma_{\text{max}}$ equals $h'\sigma_2$, where $h'$ is an integer, the steps can be increased in size by $h'$. That is, we can eliminate from the computation all but sampling steps for which $(\Delta\sigma)_{\text{comp}} = h'(\Delta\sigma)_{\text{exp}}$. Thus, another saving in computation time may be realized.

One can record the interferogram with or without spectral filters which closely define a small band. If a broad wavenumber region is recorded, the experimentalist must, with long-pass filters, provide a reasonable $\sigma_{\text{max}}$. However, for the unknown situations, he is provided with relaxed restraints and has a faster computation method if the technique developed by Connes and Nozal [4] is used.

## CONCLUSIONS

In this chapter, the effect of taking finite optical path differences has been illustrated. It was shown that the monochromatic source after treatment by the interferometer and computer has a computed spectrum ILS which is given by $2L\,\text{sinc}[2\pi(\sigma - \sigma_1)L]$. The sinc function has large sidelobes, even negative values, and is often an unacceptable line shape function. The sidelobes can be greatly reduced by apodizing. The method of apodizing was demonstrated with the triangular function and the effect on resolution was pointed out. Several other acceptable apodizing functions were mentioned along with their effect on resolution. Then, a general procedure for apodizing was given in Eq. (5-12).

Mathematical filtering was extended to illustrate the different spectra to be found from scans of infinite and of finite optical path differences. The spectra for the two cases were shown to be related with the use of the convolution integral.

## *References*

*1.* A. E. Martin, "Infra Red Instrumentation and Techniques," Elsevier, New York, 1966.

*2.* J. Connes, Computing Problems in Fourier Spectroscopy, *Aspen Int. Conf. on Fourier Spectrosc., 1970*, p. 83. (G. A. Vanasse, A. T. Stair, Jr., and D. J. Baker, eds.). AFCRL-71-0019, 5 Jan. 1971, Spec. Rep. No. 114.

*3.* W. H. Steel, "Interferometry," p. 192. Cambridge Univ. Press, London and New York, 1967.

*4.* J. Connes and V. Nozal, Mathematical Filtering in Spectroscopy by the Fourier Transformation, *J. Phys. Radium* **22**, 359 (1961).

*Note added in proof:* An excellent reference giving many apodization functions and their effect on the spectrum is K. Rasmussen, Conformations and Vibrational Spectra of TRIS-(DIAMINE) Chelate Complexes, Chem. Dept., The Technical Univ. of Denmark, 2800 Lyngby (1970).

CHAPTER SIX

# Resolution

## INTRODUCTION

We shall show in this chapter that the resolution of a Fourier transform spectrometer varies inversely as the maximum optical path difference used to obtain the interferogram. Detailed calculations of the resolution under different conditions are presented to illustrate the point. Finally, a practical limitation on resolution is pointed out.

## INSTRUMENT BROADENING OF A LINE (WITHOUT AND WITH APODIZATION)

One might wish to use the term "the resolution of the interferometer" to describe the increase in linewidth of a very narrow line introduced by the instrument and/or measuring technique. For instance, a source, such as a laser, may emit a very monochromatic line and after the analysis of the line by the instrument and the computer, the measured line will be found to have an increased width. This ever-present situation is very important from an experimental viewpoint. As we saw in the last chapter, an unapodized monochromatic line yields an instrumental line shape given by sinc $z$, where $z = 2\pi(\sigma - \sigma_0)L$, with the monochromatic wave number $\sigma_0$ and the maximum optical path difference $L$. The function sinc $z$ is called the instrumental line shape function because it is the line shape given by the interferometer for a perfectly monochromatic source. The full-width at one-half peak intensity for the sinc $z$ instrumental line shape function is given by $\delta z = 1.21\pi$, as can be seen from Fig. 5-2. In terms of wave number, one has

$$\delta\sigma = 1.21/2L. \tag{6-1}$$

[Thus, Eq. (6-1) gives the full-width at one-half the peak intensity due to the instrument.] If the instrument function were $\text{sinc}^2(z/2)$, the linewidth would be $0.90/L$, as can also be seen in Fig. 5-2, for triangular apodization.

## SEPARATION OF RESONANCES WITH APODIZATION

We now wish to discuss resolution in terms of the separation of two lines of equal intensity (or resonances) in a spectrum. We will assume that natural linewidths are negligibly small. One may claim that two resonances are resolved if they are separated such that the peak of one resonance falls at the first zero of the second resonance. This is the Rayleigh criterion. If the two resonances are very narrow so they have the instrumental line shapes given by $\text{sinc}^2(z/2)$, it may be shown that the distance between the peaks in terms of $z$ is given by $2\pi$; see Fig. 6-1. This is equivalent to the Rayleigh criterion,

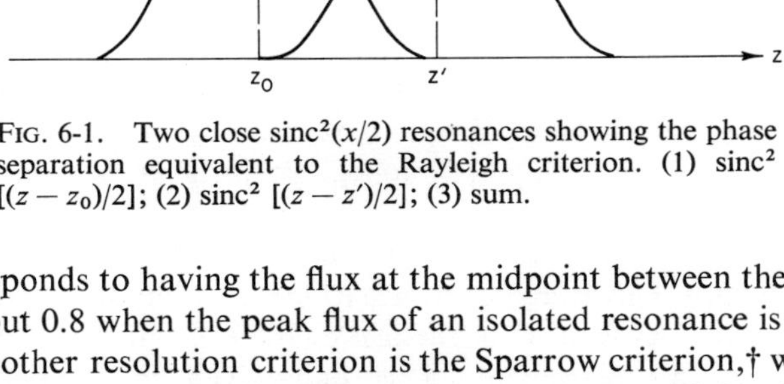


FIG. 6-1. Two close $\text{sinc}^2(x/2)$ resonances showing the phase separation equivalent to the Rayleigh criterion. (1) $\text{sinc}^2[(z - z_0)/2]$; (2) $\text{sinc}^2[(z - z')/2]$; (3) sum.

and it corresponds to having the flux at the midpoint between the resonances equal to about 0.8 when the peak flux of an isolated resonance is normalized to unity. Another resolution criterion is the Sparrow criterion,† which brings the same resonances twice as close together as the Rayleigh criterion. For

† The Sparrow criterion states that two points sources can be considered resolved if the second derivative of the resultant distribution of irradiance in the diffraction image of the two points vanishes on axis.

$\delta z = \pi$, one finds that the sum of the $\text{sinc}^2(z/2)$ fluxes is flat between the two line frequencies. If the Sparrow criterion were relaxed just a little to allow the difference in the arguments to slightly exceed $\pi$, a small dip would develop in the sum at the midpoint. Thus, if one had an infinite signal-to-noise ratio, the Sparrow criterion would yield the best resolution that could be obtained (for equal resonances). Figure 6-1 illustrates the Rayleigh criterion; the Sparrow criterion is shown by Strong [1, p. 212].

As we saw in Chapter 5, the instrumental line shape depends on the apodization of the interferogram. One has [from Eq. (5-7)]

$$B_U(\sigma) \propto \text{sinc}\, z \tag{6-2}$$

for the unapodized spectrum, and [from Eq. (5-11)]

$$B_A(\sigma) \propto \text{sinc}^2(z/2) \tag{6-3}$$

for the triangularly apodized spectrum. Both Eqs. (6-2) and (6-3) assume that the source is monochromatic. From the discussion in Chapter 5 and Table 5-1, one can see that the resolving power depends on the apodizing function used. The triangular apodization was used to derive Eq. (6-3).

If one concludes that the best and most reliable spectrum he can get is from the apodized data, then, by applying the Rayleigh criterion to Eq. (6-3), one has

$$2\pi = (z_0 - z'). \tag{6-4}$$

In Eq. (6-4), $z_0 \equiv [2\pi(\sigma - \sigma_0)L]$ and $z' \equiv [2\pi(\sigma - \sigma')L]$ for resonance wave numbers of $\sigma_0$ and $\sigma'$. The maximum optical path difference is $L$. Substituting the definitions of $z_0$ and $z'$ in Eq. (6-4), the resolution is found to be

$$(\sigma' - \sigma_0) \equiv (\delta\sigma)_A = 1/L. \tag{6-5}$$

So, the resolution given by Eq. (6-5) is for (1) a triangularly apodized interferogram, and (2) the Rayleigh criterion with a phase difference of $2\pi$.

## SEPARATION OF RESONANCES WITHOUT APODIZATION

If the Rayleigh criterion is applied to two ILS functions, the resolution becomes one-half of the resolution for the apodized spectrum. Remembering that the Rayleigh criterion separates the peaks such that the maximum of one resonance falls at the *zero* of the other resonance, $\delta z$ becomes $\pi$, as can be seen by observing Fig. 5-2. So the resolution as dictated by the instrumental line shape is

$$(\delta\sigma)_{\mathrm{ILS}} = 1/2L. \tag{6-6}$$

Thus, the resolution given by Eq. (6-6) is for (1) an unapodized interferogram, and (2) the Rayleigh criterion with a phase difference of $\pi$.

## COMPARISON OF LINE BROADENING AND THE SEPARATION OF RESONANCES

By comparing Eqs. (6-1) and (6-6), one sees that the full-width at one-half the peak intensity of the ILS is given by $1.21/2L$ and the Rayleigh criterion separates two equal ILS functions by $1/2L$. Thus, the two frequency spreads are equal to each other to within about 20%. We shall use these two values interchangeably for resolution.

Equation (6-5) gives a line separation of $1/L$ for two $\mathrm{sinc}^2(z/2)$ functions (obtained in triangular apodization) to be resolved by the Rayleigh criterion. As mentioned before, the full-width at one-half the peak intensity of the same function is $0.90/L$. Thus, the two frequency differences are equal to within about 10%.

For both cases, the Sparrow criterion (found for the limit of an infinite signal-to-noise ratio) is realized by halving the wavenumber spreads. However, the experimentalist is usually very happy if he achieves resolution that is as good as that demanded by that Rayleigh criterion.

## COUNTING FRINGES AND RESOLUTION

Another way of obtaining Eq. (6-5) is to consider the interference pattern in the interferogram from a Michelson interferometer. If in the arm scan to $L/2$, one resonance produces one more cycle of interference than the other, nearby resonance, the experimentalist should be able to separate the resonances. That is, the resonances should be resolved. If the scan has been terminated with the total optical path difference $L$, one might have $m$ constructive interference peaks for the resonance peak at $\sigma$. Thus, we may write

$$L = m/\sigma. \tag{6-7}$$

The next resolvable resonance at $\sigma + \delta\sigma$ will have $m + 1$ positions of constructive interference in the interferogram. So, for the second resonance, we have

$$L = (m + 1)/(\sigma + \delta\sigma). \tag{6-8}$$

Equating Eqs. (6-7) and (6-8), solving for $\delta\sigma$, and eliminating $m$ using Eq. (6-7), the resolution is

$$\delta\sigma = 1/L. \tag{6-9}$$

This interferometric viewpoint yields the same results as the apodized spectrum whose resolution is given in Eq. (6-5).

## CONCLUSIONS AND GENERAL COMMENTS

Aside from the preceding comments, one should not be overwhelmed by subjective definitions such as the Rayleigh and Sparrow criteria. They are convenient and internationally agreed upon, but they are unsophisticated concepts. The signal-to-noise ratio, the relative magnitudes of the peaks, and other experimental parameters can drastically alter any conclusions offered so far. Deciding which criterion is to be used is very much the experimentalist's choice—provided he outlines what he is doing with reproducible definitions. (In this chapter, the Rayleigh criterion for any resonance function shall be that the peak of one resonance lies on the first zero of the next resonance.)

In conclusion, we have expressed the resolution in terms of the separation of two lines and also in terms of half-widths. We related the resolution to specific functional line shapes. Example calculations were given for the resolution of an apodized spectrum in Eq. (6-5), of the instrumental line shape function in Eq. (6-1), and of an interferometer viewed in terms of the interferogram in Eq. (6-9). The different functions all illustrate the most important point: *the resolution of an interferometer varies inversely as the maximum optical path difference.*

For a more sophisticated treatment of resolution, the reader should study Williams and Chang [2].

In Chapter 11 on field of view and its effect on resolution, the size of the source is considered. It is found that the source size puts what are usually weak restrictions on the experimental resolution. Nevertheless, the size of the source can, under certain conditions, furnish the limiting factor to the resolution.

Since the interferogram is decreasing as the arm displacement is increasing, there is a path difference at which the *interferogram* signal-to-noise ratio falls below unity. Recording the interferogram for larger path differences only increases the noise in the computed spectrum and better resolution is not achieved. Thus, the strength of the signal can determine the $L$ and, thus, the resolution.

## References

*1.* J. D. Strong, "Concepts of Classical Optics," p. 212. Freeman, San Francisco, California, 1958.

*2.* R. A. Williams and W. S. C. Chang, Resolution and Noise in Fourier-Transform Spectroscopy, *J. Opt. Soc. Amer.* **56**, 167 (1966).

CHAPTER SEVEN

# *Sampling Intervals*

## INTRODUCTION

In this chapter, it will be shown why it is often a good procedure to sample the interferogram at equally separated small path differences $\Delta\delta$. The effects on the spectrum of using a sampled interferogram will be exhibited, and the restrictions on the maximum size of the sampling interval $\Delta\delta$ that can be employed will be given. Finally, an experimental technique for improving the signal-to-noise ratio in a sampled interferogram will be discussed.

## WHY SAMPLE?

If one is going to record an interferogram to be Fourier-analyzed for computation of the spectrum, he must somehow digitize the data for insertion into the computer. One can pick points off a strip-chart recording of the interferogram or record the interferogram automatically at equal increments of optical path difference $\Delta\delta$. If the interferogram is recorded at *only* the equally spaced intervals, there can be considerable lost recording time in driving the movable mirror continuously. In fact, with a continuous mirror drive, more time can be spent between recording increments than during the actual time of signal averaging. However, if the movable mirror is moved in steps, the wasted time can be minimized. In the stepping procedure, the mirror movement is made as rapidly as possible, and the recording of signal at the chosen interval is made as long as desired for a given signal-to-noise ratio.

For instance, one might have a stepping motor arrangement that moves the mirror a distance $\Delta\delta/2$ in a few milliseconds and lets it sit in position for seconds (or even minutes) while the signal is averaged for that particular arm displacement.

There is a drawback to stepping or sampling that must be mentioned. It is not necessary for two-sided interferograms to have the first step fall exactly at the origin, as will be discussed in Chapter 12. However, if a step does nearly coincide with the origin (sometimes recorded on a strip-chart recorder in parallel with the automatic data acquisition or computer system), the interferogram will be nearly symmetric. With a symmetric interferogram, one can quickly detect unusual electronic noise problems and any other departures from symmetry. On the other hand, if a one-sided interferogram is recorded and a step does not fall exactly at the origin, care must be exercised in computing the spectrum. Missing the origin introduces phase errors in the interferogram which are mentioned in Chapter 8 and discussed in Chapter 12.

Finally, by recording the interferogram in equal intervals, one is able to use the Cooley–Tukey algorithm for the Fourier analysis. The Cooley–Tukey algorithm is discussed in detail in Chapters 16–18 and Appendix B. The algorithm can reduce computation time by orders of magnitude, but one of the requirements for its use is that the data be recorded in equal increments.

In Appendix D, the new rapid continuous scan technique is outlined.

## SHAH FUNCTION

If one has a continuous or complete interferogram $F_c(\delta) \equiv [I_R(\delta) - I_R(\infty)]_c$ and wants to relate it mathematically to a sampled interferogram $F_s(\delta)$, the shah function is convenient. The sampled interferogram, $F_s(\delta) \equiv [I_R(\delta) - I_R(\infty)]_s$, is recorded *only* at steps of path difference $\Delta\delta$ apart, with one step falling exactly at the origin. We shall need the shah function $\text{Ш}(x)$ as described by Bracewell [1], to relate the sampled and the complete (unsampled) interferograms by

$$F_s(\delta) \equiv \text{Ш}(\delta/\Delta\delta)F_c(\delta). \tag{7-1}$$

The shah function must be a comb function to perform the operation in Eq. (7-1). That is, it must pick out only those values of the interferogram such that the interferogram is recorded at $\delta = 0, \pm\Delta\delta, \pm 2\Delta\delta, \pm 3\Delta\delta, \ldots$. Mathematically, the shah function is defined as

$$\text{Ш}(x) \equiv \sum_{n=-\infty}^{\infty} \delta'(x - n), \tag{7-2}$$

where $\delta'(x - n)$ is a Dirac delta function and $n$ is an integer. So, from Eq. (7-2), one can see that the shah function will allow nonzero values only for $x$ equal to integers (positive and negative). Then, $\text{Ш}(\delta/\Delta\delta)$ in Eq. (7-1) will allow in $F_s(\delta)$ only those values of the interferogram for which $\delta = 0, \pm\Delta\delta, \pm 2\Delta\delta, \pm 3\Delta\delta, \ldots$.

Thus, the shah function mathematically relates the sampled interferogram and the complete interferogram. Before we can relate the resultant sampled spectrum $B_s(\sigma)$ to the complete spectrum $B_c(\sigma)$, we need to investigate some of the mathematical properties of the shah function as defined in Eq. (7-2).

The shah function is periodic to any integer change in $x$; so,

$$\text{Ш}(x + m) = \text{Ш}(x), \tag{7-3}$$

where $m$ is any integer. This property, in part, depends on the fact that the limits of the sum in Eq. (7-2) are infinite.

We may also derive a scale rule for the shah function in terms of the Dirac delta function. Suppose one has a shah function defined by

$$\text{Ш}(ax) \equiv \sum_{n=-\infty}^{\infty} \delta'(ax - n), \tag{7-4}$$

where $a$ is some number. We want to change the variable in the Dirac delta function from $(ax - n)$ to $x - (n/a)$. If we consider the definition of the Dirac delta function for any function $f(x)$, then we can have

$$\int_{-\infty}^{\infty} \delta'(x - [n/a]) f(x)\, dx = f(n/a), \tag{7-5}$$

and

$$\begin{aligned}\int_{-\infty}^{\infty} \delta'(ax - n) f(x)\, dx &= (1/|a|) \int_{-\infty}^{\infty} \delta'(y) f([y + n]/a)\, dy \\ &= (1/|a|) f(n/a).\end{aligned} \tag{7-6}$$

Comparing Eqs. (7-5) and (7-6), one has

$$\delta'(ax - n) = (1/|a|)\, \delta'(x - [n/a]); \tag{7-7}$$

thus

$$\text{Ш}(ax) = (1/|a|) \sum_{n=-\infty}^{\infty} \delta'(x - [n/a]). \tag{7-8}$$

Using the scale change expression in Eq. (7-8), we would like to inquire about the inverse Fourier transform of a shah function. The inverse Fourier transform of $\text{Ш}(ax)$ is [using Eq. (7-8)]

$$\mathscr{F}^{-1}\{\text{Ш}(ax)\} = (1/|a|) \sum_{n=-\infty}^{\infty} \int_{-\infty}^{\infty} \delta'(x - [n/a]) e^{-i2\pi\sigma x}\, dx. \tag{7-9}$$

Employing the definition of the Dirac delta function, one has

$$\mathscr{F}^{-1}\{\text{Ш}(ax)\} = (1/|a|) \sum_{n=-\infty}^{\infty} e^{-i2\pi(\sigma/a)n}, \tag{7-10}$$

where the term $\sigma/a$ has been formed for convenience. Using Euler's formula, one can write

$$\sum_{n=-\infty}^{\infty} e^{-i2\pi(\sigma/a)n} = \sum_{n=-\infty}^{\infty} \cos[2\pi(\sigma/a)n] - i \sum_{n=-\infty}^{\infty} \sin[2\pi(\sigma/a)n]. \tag{7-11}$$

For all $\sigma/a$, the sum from $n = -\infty$ to $\infty$ for the sine function (odd function) is zero. If $\sigma/a$ is anything but an integer, the sum from zero to infinity (or minus infinity to zero) vanishes for the cosine function as $\sigma/a$ is varied. That is, in the sum involving cosine terms, one would be summing over so many terms (an infinite number) that when $\sigma/a$ is not equal an integer, the cosines would essentially be adding with random phase and amplitude. As $\sigma/a$ is varied, everytime this ratio equals an integer, the sum over $n$ would yield an infinite, positive value. But all this is a description of the shah function in terms of the Dirac delta function. So Eq. (7-10) becomes

$$\mathscr{F}^{-1}\{\text{Ш}(ax)\} = (1/|a|) \sum_{n=-\infty}^{\infty} \delta'([\sigma/a] - m) \tag{7-12}$$

or

$$\mathscr{F}^{-1}\{\text{Ш}(ax)\} = (1/|a|)\text{Ш}(\sigma/a). \tag{7-13}$$

That is, the inverse Fourier transform of a shah function is a shah function.

## RELATING THE SAMPLED AND THE COMPLETE SPECTRA

In Eq. (7-1), we gave the relation between the sampled and the complete interferogram. In order to compute spectra from interferograms. one calculates the inverse Fourier transform of the interferogram. Recall that the sampled spectrum $B_s(\sigma)$ is defined as

$$B_s(\sigma) \equiv \mathscr{F}^{-1}\{F_s(\delta)\} \tag{7-14}$$

and that the complete spectrum is defined as

$$B_c(\sigma) \equiv \mathscr{F}^{-1}\{F_c(\delta)\}. \tag{7-15}$$

Taking the inverse Fourier transform in Eq. (7-1) and using Eqs. (7-14) and (7-15) and the convolution theorem, one has

$$B_s(\sigma) = \mathscr{F}^{-1}\{\text{Ш}(\delta/\Delta\delta)\} \star B_c(\sigma). \tag{7-16}$$

Using the results of Eq. (7-13), one may write

$$B_s(\sigma) = (\Delta\delta)\text{Ш}(\sigma\, \Delta\delta) \star B_c(\sigma). \tag{7-17}$$

Using the definition of the shah function, Eq. (7-2), the change of variable statement in Eq. (7-7), and the definitions of the convolution integral and the Dirac delta function, one has

$$\boxed{\mathscr{F}^{-1}\{F_s(\delta)\} \equiv B_s(\sigma) = \sum_{n=-\infty}^{\infty} B_c[\sigma - n(\Delta\sigma)],} \tag{7-18}$$

where

$$\Delta\sigma = 1/\Delta\delta. \tag{7-19}$$

That is, if we compute the inverse Fourier transform of the sampled interferogram, we obtain the complete spectrum *every time* $\sigma$ equals $n(\Delta\sigma)$ for all integers $n$. That is, we have a duplicate spectrum starting at $n\ \Delta\sigma$.

If the repeated spectra do not overlap, there is no problem because one could identify the region of the spectrum from zero to $\sigma_{max}$. Whether or not the repeated spectra overlap depends on the magnitude of $\Delta\sigma$. If $\Delta\delta$ is very small and $\Delta\sigma$ is, therefore, very large, the spectra will be separated. If the reverse is true, there will be bad overlap (*aliasing*) and one could not determine the true spectrum from the spectrum computed from the sampled interferogram. Figures 7-1 and 7-2 present a sequence of interferograms and spectra which relate to the problem of the overlap of spectra. In Fig. 7-1(a), an ordinary interferogram is presented and in Fig. 7-1(b), the complete inverse Fourier transform of the interferogram is given. As was indicated in Chapter 5, the computed spectrum from the complete interferogram has values for positive *and* negative frequencies. Ordinarily, one would just reject the mirror spectrum for negative frequencies. However, for the sampled interferogram, we must look further into the problem. In Fig. 7-2(a), the interferogram that was shown in Fig. 7-1(a) is shown as it would be when sampled. From Eq. (7-18), we would get the spectra plotted in Fig. 7-2(b). That is the sampled spectrum is completely symmetric about $\sigma = 0, \pm\Delta\sigma, \pm 2\,\Delta\sigma, \ldots$. The dashed lines are used to indicate the contribution from what were negative frequencies in Fig. 7-1(a). Thus, the total spectrum, Fig. 7-2(c), which would result

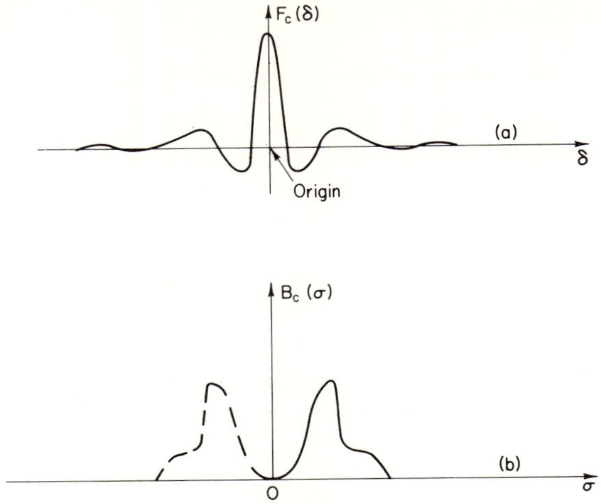


FIG. 7-1. (a) Interferogram and (b) computed spectrum from a continuous, complete scan. The inverse Fourier transform of the interferogram in curve (a) would yield the two-sided spectrum in curve (b). One usually disregards the negative $\sigma$ spectrum, but in the sampling theory, one cannot ignore it.

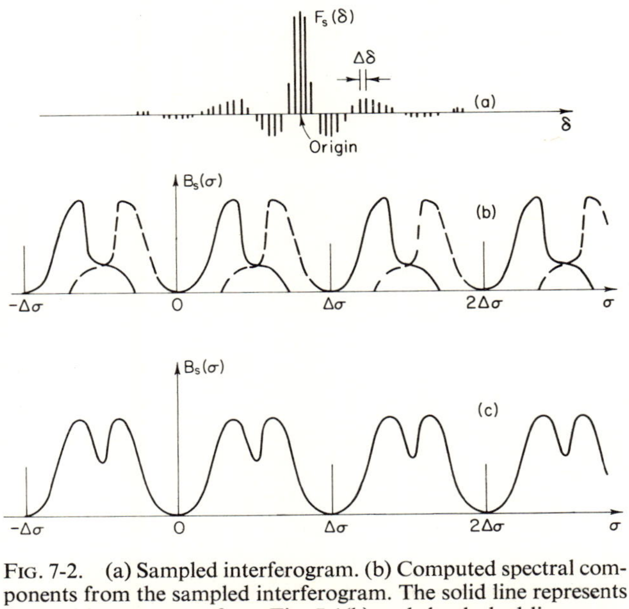


FIG. 7-2. (a) Sampled interferogram. (b) Computed spectral components from the sampled interferogram. The solid line represents the positive $\sigma$ spectra from Fig. 7-1(b) and the dashed line represents the negative $\sigma$ spectra from Fig. 7-1(b). The total spectrum as would be computed from the sampled interferogram is shown in (c). The spectrum in (c) is the sum of the spectral components in (b).

from the computation of the inverse Fourier transform of the sampled interferogram would be the sum of the solid and dashed lines. That is, the originally positive and negative frequency spectra would be overlapped or aliased, and one could not separate them.

In order to separate the spectra, we must make $\Delta\sigma$ big enough so that the maximum frequency contribution of the positive $\sigma$ spectrum does not overlap with the imaged negative $\sigma$ spectrum. This may be accomplished by requiring that

$$\Delta\sigma \geq 2\sigma_{max} \tag{7-20}$$

or

$$\boxed{\Delta\delta \leq 1/2\sigma_{max}.} \tag{7-21}$$

Thus, the stepping interval has to have an upper bound if overlapping or aliasing is to be avoided.

If the spectrum has a lower limit of $\sigma_{min}$ instead of zero, one has

$$\Delta\delta \leq 1/2(\sigma_{max} - \sigma_{min}) \tag{7-22}$$

if and only if

$$\sigma_{max} = (\text{integer})(\sigma_{max} - \sigma_{min}) \tag{7-23}$$

The condition expressed in Eq. (7-23) allows one to piece several copies of the spectrum close together without overlap.

## EXPERIMENTAL COMMENTS

Stepping may be controlled with a timing circuit that activates a stepping motor and a data acquisition system (or computer) almost simultaneously. The step size desired from Eq. (7-21) obviously depends on the maximum wave number being used. Of course, it behooves the experimentalist to keep $\sigma_{max}$ to a minimum so that large steps may be used. The large steps, as seen from Eq. (7-18), do not forfeit any information, but do save scan time.

Many data acquisition systems are expensive (\$5000–\$15,000) and either should be used judiciously, or the money should be applied toward the cost of a minicomputer. However, with a good data acquisition system, there is high probability that it records voltages by converting the input voltage to a proportional frequency and then actually counts cycles. By proper adjustment of the data acquisition controls (maybe internally), one can improve the accuracy of the interferogram with the considerations presented in the following paragraphs.

Ordinarily, one would record the interferogram data for a time slightly longer than the response time of the detector–amplifier system. Usually, the electronic time constant is made long, to smooth the data and improve the signal-to-noise ratio. If the stepping period is too short in comparison with the electronic time constant, the high-frequency spectral signal is not fully modulated in the interferogram and results in the spectrum being falsely depleted at high wave numbers (Chapter 14). In order to avoid these problems, an electronic *RC* time constant is often used to provide the proper signal-to-noise ratio and a longer stepping period is then picked. If the data acquisition system averages over most of the stepping period, the signal will be time-varying (rapidly near the origin of the interferogram). Thus, a signal partially dependent on the signal at the previous step and on the signal at the step in question is averaged and recorded. The larger is the change in signal between steps, the more the "recorded average" departs from the "average" signal at the particular step.

This error can be practically eliminated with stepping periods that are about 30% (or larger) greater than the electronic time constant. However, if the electronic amplifier time constant is reduced to 1–10 times the detector response time (for an Unicam Golay cell, the detector response time is about 50 msec), the data acquisition system can be used to average the signal. The signal from the detector–amplifier system is fed to the data acquisition system with the amplifier *RC* time constant minimized. Then, over the step period, the data acquisition averages the true signal at the particular step without any variation determined by the signal from the previous step position. The signal will be noisier, but the acquisition system will average the noise by counting the total number of cycles during a step period. Thus, one may arrive at a true signal average at a particular step. Particularly near the origin of the interferogram, this procedure can be used to record more accurate interferograms.

## CONCLUSIONS

In this chapter, we showed the necessity of recording an interferogram in steps. The stepping or sampling procedure produces an equally spaced set of interferogram points. It was shown that the sampled interferogram, when

Fourier analyzed, produces the spectrum in complete generality, Eq. (7-18). The only restriction placed on the step was that it not be too large, Eq. (7-21). Finally, some experimental procedures that are concerned with stepping or sampling were discussed.

## *Reference*

*1.* R. Bracewell, "The Fourier Transform and Its Applications." McGraw-Hill, New York, 1965.

*Note added in proof*: For high wavenumber interferometry, the mirror position must be known too accurately for obtaining the position only from a stepping motor system (i.e. for wave numbers greater than about 500 $cm^{-1}$). Moiré grating systems using a continuous mirror drive are good enough to use to measure the mirror position for interferometry up to wave numbers of the order of 1500 $cm^{-1}$. By using a simple Moiré system as described by C. Irslinger (Diplomarbeit, Physikalisches Institüt de Universitat Freiburg, Freiburg, W. Germany, July 1968), one can measure the mirror position well enough for use in the 10–500-$cm^{-1}$ range. Irslinger's system furnishes a voltage which is proportional to the optical path difference.

Small laser beam systems to give the optical path difference are now available from both the Hewlett–Packard Co. and the Siemans Co. The H–P laser–interferometer system, when used with a white light source to locate the origin, can give the mirror position (2000 Å accuracy) and velocity (accurate to 0.2 mm/min). Also the H–P system (5525A presently priced at $11,500) has a small corner cube retroreflector for insertion in many existing Fourier interferometers.

Thus with the Jacquinot advantage intact, Fourier spectroscopy can gain prominence in the visible wavenumber range.

CHAPTER EIGHT

# *Asymmetric Interferometers and Amplitude Spectroscopy*

## INTRODUCTION

This chapter is necessary to show how complex inverse Fourier transforms of the interferogram can be used for ordinary FTS. We also discuss a unique, major application of FTS-amplitude spectroscopy. However, if the reader only needs the justification for using the complex inverse Fourier transform of the interferogram in FTS, he can refer to the section, Complex Inverse Fourier Transforms of the Interferogram.

It is usual in Fourier transform spectroscopy to place the sample in the recombined radiation beams after the beamsplitter (power FTS). However, the sample may also be placed in the optical path of just one arm (amplitude FTS) as indicated in Fig. 8-1. In this chapter, the basic Fourier transform integral will be rederived for various types of samples in one arm of the spectrometer. The principal change in the basic Fourier transform integral found by putting the sample in one arm is that one uses a *complex* inverse Fourier transform of the interferogram. The complex inverse Fourier transform is equal to minus the complex conjugate of the *complex* reflection *amplitude* or *amplitude reflectance* (transmission amplitude) times the source spectrum $B(\sigma)$. In ordinary FTS, with the sample in the radiation beam after the beamsplitter and symmetric interferograms, the cosine Fourier transform of the interferogram yields just the spectrum $B(\sigma)$ as shown in Eq. (3-25). The complex reflection amplitude (transmission amplitude) is defined in terms of the ratio of reflected (transmitted) and incident electric fields and *not* in terms of the flux, intensity, or power. If one wishes the reflectance (transmittance), then the product of the complex reflection amplitude (transmission

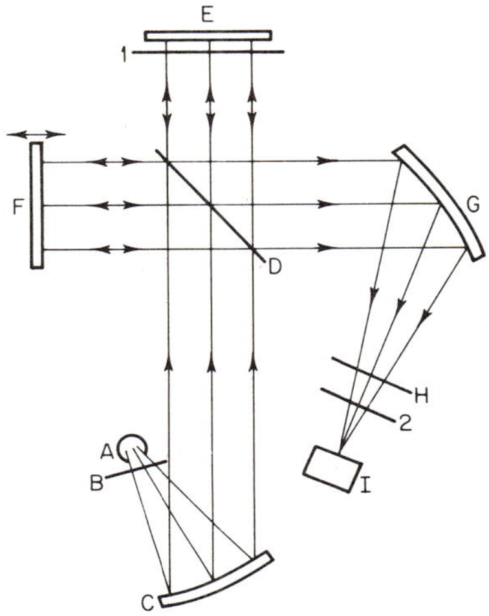


FIG. 8-1. Basic Michelson Interferometer with the sample in the combined beams for ordinary or power FTS (2) and with the sample in the optical path of one arm of the interferometer for amplitude FTS (1) as described in this chapter. A: Source. B: Chopper. C: Collimator. D: Beamsplitter. E: Fixed mirror. F: Movable mirror. G: Focusing mirror. H: Spectral filters. I: Detector.

amplitude) with its complex conjugate is calculated. Thus, putting the sample in just one radiation beam of the interferometer is referred to as amplitude spectroscopy; and amplitude spectroscopy is unique to the Michelson interferometer in its various forms.

In this chapter, the basic Fourier transform integral of the interferogram shall be derived for a reflection sample in one arm. The complex inverse Fourier transform of the interferogram is proportional to the complex conjugate of the complex reflection coefficient of the sample times $B(\sigma)$. Besides amplitude spectroscopy, this result also enables the experimentalist in ordinary Fourier transform spectroscopy to use Michelson interferometers with unbalanced reflectivities of the fixed and movable mirrors. In ordinary FTS with the sample after the beamsplitter and operating in the single-beam mode, the reflection coefficient of one mirror relative to the other will cancel in taking ratios to compute the transmittance or reflectance of the sample. Thus, one has the justification in using a *complex* Fourier transform in ordinary FTS as mentioned in Chapter 3 for Eq. (3-26). This will be discussed in detail in this chapter.

In reflection amplitude spectroscopy, the optical constants of the sample can be obtained directly from the complex reflection coefficient. The optical constants $n$ and $K$ defined by a complex index of refraction $n_c$, with

$$n_c = n + iK, \tag{8-1}$$

are found without recourse to any physical model or a Kramers–Kronig type of calculation. In Eq. (8-1), $n$ is the refractive index and $K$ is the extinction coefficient, given in terms of the absorption coefficient by

$$\alpha = 4\pi K\sigma. \tag{8-2}$$

In this chapter, the equations for obtaining the optical constants from the complex Fourier transform of the interferogram will be derived. An example of reflection amplitude spectroscopy will be presented from the data of Professor Ely E. Bell.

Since the details of how many passes the beam makes through the sample are critical to transmission amplitude spectroscopy, several cases will be studied. In transmission amplitude spectroscopy for solids, multiple internal reflections are important. In transmission amplitude spectroscopy of gases, the optical path in the sample and phase effects produced by the chamber windows must be considered. Thus, the basic equations of transmission amplitude spectroscopy will be derived for self-supporting solid samples and for gas samples in cells. In both cases, it is again possible to obtain the optical constants.

It will also be shown that amplitude spectroscopy can be used for liquid-phase studies. Usually, the interferometer is turned on edge and the sample sits on a transparent substrate in the optical beam in one arm.

In these studies by reflection and transmission amplitude spectroscopy, it is possible to make measurements in all three phases, solid, liquid, and gaseous. Transmission amplitude spectroscopy can be used to study both bulk and surface properties if the sample is reasonably transparent ($T \gtrsim 0.1\%$). However, if the sample is opaque, then reflection amplitude spectroscopy is mandatory. For gaseous studies, transmission amplitude spectroscopy is used to obtain the real part of the index of refraction $n$, and ordinary FTS is used to measure $K$. In ordinary FTS with the sample in the combined radiation beams, the absorption amplitude is not very sensitive to the sample condition or windows, and accurate absorption coefficients can be obtained. For liquid studies in transmission amplitude spectroscopy, it is possible to separate front and back surface effects from bulk phenomena with a knowledge of the interferogram.

Finally, it should be pointed out that amplitude spectroscopy is capable of measurements of the power transmittance (reflectance) to extremely small values. Since the amplitudes are measured and there is no false radiation problem, after one squares the amplitude to obtain the power, power ratios as small as $1 \times 10^{-4}$ may be obtained.

The first paper on the subject of amplitude Fourier transform spectroscopy was Refractometry in the Far-Infrared Using a Two-Beam Interferometer, by Chamberlain *et al.* [1]. Since then, the work has proceeded rapidly at the National Physics Laboratory (NPL) in England and at Ohio State University in the United States. Amplitude spectroscopy is well summarized in five articles [2–6], which, with their references, cover the subject quite well. The references cited in the review article by Chamberlain *et al.* [2] are comprehensive with regard to the NPL work and much of the Ohio State University research.

The first four articles will be extensively referred to in this chapter. The new developments exhibited in the article by Sanderson and Scott are quite significant for gaseous asymmetric interferometry instrumentation and are discussed in Appendix C. This appendix is devoted to the so-called "cat's-eye" retroreflectors, which greatly simplify alignment and the effects of yaw and pitch in the motion of the movable mirror assembly.

## GENERAL THEORY AND REFLECTION STUDIES: SOLIDS—SINGLE SURFACE

We will now derive the basic, complex Fourier transform integral for Fourier transform spectroscopy. Everything that was assumed in Chapter 3 will again be invoked: (1) the law of superposition and (2) the assumption that the electric fields are real functions. In addition, we now assume that the reflectivity of the mirror (or sample replacing the mirror) in the fixed mirror arm is different from the reflectivity of the mirror in the movable arm. We may begin with Eq. (3-11), which stated that the electric field as a function of position $y(z)$ is given by

$$y(z) = \int_{-\infty}^{\infty} \mathscr{E}(\sigma) e^{i2\pi\sigma z}\, d\sigma, \tag{8-3}$$

where $\mathscr{E}(\sigma)$ is the electric field for wave number $\sigma$. Since we are dealing with electric field *amplitudes*, the reflectivity is a reflection amplitude or reflection coefficient $r(\sigma)$. The reflectance, or ratio of reflected to incident power, $R$ is given by

$$R(\sigma) = r(\sigma) r^*(\sigma). \tag{8-4}$$

Also, it should noted that the reflection coefficient may be expressed as

$$r(\sigma) = r_0(\sigma) e^{i\varphi_r(\sigma)}, \tag{8-5}$$

where $r_0(\sigma)$ is the modulus and $\varphi_r(\sigma)$ is the phase. For a plane mirror which is a very good conductor, $\varphi_r(\sigma) = \pi$ and $r_0(\sigma) = 1$. These properties for the mirror are consistent with the assumption that traveling plane waves are incident on it of the form exp $i(2\pi\sigma z - \omega t)$. Henceforth, one cannot use exp $- i(2\pi\sigma z - \omega t)$ for the incident wave. It is usual to assume that the sign is unimportant; however, the correct statement is that one must be consistent with the sign. So, the field from the sample arm is

$$y_1(z) = \int_{-\infty}^{\infty} r(\sigma) \mathscr{E}(\sigma) e^{i2\pi\sigma z}\, d\sigma, \tag{8-6}$$

and the field from the movable mirror arm is

$$y_2(z) = \int_{-\infty}^{\infty} e^{i\pi}\mathscr{E}(\sigma)e^{i2\pi\sigma(z+\delta)}\, d\sigma. \tag{8-7}$$

Using the law of superposition, we have the resultant field $\mathscr{E}_R(\delta, \sigma)$ given by

$$\mathscr{E}_R(\delta, \sigma) = \mathscr{E}(\sigma)[r(\sigma) + e^{i(2\pi\sigma\delta+\pi)}]. \tag{8-8}$$

Within a multiplying constant, the resultant flux is

$$B(\delta, \sigma) = \mathscr{E}^2(\sigma)\{1 + R(\sigma) + r_0(\sigma)[e^{i[2\pi\sigma\delta+\pi-\varphi_r(\sigma)]} + e^{-i[2\pi\sigma\delta+\pi-\varphi_r(\sigma)]}]\} \tag{8-9}$$

In Eq. (8-9), the sample reflectance $R(\sigma)$ is $r(\sigma)r^*(\sigma) = r_0{}^2(\sigma)$. Summing over all wave numbers, the signal versus relative arm displacement is given by

$$I_R(\delta) = \int_{-\infty}^{\infty} \mathscr{E}^2(\sigma)\{1 + R(\sigma) + 2r_0(\sigma)\cos[2\pi\sigma\delta + \pi - \varphi_r(\sigma)]\}\, d\sigma. \tag{8-10}$$

Negative wave numbers are included in the integral to conserve energy between the interferogram and the spectrum as was done for Eq. (5-3). (This also preserves mathematical rigor.) For large $\delta$, the cosine term oscillates at extremely high frequencies and will average to zero, so we have the interterogram given by†

$$I_R(\delta) - I_R(\infty) = 2\int_{-\infty}^{\infty} r_0(\sigma)\mathscr{E}^2(\sigma)\cos[2\pi\sigma\delta + \pi - \varphi_r(\sigma)]\, d\sigma. \tag{8-11}$$

Expressing the cosine function as the sum of complex exponentials, we may multiply both sides of Eq. (8-11) by $\exp(-i2\pi\sigma'\delta)$, and we have on integration

$$\begin{aligned}\int_{-\infty}^{\infty} &[\exp(-i2\pi\sigma'\delta][I_R(\delta) - I_R(\infty)]\, d\delta \\ &= \int_{-\infty}^{\infty} r_0(\sigma)\mathscr{E}^2(\sigma)\left(\int_{-\infty}^{\infty} \{\exp i[2\pi(\sigma - \sigma')\,\delta + \pi - \varphi_r(\sigma)]\right. \\ &\qquad \left. + \exp -i[2\pi(\sigma + \sigma')\,\delta + \pi - \varphi_r(\sigma)]\}\, d\delta\right) d\sigma.\end{aligned} \tag{8-12}$$

The inner integrals are of the form [7]

$$\int_{-\infty}^{\infty} e^{ikx}\, dx = 2\pi\delta'(k), \tag{8-13}$$

where $\delta'(k)$ is the Dirac delta function. Upon using Eq. (8-13), employing

† By using trigonometric identities, Eq. (8-11) can be broken into cosine and sine Fourier transforms.

the definition of the Dirac delta function, and omitting the negative-wave-number part of the solution since it is simply related to the positive-wave-number solution, one has

$$r_0(\sigma')\mathscr{E}^2(\sigma') \exp i[\pi - \varphi_r(\sigma)] = \int_{-\infty}^{\infty} [I_R(\delta) - I_R(\infty)] \exp(-i2\pi\sigma'\delta)\, d\delta, \tag{8-14}$$

where multiplicative constants have been dropped from consideration.

We can eliminate the prime on $\sigma'$ since it is arbitrary, and we should recall that $B(\sigma) = \mathscr{E}^2(\sigma)$ is the spectrum. This yields the basic complex Fourier transform integral for FTS. That is, we have the spectrum in the equation

$$\boxed{-B(\sigma)r^*(\sigma) = \int_{-\infty}^{\infty} [I_R(\delta) - I_R(\infty)]e^{-i2\pi\sigma\delta}\, d\delta.} \tag{8-15}$$

The true reflection coefficient of the sample $r(\sigma)$ is given in Eq. (8-5). In comparison with Eq. (3-25), we now have a complex inverse Fourier transform which is equal to minus $(e^{-i\pi} = -1)$ the product of the reflection coefficient of the sample and the spectrum. By replacing the sample with a perfect mirror and recording the interferogram, one may obtain $B(\sigma)$. Therefore, by taking the ratio of the complex inverse Fourier transforms of the interferograms, we may write

$$\begin{aligned} r^*(\sigma) &= r_0(\sigma)e^{-i\varphi_r(\sigma)} \\ &= -\mathscr{F}^{-1}\{[I_R(\delta) - I_R(\infty)]_{\text{sample}}\}/\mathscr{F}^{-1}\{[IR(\delta) - I_R(\infty)]_{\text{mirror}}\}. \end{aligned} \tag{8-16}$$

Equation (8-16) can be used to determine completely the reflection modulus and phase; however, the use of Eq. (8-16) for such calculations is not as convenient as the procedure presented in the following.

At normal incidence, the reflection coefficient may be given as

$$r(\sigma) \equiv (1 - n_c)/(1 + n_c) = r_0(\sigma) \cos \varphi_r(\sigma) + ir_0(\sigma) \sin \varphi_r(\sigma). \tag{8-17}$$

From Eqs. (8-16) and (8-17), we are able to compute the optical constants where the complex index of refraction is given by Eq. (8-1). Substituting Eq. (8-1) in Eq. (8-17) and equating the real and imaginary parts, we have

$$r_0(\sigma) \cos \varphi_r(\sigma) = (1 - n^2 - k^2)/[(1 + n)^2 + K^2] \tag{8-18}$$

and

$$r_0(\sigma) \sin \varphi_r(\sigma) = -2K/[(1 + n)^2 + K^2]. \tag{8-19}$$

Solving Eq. (8-18) for $K$, substituting for $K$ in Eq. (8-19), simplifying, and solving the resulting equation for $n$, we obtain

$$n(\sigma) = [1 - r_0{}^2(\sigma)]/[1 + 2r_0(\sigma)\cos\varphi_r(\sigma) + r_0{}^2(\sigma)]. \tag{8-20}$$

Knowing $n(\sigma)$ we may compute $K(\sigma)$, and find that

$$K(\sigma) = -[2r_0(\sigma)\sin\varphi_r(\sigma)]/[1 + 2r_0(\sigma)\cos\varphi_r(\sigma) + r_0{}^2(\sigma)]. \tag{8-21}$$

In terms of a complex dielectric constant defined as $\varepsilon_c \equiv n_c{}^2 \equiv \varepsilon' + i\varepsilon''$, we have $\varepsilon' = n^2 - K^2$ and $\varepsilon'' = 2nK$. So one can compute the spectrum of the complex refractive index from Eqs. (8-20) and (8-21).† The easiest way to compute $\varphi_r(\sigma)$ is from Eq. (8-15). If we define

$$\int_{-\infty}^{\infty} [I_R(\delta) - I_R(\infty)]e^{-i2\pi\sigma\delta}\, d\delta \equiv P(\sigma) - iQ(\sigma), \tag{8-22}$$

we have from Eq. (8-15) that

$$B(\sigma)r_0(\sigma)\cos[\varphi_r(\sigma) - \pi] = P(\sigma) \tag{8-23}$$

and

$$B(\sigma)r_0(\sigma)\sin[\varphi_r(\sigma) - \pi] = Q(\sigma). \tag{8-24}$$

Use has been made of the fact that the spectrum $B(\sigma)$ is real. We have defined the cosine and sine Fourier transforms on the interferogram as

$$P(\sigma) \equiv \int_{-\infty}^{\infty} [I_R(\delta) - I_R(\infty)]\cos(2\pi\sigma\delta)\, d\delta \tag{8-25a}$$

and

$$Q(\sigma) \equiv \int_{-\infty}^{\infty} [I_R(\delta) - I_R(\infty)]\sin(2\pi\sigma\delta)\, d\delta, \tag{8-25b}$$

and we note that they are quantities which can be computed from the interferogram. Taking a ratio with Eqs. (8-23) and (8-24), the sample phase is given by

$$\varphi_r(\sigma) = \pi + \arctan[Q(\sigma)/P(\sigma)]. \tag{8-26}$$

Thus, the phase can be determined by one experimental scan with the sample and with no background scan if the instrument is perfect enough to give a

† By beginning with Eqs. (8-18) and (8-19), one can show that

$$\tan\varphi_r(\sigma) = 2K/(n^2 + K^2 - 1)$$

and

$$r_0(\sigma) = \{[(1-n)^2 + K^2]/[(1+n)^2 + K^2]\}^{1/2}.$$

zero-phase spectrum in the background. Since $K(\sigma)$ approaches zero for $\sigma$ approaching zero, one has $\varphi(\sigma = 0) = 0$.

Also, using Eqs. (8-23) and (8-24), the modulus of the reflection coefficient of the sample is given by

$$r_0(\sigma) = [P^2(\sigma) + Q^2(\sigma)]/[P_M{}^2(\sigma) + Q_M{}^2(\sigma)]^{1/2} \tag{8-27}$$

where the subscript M refers to spectra taken with a mirror replacing the sample. Taking the ratio in Eq. (8-27) eliminated the spectrum $B(\sigma)$.

In order to summarize the calculation procedure to obtain $n(\sigma)$ and $K(\sigma)$, consider the following sequences:

1. Put the flat sample in place of the fixed mirror, record the interferogram, and compute the complex inverse Fourier transform of the interferogram with a real part $P(\sigma)$ and an imaginary part $-Q(\sigma)$.
2. Replace the sample with a mirror, record the interferogram, and compute the complex inverse Fourier transform with a real part $P_M(\sigma)$ and an imaginary part $-Q_M(\sigma)$.
3. The reflection coefficient modulus $r_0(\sigma)$ is given by Eq. (8-27). The reflectance is given by $r_0{}^2(\sigma)$.
4. The reflection phase angle $\varphi_r(\sigma)$ for the sample is given by Eq. (8-26) if the mirror has zero phase factors.
5. The real part of the index of refraction $n$ may be computed by substituting $r_0(\sigma)$ and $\varphi_r(\sigma)$ in Eq. (8-20).
6. The extinction index $K$ may be computed by substituting $r_0(\sigma)$ and $\varphi_r(\sigma)$ in Eq. (8-21).

All of these steps are straightforward computer problems. Experimentally, one usually has difficulty getting flat samples which are large enough to replace the fixed mirror. Also, it is not an easy matter to replace the fixed mirror by an exactly aligned sample. Alignment errors of very small dimensions can introduce grotesque errors in the phase. Bell [3] of Ohio State University has solved the problem with the interferometer shown in Fig. 8-2. Another Michelson interferometer which can study small samples in the beam in one arm was designed by Chamberlain *et al.* [1]. Bell points out the following main features of his asymmetric interferometer: (1) the sample is placed in one arm of the interferometer; (2) in order to use small samples, the radiation beam is not collimated in the interferometer arms, but instead is imaged on the sample in one arm; (3) samples can be measured either in transmission or reflection; (4) the arm without the sample contains the movable mirror for scanning the interferogram; and (5) the radiation signal is chopped in one arm of the interferometer and the backs of the chopper blades are used to reflect a compensating signal into the path leading to the detector. The

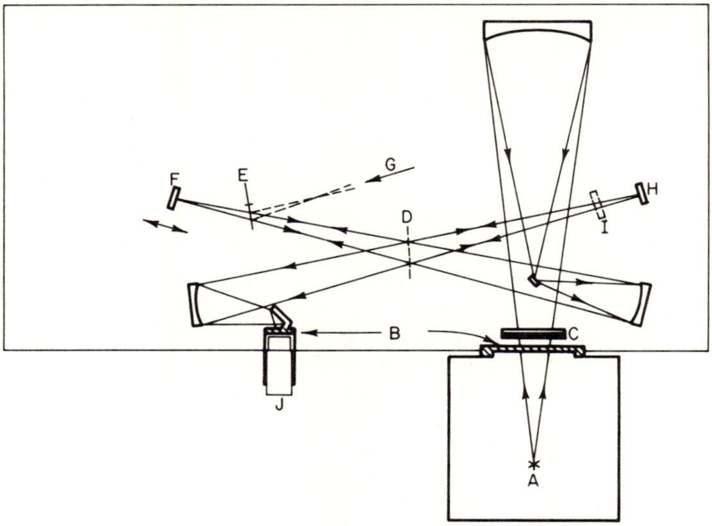


FIG. 8-2. The optical plan of a vacuum-far-infrared Michelson interferometer used in the asymmetric mode for amplitude spectroscopy. (After Russell and Bell [4]). A: Mercury arc. B: Filters. C: Polarizer. D: Beamsplitter. E: Chopper. F: Movable mirror. G: Compensating beam. H: Mirror or reflection sample. I: Transmission sample. J: Golay detector.

compensating signal is adjusted to reduce $I_R(\infty)$ to zero. As will be mentioned later, for transmission studies, the radiation beam passes *only once* through the sample, The beam transmitted through the sample does not retrace the incident beam path, but returns to the beamsplitter above the plane of the diagram.

The asymmetric interferometer is particularly well suited for reflection coefficient measurements because it only requires small samples and the samples can be aligned easily. For reflection studies, Professor Bell has the sample and reference mirror mounted over holes near the periphery of a *horizontal* disk.

The power reflectance $R = r_0^2(\sigma)$ and the phase $[\varphi_r(\sigma) - \pi]$ for potassium bromide are shown in Fig. 8-3. These data were obtained on Bell's asymmetric Michelson interferometer using amplitude spectroscopy. Note the very small error bars, which will be discussed later in the chapter. A major achievement of amplitude spectroscopy is exhibited in Fig. 8-4, which presents the optical constants of potassium bromide, as determined by amplitude spectroscopy, using no physical model of potassium bromide, nor any Kramers–Kronig type of calculation! It is truly a major advance in optics to be able to measure and compute by simple means the complex refractive index spectra of materials.

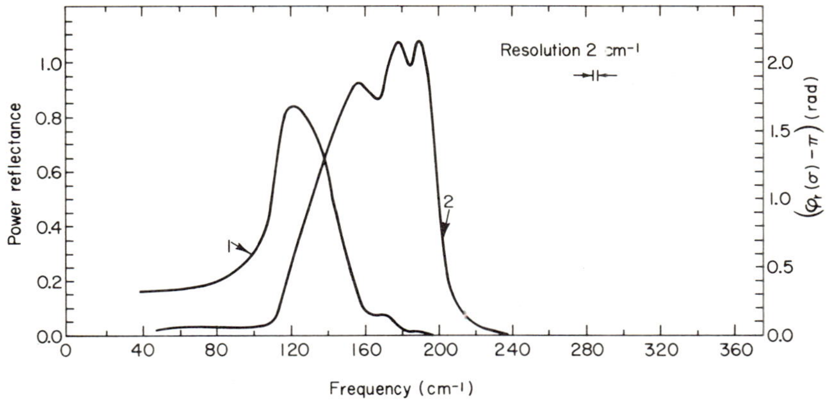


FIG. 8-3. (1) The power reflectance $r_0{}^2(\sigma)$ and (2) the phase $\varphi_r(\sigma)$ for potassium bromide as measured by amplitude spectroscopy. The error bars on the curves show the size of the variation in the measured values for several different sample measurements. The phase $\varphi_r(\sigma) - \pi$ is defined in Eq. (8-5). (After Bell [5].)

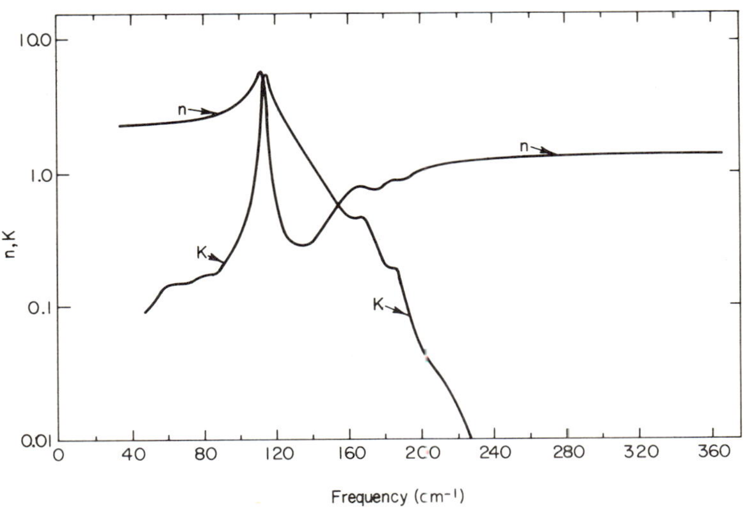


FIG. 8-4. The index of refraction $n(\sigma)$ and extinction coefficient $K(\sigma)$ of potassium bromide determined by amplitude spectroscopy. (After Bell [5].)

## COMPLEX INVERSE FOURIER TRANSFORM OF THE INTERFEROGRAM

Before leaving reflection amplitude spectroscopy, it should be pointed out that two fundamental questions concerning ordinary FTS (combined beam sampling) with Michelson interferometers have been answered. The questions are, "What effect on the spectrum or measurements is caused by not having the mirrors in the two arms identical?," and "What is the justification in using a *complex* Fourier transform of the interferogram to compute the spectrum?" Both questions have the same answer. If the complex inverse Fourier transform of the interferogram is computed, one has from Eq. (8-15) the quantity $-B(\sigma)r^*(\sigma)$. Since $r(\sigma)$ is held constant in ordinary FTS, the factor $r(\sigma)$ would cancel in taking ratios for the transmittance or reflectance of the samples. So the answer to the first question is that very slight arm asymmetries in symmetric FTS can be ignored for two-sided interferograms. To answer the second question, clearly, one is justified in using the complex inverse Fourier transform in ordinary FTS. For single-beam spectroscopy, the results will be the same as for just using the Fourier cosine transform as in Eq. (3-25). Also, as will be shown in Chapter 12, there is a distinct advantage in using the complex inverse Fourier transform of the interferogram. The advantage is that the complex inverse Fourier transform of the interferogram, when used for two-sided interferograms, eliminates any dependence in the computed spectrum on the location of the origin. Two-sided interferograms are recorded when the signal versus arm displacements are measured on both sides of the null or balanced arm position or origin. If the origin is mislocated in one-sided interferograms, serious distortion in the spectrum can result, as discussed in Chapter 12. These problems are overcome with two-sided interferograms which is also discussed in Chapter 12.

## TRANSMISSION STUDIES: SOLIDS—SINGLE PASS (NO CHANNEL SPECTRA)

If we want to use the asymmetric interferometer with *single* passes of the optical beam through the sample, we can employ Bell's instrument shown in Fig. 8-2. Consider the transmission coefficient of a beam that makes a single pass through a sample, but with part of the beam multiply internally reflected. In Fig. 8-5, an incident beam of amplitude unity on a plane, parallel

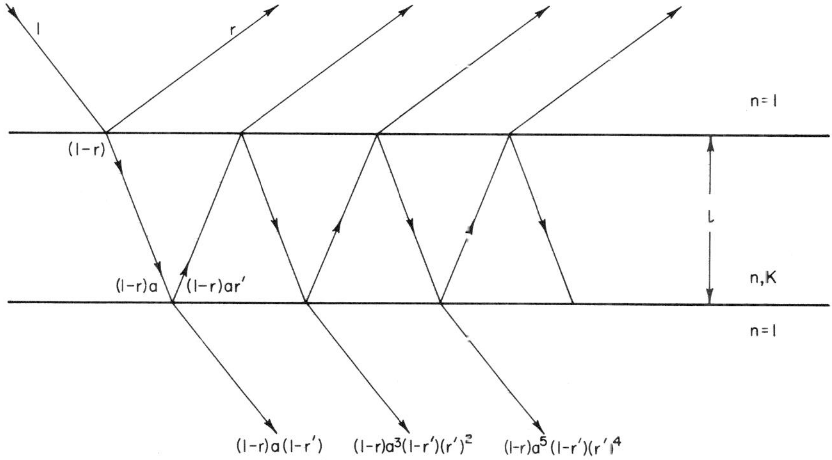


FIG. 8-5. Radiation beams through a plane, parallel slab of thickness $l$ having a complex index of refraction $n_c = n + iK$. It is understood that the electric fields in the material are reduced by the dielectric constant.

slab is shown with the complex amplitude of each emergent beam labeled. The external reflection coefficient is $r$, the transmission coefficient is $a$, and the internal reflection coefficient is $r'$. We have $r = -r'$, with $r$ given in Eq. (8-17). Assuming the beam is at normal incidence, the transmission coefficient $a$ is

$$a \equiv e^{-2\pi\sigma Kl} e^{i2\pi\sigma nl}, \tag{8-28}$$

where $n$ and $K$ are defined in Eq. (8-1). The slab thickness is $l$. Summing the transmission terms, we get the transmission amplitude $\tau$,

$$\tau = (1 - r)a(1 - r')[1 + a^2(r')^2 + a^4(r')^4 + \cdots] \tag{8-29}$$

or, noting the geometric series, $\tau$ is equivalent to

$$\tau = (1 - r)a(1 - r')/[1 - a^2(r')^2]. \tag{8-30}$$

It is more convenient to use $r = -r'$ and rewrite Eq. (8-29) as a simple power series, and we have

$$\tau = (1 - r^2)a[1 + (ar)^2 + (ar)^4 + \cdots]. \tag{8-31}$$

If the reflection coefficient is small, we can take just the first term of the series in (8-31). With Eq. (8-28) and $r^2 \approx r_0{}^2 e^{i2\pi} = r^2{}_0$, we have

$$\tau \simeq (1 - r_0{}^2)e^{-2\pi\sigma Kl} e^{i2\pi\sigma nl}. \tag{8-32}$$

From Fig. 8-5, this is equivalent to just considering the first or primary transmitted beam. We will put the solid sample in the beam of the arm with the fixed mirror. If the sample is of thickness $l$ and has index of refraction $n(\sigma)$, the optical path in the sample is $n(\sigma)l$, or the phase is $2\pi\sigma nl$. If the sample space were empty, the optical path would be $l$ and the phase $2\pi\sigma l$. Then, the increase in optical path in the beam in the fixed mirror arm due to the sample is $nl - l = (n - 1)l$; thus the phase is increased by $2\pi\sigma(n - 1)l$ due to the sample. The electric field as a function of position $z$ as it emerges from the fixed arm is given by

$$y_1(z) = \int_{-\infty}^{\infty} e^{i\pi}\tau_0(\sigma)\mathscr{E}(\sigma)e^{i2\pi\sigma[z+(n-1)l]}\, d\sigma, \tag{8-33}$$

and form the movable arm,

$$y_2(z) = \int_{-\infty}^{\infty} e^{i\pi}\mathscr{E}(\sigma)e^{i2\pi\sigma(z+\delta)}\, d\sigma, \tag{8-34}$$

where

$$\tau_0(\sigma) = (1 - r_0{}^2)e^{-2\pi\sigma Kl}. \tag{8-35}$$

Using the same procedure as used from Eqs. (8-6) and (8-7) to Eq. (8-15), we obtain

$$B(\sigma)\tau_0(\sigma)e^{i2\pi\sigma(n-1)l} = \int_{-\infty}^{\infty} [I_{\mathrm{R}}(\delta) - I_{\mathrm{R}}(\infty)]e^{-i2\pi\sigma\delta}\, d\delta. \tag{8-36}$$

Taking the Fourier transform of Eq. (8-36), we have

$$I_{\mathrm{R}}(\delta) - I_{\mathrm{R}}(\infty) = \int_{-\infty}^{\infty} B(\sigma)\tau_0(\sigma)e^{i2\pi\sigma\{\delta-[n(\sigma)-1]l\}}\, d\sigma. \tag{8-37}$$

Denote the average value of the index of refraction as $\bar{n}$. (For constant $n$, one has $n = \bar{n}$.) We note that for $\delta = (\bar{n} - 1)l$, we have an integral over the positive function $B(\sigma)\tau_0(\sigma)$. When $\delta \neq (\bar{n} - 1)l$, one has an integrand of $B(\sigma)\tau_0(\sigma)$ times an oscillating function which is always less than or equal to unity. Therefore, the maximum of the integral always occurs near or at $\delta = (\bar{n} - 1)l$. That is, the grand maximum of the interferogram will *not* occur at $\delta = 0$ but will be at or near $\delta = (\bar{n} - 1)l$. As will be shown later, it is convenient in transmission studies in amplitude spectroscopy to locate the path difference at the grand maximum or peak intensity in the interferogram. For nondispersive samples the grand maximum displacement from the origin is given precisely by $(\bar{n} - 1)l$. For dispersive media, the grand maximum will be close to $(\bar{n} - 1)l$, where $\bar{n}$ is some suitable average index of refraction. Thus, for computational advantages which will be discussed later, we shall define $D$ as the optical path difference *between* the origin and the position of the grand maximum, which for materials of constant index gives $D = (\bar{n} - 1)l$

for single passes through the sample. In Fig. 8-6, the interferograms for a background and two samples are shown with $\delta$ explicitly given at the bottom. To determine approximately the index of refraction of a material with a constant index of refraction such as polyethylene, we can measure $D$ and compute $\bar{n}$ if we know the thickness of the sample. Thus, for samples with constant indices of refraction, the analysis for the real part of the index of refraction is exceedingly simple. For the polyethylene example in Fig. 8-6, one can see that $D = 0.717$ mm and with $l = 1.510$ mm, $\bar{n}$ is computed as 1.47 (accurately, 1.52).

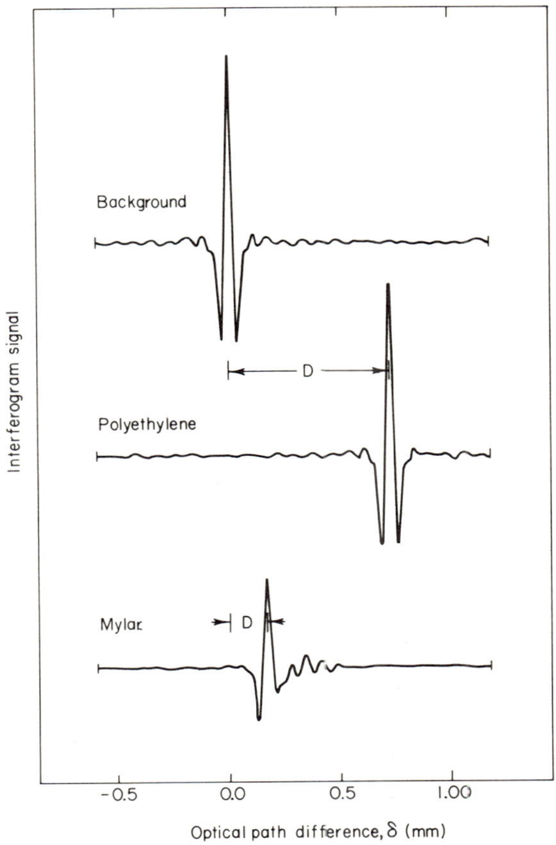


FIG. 8-6. The central portions of a background interferogram, a polyethylene sample interferogram, and a Mylar sample interferogram, to show the displacement and distortion of the central structure in amplitude spectroscopy. The thickness of the polyethylene was 1.510 mm; the thickness of the Mylar was 0.249 mm. (After Bell [5].)

To measure the small $\alpha(\sigma)$ or $K(\sigma)$ as given in Eq. (8-2), it is also possible to place the sample in the combined beams and perform an ordinary transmission experiment on a very thick sample.

We could have completed the analysis of Eq. (8-36) to obtain $n(\sigma)$ and $K(\sigma)$ when $n(\sigma) \neq \bar{n}$. Let us define

$$P(\sigma) - iQ(\sigma) \equiv \int_{-\infty}^{\infty} [I_{\mathrm{R}}(\delta) - I_{\mathrm{R}}(\infty)]e^{i2\pi\sigma\delta}\, d\delta \tag{8-38}$$

or

$$P(\sigma) - iQ(\sigma) = B(\sigma)\tau_0(\sigma)e^{-i\varphi_\tau(\sigma)}. \tag{8-39}$$

That is, let us separate the Fourier transform of the interferogram into its real (cosine) and imaginary (sine) transforms. The modulus is given by $B(\sigma)\tau_0(\sigma)$ and the transmission phase by

$$\varphi_\tau(\sigma) = 2\pi\sigma[n(\sigma) - 1]l. \tag{8-40}$$

Thus, we may write

$$\varphi_\tau(\sigma) = \arctan[Q(\sigma)/P(\sigma)] \tag{8-41}$$

and

$$B(\sigma)\tau_0(\sigma) = \{P^2(\sigma) + Q^2(\sigma)\}^{1/2}. \tag{8-42}$$

As discussed in the next section, $\varphi_\tau(\sigma)$ using Eqs. (8-40) and (8-41), is accurate to only within constant multiples of $\pm 2\pi m$, where $m$ is an integer assuming that the interferometer is perfectly symmetric.

Thus, if $n$ and $K$ depend on the wave number, we can use Eqs. (8-40) and (8-41) to obtain $n$. By taking a background scan to obtain $B(\sigma)$ and using Eqs. (8-35) and Eq. (8-42) with $r_0{}^2 = R = (n-1)^2/(n+1)^2$, we can compute $K(\sigma)$ from

$$\tau_0(\sigma) = \{[P^2(\sigma) + Q^2(\sigma)]/[P_{\mathrm{BG}}^2(\sigma) + Q_{\mathrm{BG}}^2(\sigma)]\}^{1/2}. \tag{8-43}$$

The subscript BG refers to the background scan used to measure $B(\sigma)$. Again, $K(\sigma)$ is most accurately obtained from an ordinary FTS transmission experiment with a thick sample in the combined beams.

## PHASE ERRORS OF $\pm 2\pi$(INTEGER)

In Eq. (8-41), it is important to note that principal values of the phase have been used. The phase $\varphi_\tau(\sigma)$ as given by Eq. (8-41) might be in error by a term of $\pm 2\pi m$, where $m$ is an integer. If there is a $2\pi$ discontinuity in the phase as a

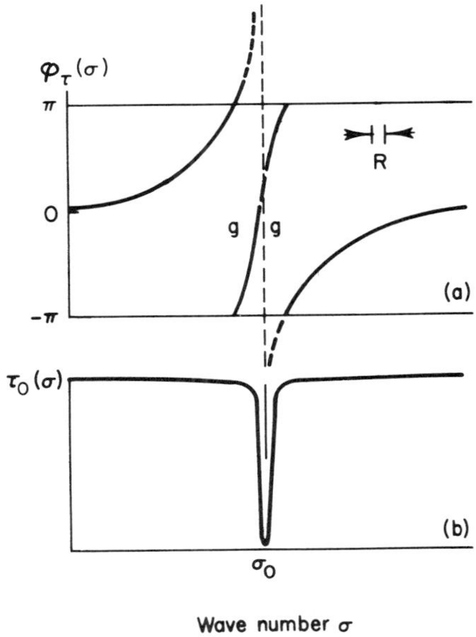


FIG. 8-7. (a) The phase spectrum and (b) the transmission spectrum of an unresolved absorption centered at $\sigma_0$ such as for a gas. The solid lines in (a) show the principal value of $\varphi_\tau(\sigma)$. The discontinuities at $g$ may be eliminated by the suitable addition of $\pm 2\pi$ to $\varphi_\tau(\sigma)$. The true phase, indicated by the shorter dashed lines, is obtained by the addition of $2\pi$ where $d\tau_0/d\sigma < 0$, the subtraction of $2\pi$ where $d\tau_0/d\sigma > 0$, and the retention of a discontinuity where $d\tau_0/d\sigma = 0$. (After J. E. Chamberlain *et al.* [2].)

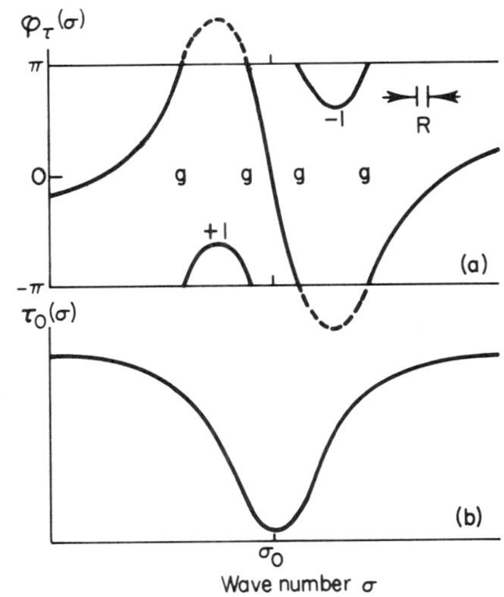


FIG. 8-8. (a) The phase spectrum and (b) the transmission spectrum of a resolved absorption centered at $\sigma_0$ such as for a solid or liquid. The solid lines in (a) show the principal value of $\varphi_\tau(\tau)$. The discontinuities at $g$ may be eliminated by the suitable addition of $+2\pi$ or $-2\pi$ to $\varphi_\tau(\sigma)$. The true phase, indicated by the dashed lines, is obtained by the addition of $2\pi$ where $d\tau_0/d\sigma < 0$ and the subtraction of $2\pi$ where $d\tau_0/d\sigma > 0$. (After Chamberlain *et al.* [2].)

function of $\sigma$, one can frequently surmise how to manipulate the factor of $\pm 2\pi m$ correctly. At very low wave numbers away from a resonance, $\varphi_\tau(\sigma)$ should approach zero since Eq. (8-40) indicates $\varphi_\tau(\sigma)$ would be approximately proportional to $\sigma$. This point is illustrated in Figs. 8-7 and 8-8 for gaseous or liquid and solid samples. The solid lines are the phase angles for various wave numbers as computed by Eqs. (8-40) and (8-41). Noting the discontinuities in the phase at $g$, we see that the phase curves can be corrected by shifting the curves near $\sigma_0$ by $2\pi$ and $-2\pi$. The shifted portions are indicated by the dashed lines. From the transmission curves (b), we note that the absorption is near zero for small $\sigma$, and from the phase curve, we have $\varphi_\tau(\sigma)$ near zero for small $\sigma$, as it should be.

Another way to determine the phase shifts of $2\pi m$ is to vary $l$, the thickness of the sample. As $l$ approaches zero, from Eq. (8-40), one can see that $\varphi_\tau(\sigma)$ should also approach zero.

## SHIFTING THE COMPUTATION ORIGIN TO THE GRAND MAXIMUM POSITION

As mentioned earlier, the position of the grand maximum is very close to the path difference of $\delta = (\bar{n} - 1)l$ for single passes of transmitted beams through the sample. Using the Fourier analysis shift theorem, it it easy to shift the center of the numerical analysis of the interferogram to the center of the grand maximum. Let us define a new coordinate $\delta_D$ given by

$$\delta_D \equiv \delta - D, \tag{8-44}$$

where the origin of the new coordinate system is at the grand maximum. Then, by the shift theorem,

$$\mathscr{F}^{-1}\{I_R(\delta) - I_R(\infty)\} = e^{i2\pi\sigma D}\mathscr{F}^{-1}\{I_R(\delta_D) - I_R(\infty)\}, \tag{8-45}$$

where

$$\mathscr{F}^{-1}\{I_R(\delta_D) - I_R(\infty)\} \equiv P_D(\sigma) - iQ_D(\sigma). \tag{8-46}$$

Using Eq. (8-45) on the Fourier transform of the interferogram in Eq. (8-36) for a sample in single pass, we have

$$B(\sigma)\tau_0(\sigma)e^{-i2\pi\sigma[(n-1)l-D]} = \int_{-\infty}^{\infty} [I_R(\delta_D) - I_R(\infty)]e^{-i2\pi\sigma\delta_D}\, d\delta_D \tag{8-47}$$

and the new phase $\varphi_{\tau D}(\sigma)$ would be

$$\varphi_{\tau D}(\sigma) = \arctan[Q_D(\sigma)/P_D(\sigma)] = 2\pi\sigma\{[n(\sigma) - 1]l - D\}. \tag{8-48}$$

The index can be computed from Eq. (8-48) in a straightforward manner. The new phase $\varphi_{\tau,D}(\sigma)$ is a function that does not depart very far from $2\pi\sigma[(n-1)l - D] \simeq 2\pi\sigma(n - \bar{n})l$. Thus, the shift in the center of the computation origin to $\delta_D = 0$ reduces the oscillations in phase.

Henceforth, after the inverse Fourier transform of the interferogram has been found, it will be shifted by $D$ so that numerical transforms will be performed with the variable $\delta_D$ as defined in Eq. (8-46). It has been shown by Chamberlain *et al.* that shifting the numerical calculation to have the origin at $\delta_D = 0$ reduces the noise in the computed spectrum. The reduction in noise by the shift can amount to factors of two or more. In more proper terms, by not using the computation origin at $\delta_D = 0$, noise is falsely introduced into the computed spectrum.

There is another problem associated with shifting the computational origin to the grand maximum. For *each sample*, the grand maximum must be located very accurately. When one records an interferogram in steps, he may not actually have a step precisely at the grand maximum. Moving the mirror in one arm in discrete steps has distinct advantages as discussed in Chapter 7. Since the interferogram is stepped in equal increments and a step does not necessarily fall precisely at the grand maximum, it is necessary to investigate the effect. Suppose one had a grand maximum as shown in Fig. 8-9 and that

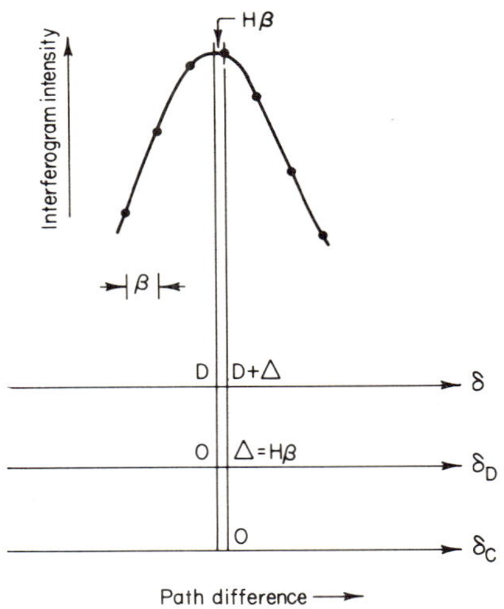


FIG. 8-9. Interferogram intensity versus path difference near the grand maximum. The variables for the path difference are shown. The top axis for $\delta$ has the origin located when the arms are in balance with no sample. The middle axis for $\delta_D$ has the origin located at the grand maximum. The bottom axis for $\delta_C$ has the origin located at the step for which the peak intensity is recorded.

the steps fell at the dots which are a path difference $\beta$ apart. The true path difference is indicated by $\delta$, the grand maximum is at the origin of $\delta_D$, and the recorded data would have a maximum at the origin of $\delta_c$ which would be used in the computation. One would, in using $\delta_c$, have missed the grand maximum by a path difference of $\Delta = H\beta$, where $H$ is the fraction of the path step $\beta$. One would locate the true maximum either with a parabolic fit of the data points, or by scanning the grand maximum with very small steps. Thus, $\Delta$ and then $H$ could be determined for a particular optical path step used in recording the interferogram with a particular $\beta$. The sample data could have the maximum missed by $H\beta$ and the background scan could have the grand maximum missed by $H_0\beta$. One may designate $n_{\text{comp}}$ as the index of refraction computed from the interferogram with the grand maximum mistakenly located at $\delta_c = 0$. Then, Chamberlain *et al.* [2] have shown that the true index of refraction $n_{\text{true}}$ is given by

$$n_{\text{true}}(\sigma) = n_{\text{comp}}(\sigma) + [(H - H_0)\beta/l]. \tag{8-49}$$

Thus, the phase error in missing the origin is a constant that introduces a constant error in the index of refraction. Usually these errors in $n$ are of the order of one part in $10^3$, but for accurate measurements they are important.

## TRANSMISSION STUDIES: SOLIDS—SINGLE PASS (WITH CHANNEL SPECTRA)

We have already calculated the modulus and phase for the transmission amplitude through a sample with a single pass of the beam through the sample. Now, we wish to extend the example to include the effect of multiple internal reflections of the beam inside the sample. In this section, we shall be content with including in the transmission amplitude the amplitude of the primary partial wave $[(1 - r^2)a]$ and the amplitude of the secondary partial wave $[(1 - r^2)a(ar)^2]$. These expressions may be verified by setting $r = -r'$ in Fig. 8-5. So, the secondary beam is just internally reflected once, and it will be the strongest internally reflected beam.

A treatment similar to that carried out for the transmission studies without channel spectra can be repeated if extra terms in the transmission coefficient are put in polar form. That is, letting

$$1 + [a(\sigma)r_0(\sigma)]^2 \equiv C_0(\sigma)e^{i\varphi_c(\sigma)} \tag{8-50}$$

for $(a_0 r_0)^2 \ll 1$ yields

$$\varphi_c(\sigma) \simeq \arcsin\{[a_0(\sigma)r_0{}^2(\sigma)] \sin[4\pi\sigma n(\sigma)l]\}. \tag{8-51}$$

Completing the analogous treatment, we have

$$n(\sigma) - 1 = (D/l) + (1/2\pi\sigma l)[\varphi_{\tau,D}(\sigma) - \varphi_c(\sigma)], \tag{8-52}$$

where

$$\varphi_{\tau,D}(\sigma) \equiv 2\pi\sigma\{[n(\sigma) - 1]l + [\varphi_c(\sigma)/2\pi\sigma D] - D\}. \tag{8-53}$$

However, from Eq. (8-51), we note that $\varphi_c(\sigma)$ is a function of $n(\sigma)$ not only through the argument of the sine function, but through the fact that for small absorption, one has

$$r_0^2(\sigma) \simeq \{[1 - n(\sigma)]/[1 + n(\sigma)]\}^2 \tag{8-54}$$

To overcome the problem of solving the transcendental equation (8-53) for $n(\sigma)$, we use a reiterative calculation which converges fairly rapidly. One notes that there are oscillations in the phase about the mean, caused by channel spectra. These oscillations come from the term for $\varphi_c(\sigma)$—as can be seen in Eq. (8-51). As $\sigma$ varies in Eq. (8-51), one can see that for nearly constant $n(\sigma)$, $\varphi_c(\sigma)$ will only oscillate in a minor way.

## TRANSMISSION STUDIES: GASES—SINGLE PASS (BELL'S INTERFEROMETER)

Gases may be studies in amplitude spectroscopy by placing a sample cell in the radiation beam in the fixed-mirror arm of the interferometer shown in Fig. 8-2. This would be for a single pass of the radiation beam through the sample. Double passes of the beam through the sample in amplitude spectroscopy can be accomplished with a nearly conventional Michelson interferometer as described in the next section.

In this case, we are not particularly concerned with reflection from the gas, but we must be careful about the windows. One would, of course, try to use nondispersive or nonabsorbing windows. Nevertheless, the reflection, absorption, and phase shifts of the windows can be included by defining the transmission coefficient of the windows as

$$\tau_w(\sigma) \equiv \tau_{0,w}(\sigma)e^{i\varphi_w(\sigma)}. \tag{8-55}$$

In the analogous treatment, the index of refraction is given by

$$n(\sigma) = 1 + [(D - D_w)/l] + (1/2\pi\sigma l) \times \{\arctan[Q_D(\sigma)/P_D(\sigma)] - \arctan[Q_{D,w}(\sigma)/P_{D,w}(\sigma)]\} \tag{8-56}$$

with

$$P_D(\sigma) - iQ_D(\sigma) = \int_{-\infty}^{\infty} [I_R(\delta_D) - I_R(\infty)]e^{-i2\pi\sigma\delta_D}\, d\delta_D. \tag{8-57}$$

For the window experiment (window only),

$$P_{D,w}(\sigma) - iQ_{D,w}(\sigma) = \int_{-\infty}^{\infty} [I_R(\delta_{Dw}) - I_R(\infty)]e^{-i2\pi\sigma\delta_{Dw}}\, d\delta_{Dw}. \tag{8-58}$$

One has $D_w$ located as the position of the grand maximum in a transmission interferogram with just the sample cell windows in the radiation beam, and $D$ specifying the grand maximum in the complete experiment.

Equation (8-56) can be used to compute the index of refraction of the gas. Again, one must remember that the principal values of the phases were used, and the phases expressed by the arctangents can be in error by $\pm[2\pi(\text{integer})]$. These phase shifts have already been illustrated in Fig. 8-7 for gases. If one wishes, the index can be obtained in the wavenumber range of anomalous dispersion introduced by rotational spectra.

The absorption coefficient can also be determined, and, on assuming for a gas that $r_0{}^2 = 0$ in Eq. (8-35), one has

$$\alpha(\sigma) \simeq (1/l)\ln\{[P_{Dw}^2(\sigma) + Q_{Dw}^2(\sigma)]/[P_D{}^2(\sigma) + Q_D{}^2(\sigma)]\}^{1/2}. \qquad (8\text{-}59)$$

## TRANSMISSION STUDIES: GASES—DOUBLE PASS (ORDINARY MICHELSON INTERFEROMETER)

It is possible to perform amplitude spectroscopy with an ordinary Michelson interferometer. For gas studies, the fixed mirror forms the back of the sample cell and a plane, parallel window is used in front to contain the gas in a cylinder jointed to the window and fixed mirror. This type of sample cell has the advantage of keeping the number of windows to a minimum. If the fixed mirror is made of polished stainless steel, even corrosive gases may be studied. The radiation beam in the fixed-mirror arm comes through the window, goes through the gas, is reflected off the fixed mirror, retraverses the gas, and passes back through the window to the beamsplitter. If one prefers, a compensating window can be placed in the radiation beam in the movable-mirror arm, but this is not always necessary.

We can express the overall window effects as $\tau_w(\sigma)\exp\varphi_w(\sigma)$ as we did in the last section. In fact, the only mathematical change from the last section for a single pass is that $(n-1)l$ is replaced by $2(n-1)l$ [not $(2n-1)l$]. The similar approach yields an index of refraction given by

$$n(\sigma) = 1 + [(D - D_w)/2l] + \{[\phi_\tau(\sigma) - \phi_{w,D}(\sigma)]/4\pi\sigma l\}, \qquad (8\text{-}60)$$

where, from the sample scan, we have

$$\varphi_\tau(\sigma) \equiv \arctan[Q_D(\sigma)/P_D(\sigma)], \qquad (8\text{-}61)$$

and from the background scan, one has

$$Q_{D,w}(\sigma) \equiv \arctan[Q_{Dw}(\sigma)/P_{Dw}(\sigma)]. \qquad (8\text{-}62)$$

The absorption coefficient is given by

$$\alpha(\sigma) = (1/l)\ln\{[P_{Dw}^2(\sigma) + Q_{Dw}^2(\sigma)]/[P_D{}^2(\sigma) + Q_D{}^2(\sigma)]\}^{1/2}, \qquad (8\text{-}63)$$

where $\alpha(\sigma)$ is defined in Eq. (8-2) and for gases we assumed $r_0{}^2 \simeq 0$. The subscript $D$ refers to the complete experiment and $Dw$ implies the "window-only" conditions were used. The corrections for the errors of $\pm 2\pi m$ were discussed near the end of the section, Phase Errors of $\pm[2\pi(\text{Integer})]$. The use of principal values also places restrictions on the sample thickness as will be discussed in the section, Transmission Studies—Limits on Sample Thickness.

Sample data would be similar to that found in the preceding section, Transmission Studies: Gases—Single Pass (Bell's Interferometer). The main advantage of the double-pass technique is that an ordinary Michelson interferometer can be used in amplitude FTS. Doubling the path length is of some advantage to some experimentalists.

## INTERFEROGRAMS FOR TRANSMISSION STUDIES

We shall now point out a dominant feature of the interferograms in Fig. 8-6. In the background interferogram, the signal features were symmetric about the origin; however, in the sample interferograms, the signal is nearly flat to the left of $\delta = 0$ regardless of where the grand maximum is located. As shown by Toll, causality dictates that the delay of the signal through the sample with $n \geq 1$ will allow transmission amplitude features only for $\delta \geq 0$. For negative $\delta$, the interferogram contains no sample imformation in transmission studies.

We can prove the fact that the transmission sample interferogram will be zero for $\delta < 0$. For single-pass transmission experiments,

$$\mathscr{F}^{-1}\{[I_R(\delta) - I_R(\infty)]_s\} = \mathscr{F}^{-1}\{[I_R(\delta) - I_R(\infty)]_{BG}\}\tau_0(\sigma)e^{-i2\pi\sigma[n(\sigma)-1]l}, \qquad (8\text{-}64)$$

where the Fourier transforms are of the sample(s) and windows or background (BG) interferograms. We can see from Eqs. (8-35) and (8-1) that (for $r_0 = 0$) one has

$$\tau_0(\sigma)\exp\{-i2\pi\sigma[n(\sigma) - 1]l\} = \exp\{-i2\pi\sigma[n_c{}^*(\sigma) - 1]l\}. \qquad (8\text{-}65)$$

Let us define $V(\delta)$ as the Fourier transform of $\exp[-i2\pi\sigma(n_c{}^* - 1)l]$. Since the inverse Fourier transform of the Fourier transform of a function is the function, we have

$$\mathscr{F}^{-1}\{V(\delta)\} = \exp\{-i2\pi\sigma[n_c{}^*(\sigma) - 1]l\} \qquad (8\text{-}66)$$

So, substituting Eq. (8-66) in Eq. (8-65) for use in Eq. (8-64), we have

$$\mathscr{F}^{-1}\{[I_R(\delta) - I_R(\infty)]_s\} = \mathscr{F}^{-1}\{[I_R(\delta) - I_R(\infty)]_{BG}\}\mathscr{F}^{-1}\{V(\delta)\}. \quad (8\text{-}67)$$

The product of the inverse Fourier transforms of two functions is the inverse Fourier transform of the convolution of the two functions. With the convolution theorem and taking the Fourier transform of the results, we have

$$[I_R(\delta) - I_R(\infty)]_s = [I_R(\delta) - I_R(\infty)]_{BG} \star V(\delta), \quad (8\text{-}68)$$

where

$$V(\delta) \equiv \int_{-\infty}^{\infty} \exp\{-i2\pi\sigma[n_c{}^*(\sigma) - 1]l\}\exp(i2\pi\sigma\delta)\, d\sigma \quad (8\text{-}69)$$

We will now use a theorem of Fourier analysis due to Titchmarsh to get $V(\delta) = 0$ for $\delta < 0$. Since $n_c$ is regular in the upper half-plane,† then $\exp[-i2\pi\sigma(n_c{}^* - 1)l]$ is regular in the lower half-plane: and also the exponential is bounded and square-integrable. With these conditions and Titchmarsh's theorem, one has $V(\delta) = 0$ for $\delta < 0$. Therefore, the transmission sample interferogram is zero for $\delta = 0$. Those small-signal features that are present in the sample interferograms for $\delta < 0$ are caused by reflections from the front surface of the sample or windows. These signals or signatures to the left of the grand maximum will be discussed in more detail in the section on liquid studies. Between $\delta = 0$ and the grand maximum, there can be transmission signal features in the interferogram, as there are to the right of the grand maximum; but for $\delta < 0$, there is no sample information in pure transmission interferograms. These remarks apply equally well to the double-pass transmission interferograms.

## TRANSMISSION STUDIES: LIMITS ON SAMPLE THICKNESS

We will make these remarks relative to the transmission studies for double-pass experiments and the reader can make the necessary modifications for single-pass studies. Note that the principal value of $\varphi_\tau(\sigma)$ was used; so we have $-\pi < \varphi_\tau(\sigma) < \pi$, or equivalently,

$$|2\pi\sigma\{2[n(\sigma) - 1]l + [\varphi_w(\sigma)/2\pi\sigma] - D\}| < \pi \quad (8\text{-}70)$$

Ignoring window effects by setting $\varphi_w(\sigma) = 0$ and replacing $D$ by its approximate value $2(\bar{n} - 1)l$, we can see that

$$|nl - \bar{n}l| < 1/4\sigma = \lambda/4 \quad (8\text{-}71)$$

† The function $n - iK$ is regular in the lower half-plane and not in the upper half-plane as sometimes stated.

That is, the variation of $n(\sigma)$ from $\bar{n}$, times the sample cell length (one-half total path in the gas) must be less than a quarter wavelength. This is a lesser restriction that one would have had if the Fourier transform computational origin had not been shifted to the grand maximum at $\delta_D = 0$.

One can also interpret Eq. (8-71) in terms of window bulge. If the gas pressure allows the windows to deform, then phase errors can be introduced and Eq. (8-71) can be interpreted in terms of new cell lengths different from $l$. These errors can also be found in solid samples with a thickness variation. Figure 8-10 shows how serious the problem can be for a "black" polyethylene sample.

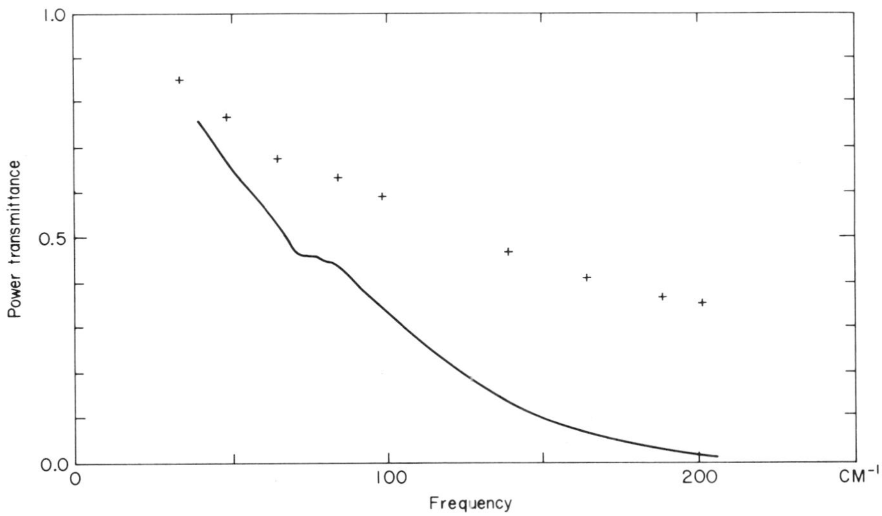


FIG. 8-10. The crosses show the power transmittance of a sample of "black" polyethylene measured with a spectrometer. The solid curve shows the measurement results for the same sample by amplitude spectroscopy. The amplitude spectroscopy results are spoiled by the thickness variation, 0.153 to 0.161 mm, of the sample. (After Bell [5].)

## TRANSMISSION STUDIES: SOLID—TWO PASSES

It is possible to study solid samples such as pressed disks of KBr in the ordinary Michelson interferometer by placing the sample in the radiation beam in the fixed-mirror arm of the interferometer. The beam will make two passes through the sample. The back-reflection signal can be eliminated by placing the plane, parallel sample at an angle in the radiation beam. If the sample face makes an angle with the wavefront, the effective thickness of the sample $l_{\text{eff}}$ is given by (two passes)

$$l_{\text{eff}}(\sigma) = (1 + \{[\sin^2\theta]/2n(\sigma)\})l. \tag{8-72}$$

The analysis would follow the same as in the section, Transmission Studies: Gases—Double Pass (Ordinary Michelson Interferometer). The parameter $l$ would be replaced by $l_{\text{eff}}(\sigma)$. All window effect terms could be used to express the losses from the radiation beam due to back reflection. Channel spectra are neglected if one uses the ideas expressed in the double-pass section, but they could be included by using the ideas expressed in the section, Transmission Studies: Solids—Single Pass (with Channel Spectra). One must be very careful with the factors of two involved in the total path of the radiation beam in the sample.

## TRANSMISSION STUDIES: LIQUIDS—DOUBLE PASS

Liquids may be studied in amplitude spectroscopy by turning the interferometer on its side with the fixed mirror in a precise *horizontal* position. If mechanical vibrations are well controlled, and if the viscosity and surface tension of the sample are reasonable, a plane, parallel liquid sample will be in the radiation beam. The liquid will probably have a vapor pressure which is sufficient to cause vacuum problems, or the vapor might be toxic or corrosive so a sample cell is built over the fixed mirror. If the fixed mirror is stainless steel, it will not be attacked by the sample and a vacuum seal can be easily made. Above the sample is positioned a plane parallel window, and it need *not* be in contact with the liquid. It is necessary to have enough sample so that the sample has plane, parallel faces. Sometimes, for highly absorbing samples, too much sample is required for there to be an easily measurable signal, and long scan times are required.

In Fig. 8-11, a ray diagram for a radiation beam incident on the liquid is shown. The radiation beams have $\pi$ phase shifts upon reflection from the mirror and the externally reflected radiation beam has a $\varphi_r$ phase shift. For internal reflection, there is $\varphi_{r'}$ phase shift. The strongest radiation beam from the sample is the one that enters the sample and is reflected once from the mirror with a phase shift. This beam is labeled (a) and it has traversed an excess path distance of $2(n - 1)l$. This ray will obey all of the equations in the section, Transmission Studies: Gases—Double Pass (Ordinary Michelson Interferometer). Even the window location for the gas and liquid studies are the same. So the index of refraction may be computed from Eq. (8-60) if one is considering the radiation amplitude from ray (a). However, the

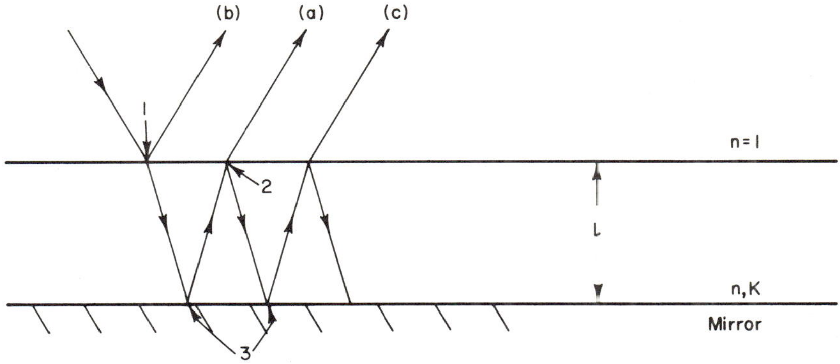


FIG. 8-11. Ray diagram for a beam incident on a plane, parallel liquid sample sitting on a mirror. Phase shifts for the reflected rays: (1) $\varphi_r$, (2) $\varphi_{r'}$, (3) $\pi$.

reflected ray, labeled (b), will reach the detector sooner than the ray (a). Since the ray (b) travels a shorter path distance ($2l$) than required to equalize the path difference between the beams in the movable-mirror arm and fixed-mirror arm, one has the signature of this ray path distance $2l$ before the origin. In Fig. 8-12, the interferogram from a liquid tetrabromoethane sample on a mirror is shown. The strongest beam (a), which has traversed the sample twice, is displaced a distance $D$. The first reflected radiation beam (b) is displaced to negative paths by a distance $2l$. The next ray, labeled (c) in Fig. 8-12, is due to a ray that has been internally reflected once. It has traversed the sample four times and has undergone a total phase shift of $\varphi_{r'}$, and, hence, it is displaced well to the right of the grand maximum.

The sample thickness can be determined from the interferogram in Fig. 8-12 since the signature at (b) is displaced from the origin a distance $2l$. For this particular sample, the thickness would be 0.068 cm. By reading the interferogram, the sample thickness can be measured with great accuracy.

Often, the signatures for the radiation beams (b) and (c) are separated almost completely from the signal near the grand maximum at (a). When this happens for thick enough samples such as shown in Fig. 8-12, one may edit the data tape for the interferogram. After reading $2l$ from the position of (b), one may replace the signal with constants equal to $I_R(\infty)$ at both (b) and (c). This editing then makes valid all of the equations in the section, Transmission Studies: Gases—Double Pass (Ordinary Michelson Interferometer), except that $r_0 \neq 0$ in obtaining Eq. (8-63). Thus, the index of refraction can be computed from Eq. (8-60) and the absorption coefficient can be computed from

$$4\pi\sigma K(\sigma) = \alpha(\sigma) = l^{-1} \ln((1 - r_0{}^2)\{[P_{D\text{w}}^2(\sigma) + Q_{D\text{w}}^2(\sigma)]/[P_D{}^2(\sigma) + Q_D{}^2(\sigma)]\}^{1/2}). \tag{8-73}$$

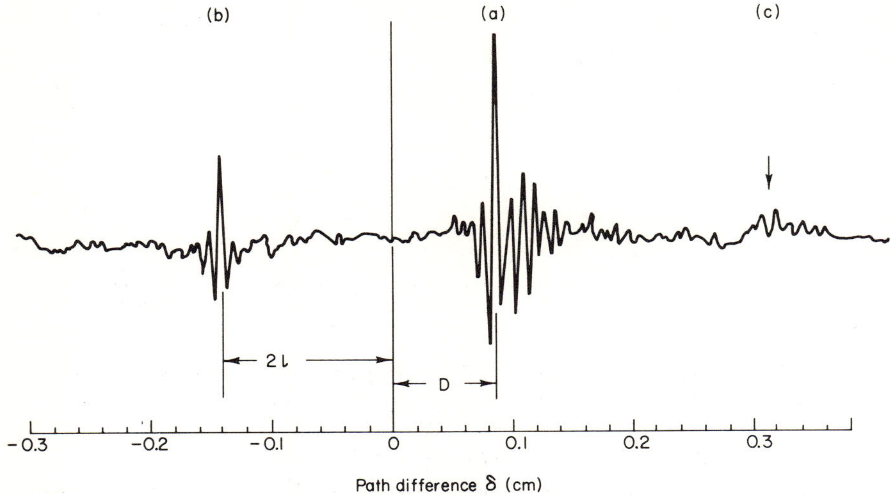


FIG. 8-12. An asymmetric mode interferogram measured for a thickness of 0.68 mm of liquid tetrabromoethane on a mirror. The signature (b) is due to the partial wave reflected from the top surface of the liquid; the signature (a) is due to the partial wave transmitted twice through the liquid layer; the signature (c) is due to the partial wave which is reflected internally once from the top liquid surface, reflected twice by the mirror, and transmitted four times through the liquid layer. (After Chamberlain *et al.* [9].)

Equations (8-60) and (8-73) for $n$ and $K$ apply only to the edited interferogram.

Sometimes, one cannot logically edit the signature at (c) due to the internal reflection but *can edit* (b), as might be the case for liquid tetrabromoethane as shown in Fig. 8-12.

With the displacement of the grand maximum in the background scan due to the windows given by $D_w$, the index of refraction is found to be

$$n(\sigma) = 1 + [(D - D_w)/2l] + \{[\varphi_{\tau D}(\sigma) - \varphi_{wD}(\sigma)]/4\pi\sigma l\}, \tag{8-74}$$

where (sample and window scan)

$$\varphi_{\tau D}(\sigma) = \arctan[Q_D(\sigma)/P_D(\sigma)] \tag{8-75}$$

and (window scan)

$$\varphi_{wD}(\sigma) = \arctan[Q_{Dw}(\sigma)/{}_{Dw}(\sigma)]. \tag{8-76}$$

Principal values have been used in Eqs. (8-75) and (8-76), and so errors of $\pm[2\pi(\text{integer})]$ will have to be compensated for as discussed in the section, Phase Errors of $\pm[2\pi(\text{Integer})]$. By editing the tape to eliminate some of the signature at (c), one may have both the signatures at (b) and (c) removed. Then, $n(\sigma)$ may be estimated as outlined earlier in this section. The absorption coefficient, and hence $K(\sigma)$, can be easily estimated from a two-beam measurement on the sample.

## ACCURATE LOW-TRANSMITTANCE MEASUREMENTS

In FTS, there is no false radiation problem (unless the radiation beam misses the sample) because all waves will produce a characteristic interference pattern. With amplitude spectroscopy, it is possible to measure the transmission amplitude $\tau(\sigma)$ of the sample so one can compute the transmittance $T$ given by $\tau(\sigma)\tau^*(\sigma) = \tau_0{}^2(\sigma)$. Suppose one measures the transmission coefficient $\tau_0(\sigma)$ with an error of $\pm\Delta\tau_0(\sigma)$. Then, the experimental transmittance $T_{\text{exp}}(\sigma)$ would be

$$T_{\text{exp}}(\sigma) \simeq T_{\text{M}}(\sigma) \pm 2\tau_0(\sigma)[\Delta\tau_0(\sigma)], \tag{8-77}$$

where $T_{\text{M}}(\sigma)$ is the mean transmittance. If $\tau_0(\sigma) = 0.1$ (10%) and $\Delta\tau_0(\sigma)$ is estimated at 0.01 (1%), then the transmittance is

$$T_{\text{exp}} \simeq 0.01 \pm 0.002.$$

That is, one has measured a transmittance of 1% with an accuracy of +0.2%! For very opaque samples, amplitude spectroscopy has a very large advantage. If the above sample had been studied in the combined beams, the signal-to-noise ratio would have been unity instead of five as it is for amplitude spectroscopy. The capability of amplitude spectroscopy is completely lost if the sample does not have high-quality, plane, parallel faces. A "pseudocoherence" effect exists in amplitude spectroscopy that does not exist in power spectroscopy. This effect reduces the magnitude of the amplitude transmittance for samples having low geometric quality.

In order to illustrate that amplitude spectroscopy is able to make accurate transmittance measurements at least as low as 0.05%, consider Fig. 8-13. The figure contains a plot of the experimental transmittance (solid line) of an infrared filter that was measured in amplitude spectroscopy using Professor E. E. Bell's interferometer shown in Fig. 8-2. The measurements were made by Ulrich [10], and at a later date, the theory for these filters was worked out by R. J. Bell and Romero [11]. The theoretical results are indicated by the dashed line. The approximations in the theory are best for the region below 100 cm$^{-1}$, and in that region one can see in Fig. 8-14 that the agreement for transmittances as low as 0.05% is excellent.

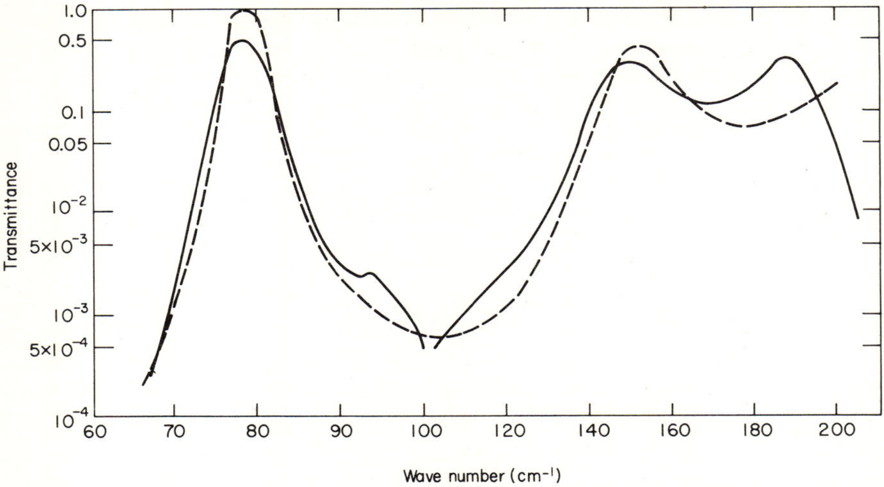


FIG. 8-13. Transmittance of a multielement grid filter in the far-infrared. The solid line is from an amplitude spectroscopy experiment and the dashed line is from theory. The theoretical approximations are weak above 100 $cm^{-1}$. (After Bell *et al.* [11].)

## CONCLUSIONS

In this chapter, we have discussed amplitude spectroscopy and have shown how it can be used to obtain the complex indices of a sample. The basic equations for gaseous, liquid, and solid sample phase studies were presented under a wide variety of conditions. Both the asymmetric interferometer and the ordinary Michelson interferometer may be used in amplitude spectroscopy, and the examples described the use of each instrument.

Sample quality must be kept high for amplitude spectroscopy, but if the conditions are met, excellent data may be obtained for complex indices, phase angles, transmission coefficients, transmittances, etc. The complex indices are found without any physical models or a Kramers–Kroing-type calculation. This is a tremendous advantage to the experimentalist.

## *References*

*1.* J. E. Chamberlain, J. E. Gibbs, and H. A. Gebbie, Refractometry in the Far Infrared Using a Two Beam Interferometer, *Nature* **198**, 874 (1963).

*2.* J. E. Chamberlain, J. E. Gibbs, and H. A. Gebbie, The Determination of Refractive Index Spectra by Fourier Spectroscopy, *Infrared Phys.* **9**, 185 (1969).

*3.* E. E. Bell, Measurement of the Far Infrared Optical Properties of Solids with a Michelson Interferometer Used in the Asymmetric Mode: Part 1, Mathematical Formulation, *Infrared Phys.* **6**, 57 (1966).

*4.* E. E. Russell and E. E. Bell, Measurement of the Far Infrared Optical Properties of Solids with a Michelson Interferometer Used in the Asymmetric Mode: Part II, the Vacuum Interferometers, *Infrared Phys.* **6**, 75 (1966).

*5.* E. E. Bell, Amplitude Spectroscopy, *Aspen Int. Conf. on Fourier Spectrosc., 1970*, p. 71. (G. A. Vanasse, A. T. Stair, Jr., and D. J. Baker, eds.). AFCRL-71-0019, 5 Jan. 1971, Spec. Rep. No. 114.

*6.* R. B. Sanderson and H. E. Scott, A High Resolution Far Infrared Interferometer, *Appl. Opt.* **10**, 1097 (1971).

*7.* G. A. Korn and T. M. Korn, "Handbook of Mathematical Physics," p. 879. McGraw-Hill, New York, 1968, 2nd ed.

*8.* R. B. Sanderson, Measurement of Rotational Line Strengths in HCl by Asymmetric Fourier Transform Techniques, *Appl. Opt.* **6**, 1527 (1967).

*9.* J. E. Chamberlain, A. E. Costley, and H. A. Gebbie, Submillimeter Dispersion of Liquid Tetrabromoethane, *Spectrochim. Acta* **23A**, 2255 (1967).

*10.* R. Ulrich, Interference Filters for the Far Infrared, *Appl. Opt.* **7**, 1987 (1967).

*11.* R. J. Bell, H. V. Romero, and J. M. Blea, Theory and Experiments for Multielement Grid Filters in a Dielectric, *Appl. Opt.* **9**, 2350 (1970).

*Note added in proof*: At the Max Planck Institute for Solid State Research (Stuttgart), there is an unusual asymmetric interferometer. It is designed for room temperature and *liquid helium temperature* sample measurements in reflection and transmission studies. Corner cube retroreflectors, uncollimated optics in the interferometer, and horizontal sample holders are important features of the instrument. This instrument was begun at Freiburg University by L. Genzel, J. Gast, and C. Irslinger and will be completed soon in Stuttgart. An article by Genzel and Gast will soon appear in the journal *Infrared Phys.*

CHAPTER NINE

# *Beamsplitters*

## INTRODUCTION

In this chapter, we will discuss various types of beamsplitter used in Fourier transform spectroscopy. For interferometry in the past, one usually thought of semitransparent or "half-silvered" metal films on transparent substrates as beamsplitters. Partially aluminized films on plates of glass or quartz have been used as beamsplitters for visible wavelength radiation. An ideal beamsplitter transmits 50% and reflects 50% of the radiation intensity; however, when equal transmitted and reflected intensities are obtained in the visible wavelength region with a metal film, it turns out that 50% of the incident flux is absorbed. Therefore, whenever possible, other, nonabsorbing beamsplitter materials are used.

We shall discuss the ideas governing the use of beamsplitters made of self-supporting dielectric sheets and made of dielectric films on substrates. In general, dielectrics with very large indices of refraction are used. However, qualifying conditions will be presented. In the far-infrared region below 400 $cm^{-1}$, self-supporting, taut sheets of Mylar or polyethylene can be used for beamsplitters. From about 400 to 3800 $cm^{-1}$, germanium-film beamsplitters on various transparent substrates are often used. From about 2000 to 16,000 $cm^{-1}$, iron oxide ($Fe_2O_3$) films on different substrates are excellent.

There are optimum wavelength regions for a given index of refraction and film thickness for a particular beamsplitter. Information relating the various beamsplitters to the appropriate wavenumber regions will be given.

Since the radiation beams are incident at 45° to the beamsplitter, there are sometimes very strong polarization effects. It will be shown how great the polarization effects are for the main beamsplitter materials presently employed.

In addition to the dielectric beamsplitters, there are wire-screen beamsplitters. There are commercially available wire grids which can be used for wave numbers as high as about 900 $cm^{-1}$. The efficiency of this type of beamsplitter will be demonstrated with experimental data.

The lamellar grating is a beamsplitter in that it divides the wavefront. It is discussed in Chapter 15.

Excellent references for the subject of beamsplitters for Michelson interferometers may be found in an article by Chamberlain *et al.* [1], in a book by Steel [2], and in a thesis by Roland [3].

## SELF-SUPPORTING DIELECTRIC BEAMSPLITTERS

With nonabsorbing dielectric sheets, one needs only to be concerned with the reflection and transmission amplitudes and phases of the radiation beams. There are $\pi$ phase shifts at external reflection surfaces and no phase shifts for internal reflection in the dielectric beamsplitter. Some waves are multiply internally reflected, and it is important to decide which waves have sufficient magnitude to be important. For the sake of discussion, consider a Mylar† beamsplitter with an index of refraction of about 1.85. Also assume that the radiation beam is unpolarized, so that the mean reflectance at a 45° angle of incidence is about equal to the reflectance at normal incidence. This assumption is fairly good if the angle of incidence is less than the polarizing angle (Brewster's angle) of $\theta_B = \tan^{-1} n = 62°$. So, the mean reflectance at a single interface or external surface of the dielectric sheet will be approximately $(n-1)^2/(n+1)^2 \simeq 0.089$. Figure 9-1 is a optical ray diagram for a radiation beam incident on a Mylar-sheet beamsplitter. One can see in the figure that the primary transmitted radiation beam contains about 83% of the incident flux, whereas the secondary transmitted radiation beam has a negligible 0.7% of the incident flux. For the reflected radiation beams, one can see that about 8.9 and 7.4% of the incident intensity are in the reflected primary and secondary radiation beams, respectively. The intensity of the next reflected beam has fallen to 0.07%. Thus, only the primary transmitted and primary and secondary reflected beams are important for Mylar-sheet beamsplitters. Even with dielectrics having higher indices of refraction, the same general conclusions may be drawn.

† Mylar is also known as polyethylene terepthalate, or melinex or terylene. Its index of refraction in the far-infrared is given as about 1.85 by Gebbie and Stone [5] and as about 1.75 by Bell [6].

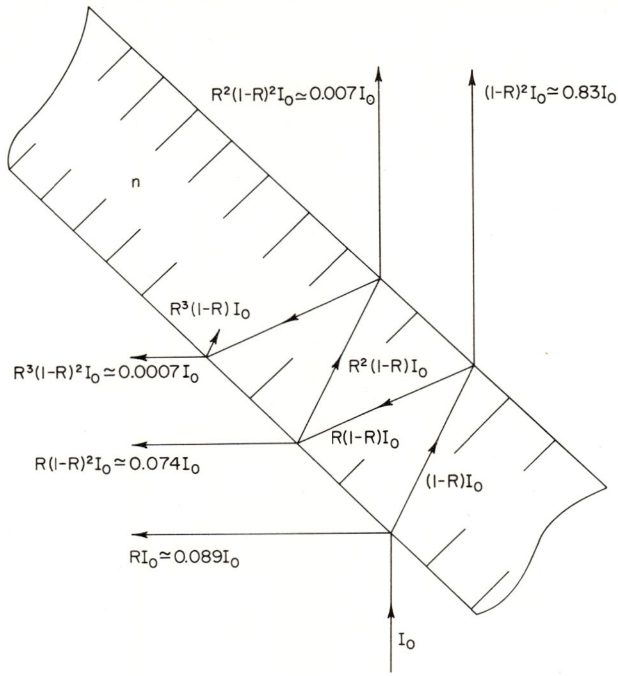


FIG. 9-1. The rays incident on, reflected from, and transmitted through a beamsplitter. Approximate (normal incident values used) flux magnitudes for each ray shown.

The transmitted beam with essentially only one component does not need to be discussed further. However, the two reflected components are nearly equal in magnitude and have a definite phase relation. Therefore, constructive and destructive interference between the reflected components is very important.

In Fig. 9-2, an optical ray diagram for a radiation beam incident on a dielectric sheet is shown. Only the optical rays with appreciable magnitude are included in the diagram, and the relevant phase shifts are indicated. We shall use Fig. 9-2 to derive the wavenumber and thickness ($l$) conditions for constructive and destructive interference between the reflected rays. One must keep in mind that the radiation beams in each of the arms of the interferometer undergo one transmission through and one reflection from the beamsplitter before reaching the detector. Whenever the reflection occurs, there will be two waves as indicated in Fig. 9-2, and then the interference between these component waves occurs.

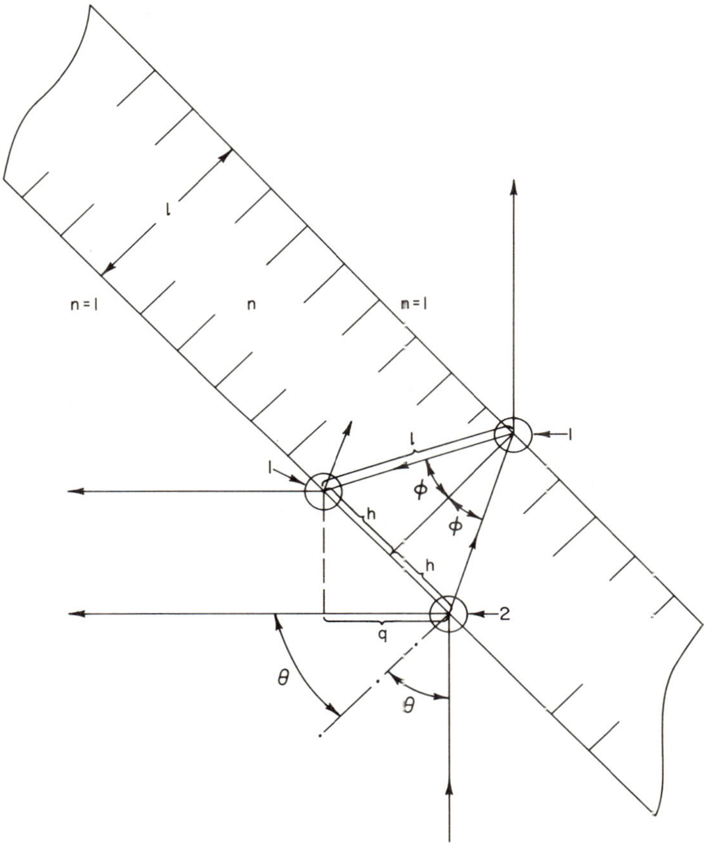


FIG. 9-2. The important rays at and in the nonabsorbing beamsplitter with dimensions defined and phase shifts as indicated: (1) no phase shift on reflection; (2) $\pi$ phase shift on reflection.

The conditions for destructive and constructive interference between the reflected waves is

$$m/\sigma = m\lambda = 2nl' - q \tag{9-1}$$

where $q$, $n$, and $l'$ are defined in Fig. 9-2. The interference conditions are

$$\begin{aligned} m &= 0, 1, 2, 3, \ldots \quad \text{for destructive interference} \\ m &= \tfrac{1}{2}, \tfrac{3}{2}, \tfrac{5}{2}, \ldots \quad \text{for constructive interference} \end{aligned} \tag{9-2}$$

The $\pi$ phase shift for external reflection and zero phase shift for internal reflection determine the values of $m$ in Eqs. (9-2). Using Snell's law, setting the angle of incidence equal to 45°, and using plane geometry and trigonometry, one has

$$m/\sigma = m\lambda = 2l(n^2 - \tfrac{1}{2})^{1/2}. \tag{9-3}$$

As a numerical example to locate the wave numbers at peak and minimal intensity, consider a 6-$\mu$m-thick Mylar sheet ($n \simeq 1.85$). The peak efficiency (constructive interference) for $m = \frac{1}{2}$ occurs at a 41 $\mu$m wavelength or 240 $\text{cm}^{-1}$. As will be seen later in this section, the efficiency oscillates from zero to some maximum value as the wave number increases. For all dielectric beamsplitters, the efficiency will be zero at $\sigma = 0$. This is because even when the optical path difference is negligible in comparison with the wavelength, there is a $\pi$ phase shift for one radiation beam and none for the other (Fig. 9-2). The two nearly equal waves will be exactly out of phase and will therefore destructively interfere. So, for this example, the efficiency is zero at $\sigma = 0$, it peaks at $\sigma = 240\ \text{cm}^{-1}$, and then it falls to zero at 490 $\text{cm}^{-1}$ ($m = 1$). The efficiency then oscillates between zero and the same maximum as predicted by the maximum and minimum conditions in Eqs. (9-2) and (9-3). In practice, there is not exact cancellation in the two components of the reflected radiation beams. As shown in Fig. 9-1, one beam has an intensity of $0.089I_0$ and the other has intensity of $0.074I_0$. Therefore, with unequal components, we must consider the following theoretical treatment as an approximation.

We would now like to obtain the efficiency curve for the beamsplitter sheet. In Fig. 9-3, an optical ray diagram is shown of a Michelson interferometer. The incident radiation beam has an intensity of unity and the overall reflectance (power reflected) and transmittance (power transmitted) of the beamsplitter are $R_0$ and $T_0$, respectively. The intensity reaching the detector is $2R_0T_0$ and the lost intensity returning to the source is $R_0^2 + T_0^2$. As is well known, the maximum efficiency $2R_0T_0$ occurs for $R_0 = T_0 = 0.50$, and is therefore equal to 50%. The other 50% of the intensity returns to the source. Unfortunately, the fact that the maximum efficiency of a perfect Michelson interferometer is only 50% is often lost in the literature. This is an important consideration when one discusses the lamellar grating beamsplitter, which can have the efficiency approach 100%. So, one must keep in mind the difference between definitions of absolute and relative efficiency when consulting the literature. (See the Added Note at the end of this chapter.)

For the Michelson interferometer, the relative efficiency (RE) will be defined as

$$(\text{RE}) \equiv 2R_0T_0/(2\text{R}_0T_0)_{\text{ideal}} = R_0T_0/0.25 = 4R_0T_0. \tag{9-4}$$

From Fig. 9-1, one can see that the transmitted beam intensity is given approximately by the primary transmitted intensity or $T_0 \simeq (1 - R)^2$. Remembering that $R$ is the *single* surface reflectance of the dielectric sheet,

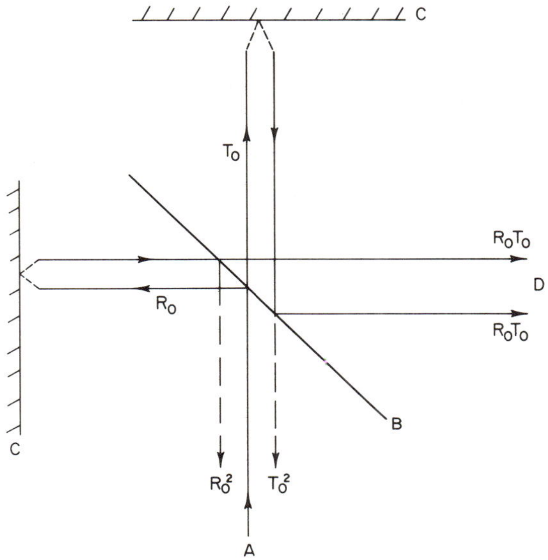


FIG. 9-3. Modified optical ray diagram of a Michelson interferometer illustrating the intensity going in various directions in the interferometer. A: Source; B: beamsplitter; C: mirror; D: detector. The optical rays are offset at the mirrors for graphic purposes only. It can be seen that for unit incident intensity, an intensity of $2R_0T_0$ reaches the detector and $R_0{}^2 + T_0{}^2$ returns to the source.

we consider the reflected components in Fig. 9-1. To a first approximation, we may consider the first two reflected waves to be of equal intensity, so that $R_0 \simeq 2R$. Higher-order approximations become important if $n$, and thus $R$, is large.

With equal amplitudes in the two reflected fields, we inquire about the interference between the two radiation beams. In elementary optics, if one has two equal, parallel waves of the same frequency, they will have a resultant intensity proportion to $\cos^2(\rho/2)$, where $\rho$ is the phase between the two waves. Considering the optical path differences and the phase shifts, the phase for the two reflected components is

$$\rho = [2\pi\sigma(\text{optical path difference}) + \pi]. \tag{9-5}$$

From Fig. 9-2, one has

$$\rho = 2\pi\sigma[2nl - q] + \pi \tag{9-6}$$

or

$$\rho = 2\pi[\sigma l(n^2 - \tfrac{1}{2})^{1/2} + \tfrac{1}{4}]. \tag{9-7}$$

Combining all these equations and mathematical statements and substituting in Eq. (9-4), we have

$$(\mathrm{RE}) \simeq 8(1 - R)^2 R \cos^2\{2\pi[\sigma l(n^2 - \tfrac{1}{2})^{1/2} + \tfrac{1}{4}]\}. \tag{9-8}$$

Using Eq. (9-8), the relative efficiencies of several Mylar beamsplitters (assuming $n = 1.85$) have been computed. The results are presented in Figs. 9-4–9-10 for Mylar thicknesses of 6, 12, 25, 51, 102, 191, and 356 $\mu$m. Curves (a) are normalized spectra (peak signal set equal to unity) taken with the beam-

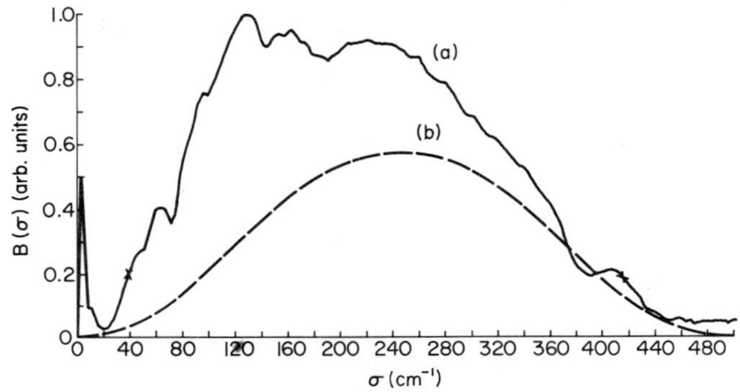


FIG. 9-4. (a) Normalized spectrum using the 25-gauge Mylar beamsplitter (6 $\mu$m thick). Mercury lamp source, Golay cell detector, and a black polyethylene lens. (b) Theoretical relative efficiency of the beamsplitter.

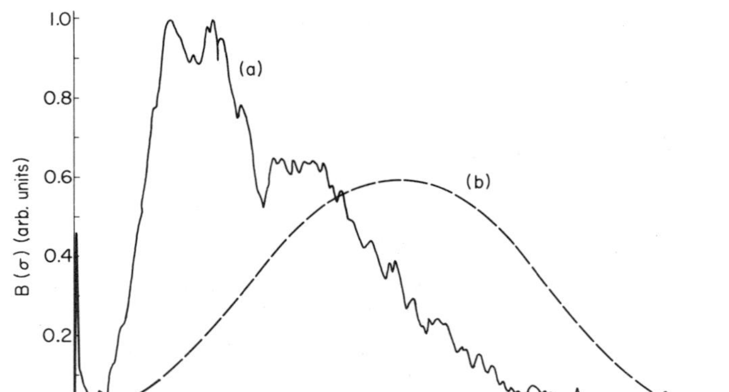


FIG. 9-5. (a) Normalized spectrum using the 50-gauge Mylar beamsplitter (12 $\mu$m thick). Mercury lamp source, Golay cell detector, black polyethylene lens, and 0.6 mm crystal quartz filter. (b) Theoretical relative efficiency of the beamsplitter.

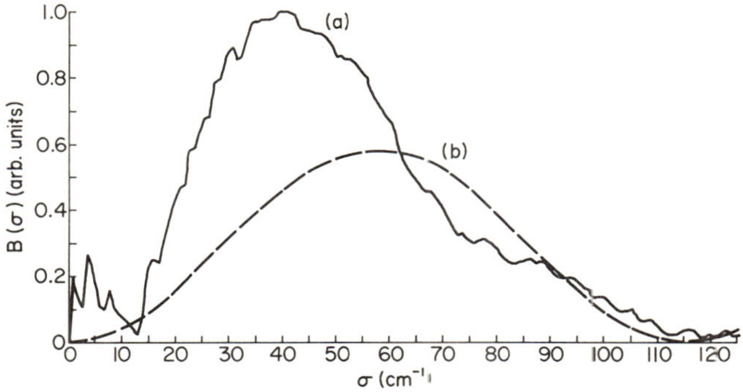


FIG. 9-6. (a) Normalized spectrum using the 100-gauge Mylar beamsplitter (25 $\mu$m thick). Mercury lamp source, Golay cell detector, black polyethylene lens, and 1/16-in. fused quartz filter. (b) Theoretical relative efficiency of the beamsplitter.

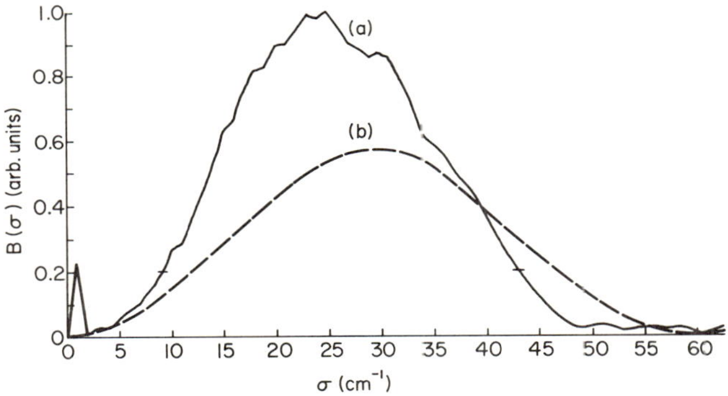


FIG. 9-7. (a) Normalized spectrum using the 200-gauge Mylar beamsplitter (51 $\mu$m thick). Mercury lamp source, Golay cell detector, black polyethylene lens, and filter "7." (b) Theoretical relative efficiency of the beamsplitter.

splitters with various filters and detectors. Curves (b) are the theoretical relative efficiencies computed using Eq. (9-8). The peak relative efficiencies are near 60%, and the intensity losses are accounted for in the flux, $R_0^2 + T_0^2$, which returns to the source. Perusing the curves, one can see that the position of the minima are closely predicted. The fact that the spectral signal curves do not equal zero at the minima is due to the fact that the two reflected components are not of exactly the same intensity, as discussed before. The best agreement in the shape of the curves (a) and (b) is in Fig. 9-9 for the 750-gauge beamsplitter. In Fig. 9-9, the spectral filters, detector sensitivity, and lamp intensity are reasonably flat between 10 and 20 cm$^{-1}$. If the spectral filters (described in Chapter 10) absorb strongly at the high frequencies as

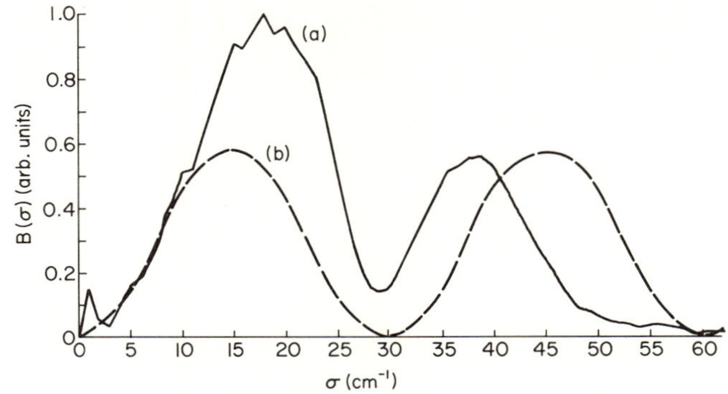


FIG. 9-8. (a) Normalized spectrum using the 400-gauge Mylar beamsplitter (102 $\mu$m thick). Mercury lamp source, Golay cell detector, black polyethylene lens, and filter "7." (b) Theoretical relative efficiency of the beamsplitter.

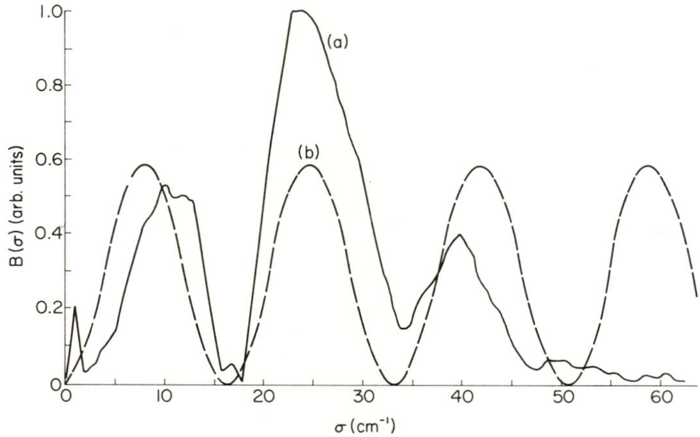


FIG. 9-9. (a) Normalized spectrum using the 750-gauge Mylar beamsplitter (191 $\mu$m thick). Mercury lamp source, gallium-doped germanium detector( 4.2 K), black polyethylene lens, and filter "7." (b) Theoretical relative efficiency of the beamsplitter.

they do in the cases exhibited by Figs. 9-4–9-7, the experimental normalized spectral peak will be shifted to lower frequencies than predicted by just the beamsplitter considerations. If the source intensity falls rapidly as it does for the mercury lamp at low wave numbers [7], the spectral signal will fall lower than predicted by the beamsplitter relative efficiency alone. The relative low spectral signal below 10 $\text{cm}^{-1}$ is shown in Figs. 9-9 and 9-10.

Also, the absorption of Mylar was neglected in the calculations of the theoretical relative efficiency of the beamsplitter. From Fig. 8-9, one can see that the index of refraction of Mylar is changing appreciably above about

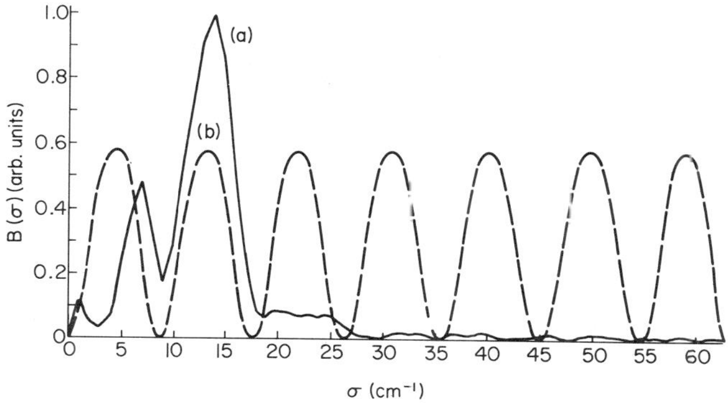


FIG. 9-10. (a) Normalized spectrum using the 1400-gauge Mylar beamsplitter (356 $\mu$m thick). Mercury lamp source, gallium-doped germanium detector (4.2 K), black polyethylene lens, filter "7," and 4.5-mm sodium chloride crystal filter. (b) Theoretical relative efficiency of the beamsplitter.

50 $\mathrm{cm}^{-1}$. As shown by Bell [6], the absorption would be most pronounced in Fig. 9-4 and would help account for the shift of the spectral peak toward lower wave numbers.

All the spectral curves show the drop to zero signal as $\sigma$ approaches zero (except for noise near $\sigma$ about zero), as predicted, because of destructive interference. The other zeros are predicted by Eqs. (9-2) and (9-3). Thus, the features of the self-supporting beamsplitters seem to be well accounted for by Eq. (9-8).

The experimental and theoretical relative efficiency curves for a 76-$\mu$m-thick Mylar beamsplitter is shown in Fig. 9-11 for (assumed) s-polarization

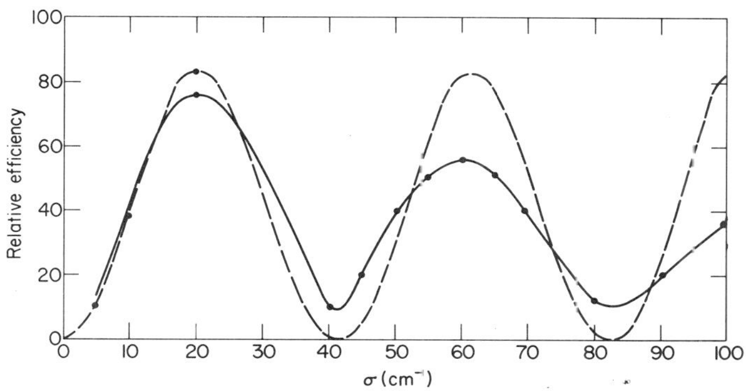


FIG. 9-11. Relative efficiency of a Mylar beamsplitter for an assumed s-polarized radiation beam. The beamsplitter was 76 $\mu$m thick (300 gauge), and for the theoretical curve (dashed line), $n$ was assumed to be 1.85. The solid line represents experimental data from Richards [8].

of the radiation beam. The agreement between theory and experiment is excellent except at large wave numbers, where the Mylar absorption becomes important.

## POLARIZATION IN DIELECTRIC-SHEET BEAMSPLITTERS

The reflectance of a dielectric depends on the angle of incidence of the radiation beam that gives the internal and external reflectances equal. The Fresnel equations for a nonabsorbing dielectric are

$$R_p = [\tan^2(\varphi - \theta)]/[\tan^2(\varphi + \theta)] \tag{9-9}$$

and

$$R_s = [\sin^2(\varphi - \theta)]/[\sin^2(\varphi + \theta)]. \tag{9-10}$$

The subscript p refers to the polarization with the electric field parallel to the plane of incidence (in the plane of the optical diagram of the Michelson interferometer), and the subscript s refers to the polarization with the electric field perpendicular to the plane of incidence (perpendicular to the plane of the optical diagram of the Michelson interferometer). The angle of incidence is $\theta$ (equal to 45°) and $\varphi$ is the angle of refraction given by Snell's law as

$$\sin \varphi = (\sin \theta)/n. \tag{9-11}$$

Referring to Fig. 9-1, the overall or total reflectance $R_0$ of the beamsplitter is given to a good approximation (even for indices of refraction of 4) by

$$R_0 \simeq R(1 - R)^2 + R = 2R[1 - R + (R^2/2)], \tag{9-12}$$

where $R$ is the *single* surface reflectance.

Using Eq. (9-12) for each polarization with $\theta = 45°$, we have computed $R_0$ for materials with different indices of refraction (see Fig. 9-12). Using $R \simeq [(R_p + R_s)/2]$ for the average reflectance of the single surface, $\bar{R}_0$ was computed for the overall reflectance of the beamsplitter for unpolarized radiation beams. The optimum beamsplitter will have an $R_0$ of 0.5, and this is indicated by a dashed line. Therefore, the optimum beamsplitter for an unpolarized beam has an index of refraction of about 3.7 (more rigorous calculations place it near 3.64). An ideal beamsplitter which is to be used with the perpendicular polarization will require an index of refraction of about 2.7 (calculations without approximations yield 2.85). For the p polarization, the ideal index of refraction is so large that only ferroelectric materials may be considered.

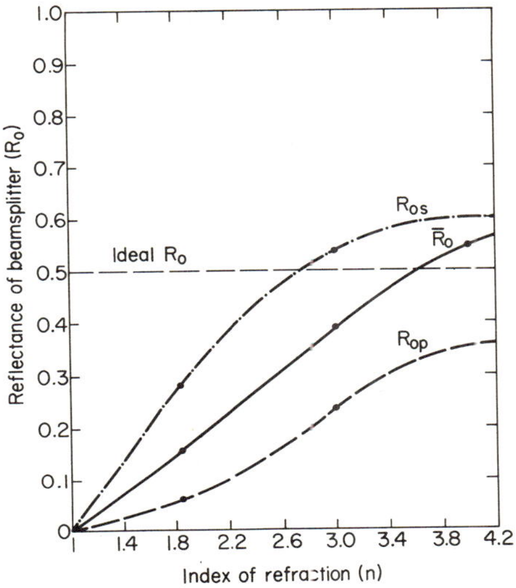


FIG. 9-12. Total reflectance of self-supporting beamsplitters for various indices of refraction. The subscript p is for parallel polarization to the plane of incidence and s is for perpendicular polarization. The unpolarized beam is represented by $\bar{R}_0$. The ideal $R_0$ is at 0.5 to keep to a minimum energy loss.

The polarization effect of having a much stronger beam in the s polarization than in the p polarization can be shown by calculating the relative efficiences $4R_0T_0 \equiv 4R_0(1 - R_0)$ of some different beamsplitters. In Table 9-1, the index of refraction $n$, the overall reflectance $R_0$, the overall transmittance $T_0$, and the relative efficiency $4R_0T_0$ of several different beamsplitter materials are presented. It must be noted, however, that germanium

TABLE 9-1

Total Reflectances $R_0$, Total Transmittances $T_0$, and Relative Efficiency $4R_0T_0$ of Some Beamsplitter Materials (if Self-Supporting) for the Different Polarizations of the Radiation Beam

| Material | Index of refraction | Unpolarized | | | p-polarization | | | s-polarization | | |
|---|---|---|---|---|---|---|---|---|---|---|
| | | $R_0$ | $T_0$ | $4R_0T_0$ | $R_0$ | $T_0$ | $4R_0T_0$ | $R_0$ | $T_0$ | $4R_0T_0$ |
| Mylar | 1.85 | 0.16 | 0.84 | 0.55 | 0.06 | 0.94 | 0.23 | 0.29 | 0.71 | 0.83 |
| $Fe_2O_3$ | 3.0 | 0.39 | 0.61 | 0.95 | 0.24 | 0.76 | 0.73 | 0.54 | 0.46 | 0.99 |
| Germanium | 4.0 | 0.55 | 0.45 | 0.99 | 0.36 | 0.64 | 0.92 | 0.62 | 0.38 | 0.94 |

and iron oxide are usually used as films on substrates because they are too thin to be self-supporting. Nonetheless, the materials with refractive indices of 1.85, 3.0, and 4.0 are on both sides of the optimum index for unpolarized and s-polarized radiation beams as seen from Fig. 9-12. By perusing the relative efficiency columns $4R_0 T_0$, one can see that the relative efficiency of the beamsplitter is larger for the s than for the p polarization. For Mylar self-supporting beamsplitters, the s polarization is 3.6 times more efficient than the p polarization. Sometimes, this places a considerable burden on some experimentalists when polarization studies are to be made if heed is not paid to this fact for Mylar beamsplitters. For materials with larger indices of refraction, the two polarizations are about equal.

We shall return to the subject of polarization for beamsplitters made of dielectric materials on substrates.

## DIELECTRIC BEAMSPLITTERS ON SUBSTRATES

As the wavelengths become shorter, it is necessary to make the dielectric beamsplitters thinner. In the near-infrared and visible spectral regions, the beamsplitters must be thinner than 10,000 Å. Thus, it is necessary to put the dielectric beamsplitters on transparent substrates for support. This is usually accomplished by vaccum-depositing the beamsplitter on some sturdy, transparent material such as quartz or alkali halide crystals.

Neglecting the reflection from the substrate, the relative efficiency of the beamsplitter for unpolarized radiation beams will be nearly described by Eq. (9-8). In Eq. (9-8), the total transmittance $T_0$ was given by $(1 - R)^2$ and overall reflectance $R_0$ was approximated by $2R$. Defining the index of refraction of the substrate as $n_2$ and keeping $n$ for the index of refraction of the beamsplitter dielectric, one has the reflectance $R_2$ at the dielectric–substrate interface given approximately by (unpolarized)

$$R_2 \simeq \{[1 - (n/n_2)]/[1 + (n/n_2)]\}^2. \tag{9-13}$$

The external reflectance of the dielectric beamsplitter will be defined as $R$, as before. By considering a simple optical ray diagram similar to the one shown in Fig. 9-1, one can obtain the overall reflectance for the two reflected components from the external surface and from the dielectric–substrate interface. The result is

$$R_0 \simeq R + R_2(1 - R)^2. \tag{9-14}$$

The total transmittance (with no absorption) is

$$T_0 \equiv 1 - R_0 . \tag{9-15}$$

Considering only beams within the dielectric with high index of refraction, the relative efficiency of the beamsplitter then becomes approximately

$$(\mathrm{RE})_s \simeq 4T_0 R_0 \cos^2\{2\pi[\sigma l(n^2 - \tfrac{1}{2})^{1/2} + \tfrac{1}{4}]\}, \tag{9-16}$$

where $R_0$ and $T_0$ are given by Eqs. (9-14) and (9-15), respectively.

If the radiation beams are polarized, then the single surface reflectance equations must be replaced with Fresnel's equation. For the outer or external reflectance, one still has the reflectance given by Eqs. (9-9) and (9-10) with $\varphi$ found from Snell's law in Eq. (9-11). This would yield the value of $R$ to use in Eq. (9-14). At the beamsplitter–substrate interface, the reflectances are given by

$$R_{2,\mathrm{p}} = [\tan^2(\varphi - \varphi')]/[\tan^2(\varphi + \varphi')] \tag{9-17}$$

and

$$R_{2,\mathrm{s}} = [\sin^2(\varphi - \varphi')]/[\sin^2(\varphi + \varphi')]. \tag{9-18}$$

In these equations, $\varphi$ can be found from Eq. (9-11) as before, but $\varphi'$ (found after double use of Snell's law) is given by

$$\sin \varphi' = (\sin \theta)/n_2 . \tag{9-19}$$

So, using Eq. (9-16) the approximate relative efficiency can be computed for a high-index-of-refraction beamsplitter on a transparent dielectric. The results would obviously be cosine-squared functions for the relative efficiency. However, if one includes the effect of the substrate and recomputes the phases, the total reflectance, and total transmittance of the beamsplitter and substrate, he gets a slightly different result. This was done by Roland, who obtained the results shown in Fig. 9-13 for germanium ($n = 4.0$) on a $CaF_2$ ($n = 1.43$) substrate. In the figure, the relative efficiency is plotted versus $\sigma/\sigma_m$. For $m = \frac{1}{2}$ from Eqs. (9-2) and (9-3), $\sigma_{1/2}$ is given (approximately) by

$$\sigma_{1/2} \simeq 1/[4l(n^2 - \tfrac{1}{2})^{1/2}]. \tag{9-20}$$

Since for germanium $n = 4.0$, one has $\sigma_{1/2} \simeq 0.063/l$. The relative efficiency curves are flatter for a large bandwidth than would be predicted by the approximations in Eq. (9-16). However, the zero-relative-efficiency points are still closely predicted by Eqs. (9-2) and (9-3). At the midband, the perpendicular polarization is still the most efficient. For beamsplitters in the spectral range from 3800 to 400 $\mathrm{cm}^{-1}$, germanium films on $CaF_2$ or KBr substrates are quite good because the relative efficiency is between about 0.7 and 0.99

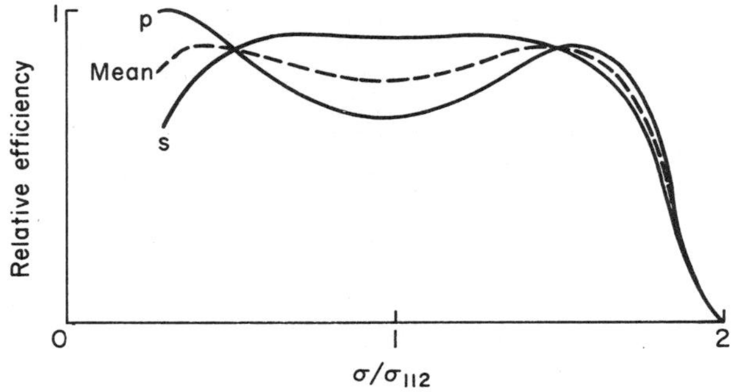


FIG. 9-13. Relative efficiency $4R_0T_0$ over a range of frequencies (or coating thicknesses) for a germanium coating ($n = 4.0$) on a $CaF_2$ substrate ($n = 1.43$). The normalizing frequency is $\sigma_{1/2} \simeq 0.063/l$, where $l$ is the thickness of the germanium film. (After Roland [3] and Steel [2].)

for both polarizations. In Fig. 9-14, a chart is given showing the wavenumber and wavelength ranges in which some substrate materials are useful. The indices of refraction of the materials are also given. If one wishes a midband wave number of 800 $cm^{-1}$, the germanium film thickness $l$, from Eqs. (9-2) and (9-3), should be about 7900 Å. Such a beamsplitter on KBr is usable between about 400 and 1500 $cm^{-1}$. The KBr can also be protected from water vapor by covering the back side of the beamsplitter with some nonhygroscopic coating which is transparent in the wavelength range of interest.

In Fig. 9-15, the experimental transmittance of a germanium film on a KBr substrate is shown. Also, assuming no absorption losses, the relative efficiency $4RT = 4T(1 - T)$ is plotted. One can see that this beamsplitter is quite useful between 400 and 2200 $cm^{-1}$.

Digilab uses a dual-thickness beamsplitter which is thicker on the outer ring than in the center section. Since the center section is much thinner than the outer ring, its transmittance maximum falls at a higher wave number than the thick film. Figure 9-15 gives data for the thick, outer ring. The radiation beam from the outer ring of the beamsplitter is passed to a pyroelectric detector (triglycine sulfate crystals, called "TGS"), which has a good low-wavenumber response. But the radiation beam from the inner ring is passed to a PbSe detector. The signals from the two detectors are individually amplified and then mixed such that a good beamsplitter efficiency is found for all wave numbers between 400 and 3800 $cm^{-1}$.

A material that is quite useful as a beamsplitter dielectric in the far-red and near-infrared is iron oxide ($Fe_2O_3$). It has an index of refraction of 3.0 and can be vacuum-plated onto crystal quartz ($n = 1.54$), fused quartz

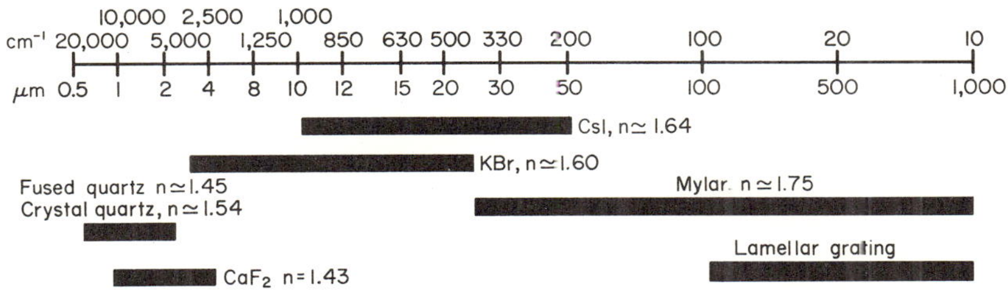


FIG. 9-14. Wavenumber and wavelength ranges covered by various beamsplitters. The quartz, $CaF_2$, KBr, and CsI materials are used as substrates for high-index-of-refraction, dielectric-film beamsplitters. The Mylar is a self-supporting dielectric-sheet beamsplitter. The lamellar grating beamsplitter is described in Chapter 15.

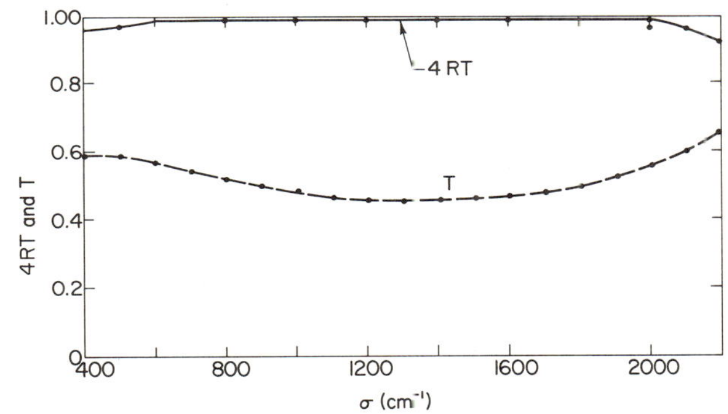


FIG. 9-15. Experimental transmittance $T$ of a germanium beamsplitter on a KBr substrate. The relative efficiency $4RT$ has been computed from the measured value of $T$ assuming no absorption losses. (After Digilab, Inc., Cambridge, Massachusetts.)

($n = 1.45$), and calcium fluoride ($n = 1.43$). In Fig. 9-14, the useful range of the quartz and $CaF_2$ substrate materials are shown. The experimental relative efficiency of an iron oxide beamsplitter on calcium fluoride with a coating thickness of 2500 Å is large enough to be quite useful between 10,000 and 3000 $cm^{-1}$.

For dielectric beamsplitters on transparent substrates, there is an optimum matching condition for the index of refraction of the coating with the index of refraction of the beamsplitter. As was indicated in Fig. 9-12, for a dielectric with no substrate, the unpolarized radiation beam required an index of refraction near 3.6 for the ideal beamsplitter. The ideal beamsplitter index of refraction for the s-polarized radiation beams was approximately 2.7. If higher-order approximations were used for $R_0$ and $T_0$, values shifted to

3.64 for the unpolarized beam and 2.85 for the s-polarized beam. The optimum beamsplitter has equal transmitted and reflected intensity as close to 50% of the incident intensity as possible. In Fig. 9-16, the optimum index of refraction of the coating or beamsplitter material versus the refractive index of the support or substrate is shown. These curves were calculated by Steel [2] using the transmittance and reflectance properties of very thin film as described in Born and Wolf [9]. There are curves in Fig. 9-16 for s-polarized and unpolarized beams. However, the p polarization is not shown because it

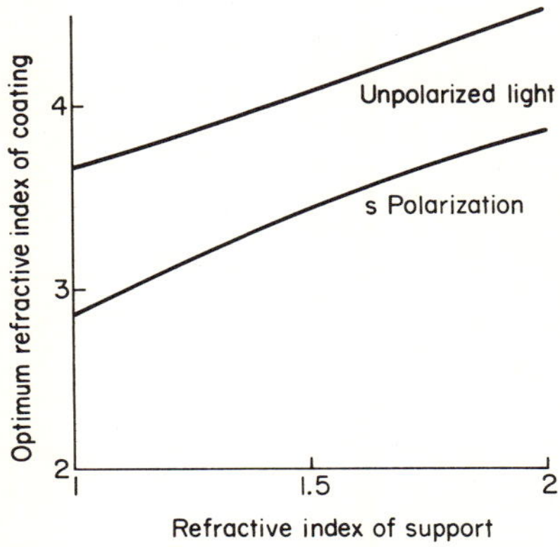


FIG. 9-16. Optimum refractive index for a single dielectric coating for use as a beamsplitter at 45°. (After Steel [2, Fig. 38].)

requires a prohibitively high index of refraction which is only available with ferroelectric materials. The only materials which can satisfy the requirements are those with indices of refraction in the range from about 3 to 4 and these materials are typically semiconductors. Some examples are germanium ($n = 4.0$) and silicon ($n = 3.5$). Also, such oxides as $Fe_2O_3$ ($n = 3.0$) are quite useful. All of the alkali halides have indices of refraction for the substrates in the range from about 1.4 to about 1.7 and are useful. With the semiconductors, one must have the energy of the radiation beam below the energy-gap value so that the material will be transparent. Also, semiconductors can have large lattice and/or free-carrier absorption which can degrade the beamsplitter function of the material. Germanium [10] has large lattice absorption at 3400 $cm^{-1}$ and silicon [10] has absorption at 1100 $cm^{-1}$ (oxygen impurities) and near 600 $cm^{-1}$.

## PHASE ERRORS DUE TO ABSORPTION

If the external reflection phase angle is not $\pi$ and the phase for reflection from the dielectric substrate interface is not zero, the situation is different from that described in the last sections. Recall from Chapter 8 that the external reflection phase angle is

$$\varphi_{\mathrm{r,E}}(\sigma) = \arctan\{2\mathrm{K}(\sigma)/[n^2(\sigma) + K^2(\sigma) - 1]\} + \pi \tag{9-21}$$

and that the internal reflection phase angle is

$$\varphi_{\mathrm{r,I}}(\sigma) = \arctan\{2K(\sigma)/[n^2(\sigma) + K^2(\sigma) - 1]\}. \tag{9-22}$$

In Eqs. (9-21) and (9-22), $n(\sigma)$ and $K(\sigma)$ are the index of refraction and the extinction index defined by the complex index of refraction $n_c \equiv n + ik$.

If absorption is present, $K(\sigma)$ is not equal to zero and the phase shifts are different from zero and $\pi$. Therefore, the phase constant in Eq. (9-5) will not be exactly $\pi$. For selenium beamsplitters, Roland [3] found phase shifts as large as $\frac{1}{2}$ radian. for $\varphi_{\mathrm{r,E}}(\sigma)$. One can use beamsplitters with absorption, but the efficiency curves can depart significantly from the ideal case because of phase changes and absorption.

## WIRE-GRID BEAMSPLITTERS

Wire grids which have been made by electroforming processes make good beamsplitters. A regular array of square openings in a metallic sheet as shown in Fig. 9-17 is a "wire grid." The thickness of the grid can be kept as small as 5–7 $\mu$m. The metal strips are of width $2a$ and the periodicity is $g$. Grids having $g$ as small as 25 $\mu$m are commercially available.†

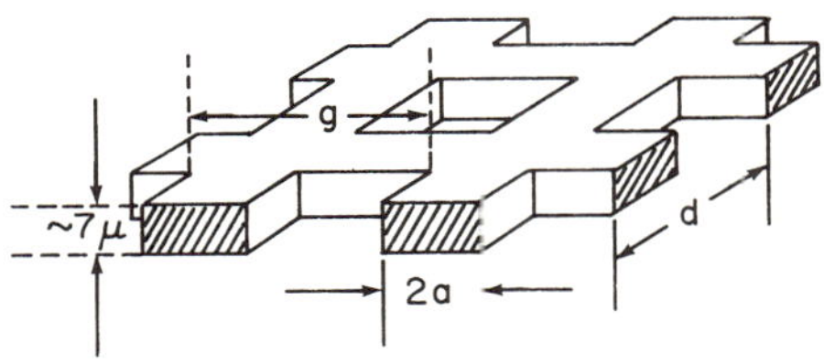


FIG. 9-17. Cross section of an electroformed metal mesh. The thickness is usually of the order of 5–7 $\mu$m, $g$ is the periodicity, and $2a$ is width of the metal between apertures. (After Vogel and Genzel [11].)

Wire-grid beamsplitters were comprehensively examined by Vogel and Genzel [11]. The orientation of the edges of the square apertures relative to the plane of incidence was used as a variable in their studies. In Fig. 9-18, the edges of the squares were in the plane of incidence, and in Fig. 9-19, the

† Buckbee-Mears Co., 245 E. Sixth St., St. Paul 1, Minnesota.

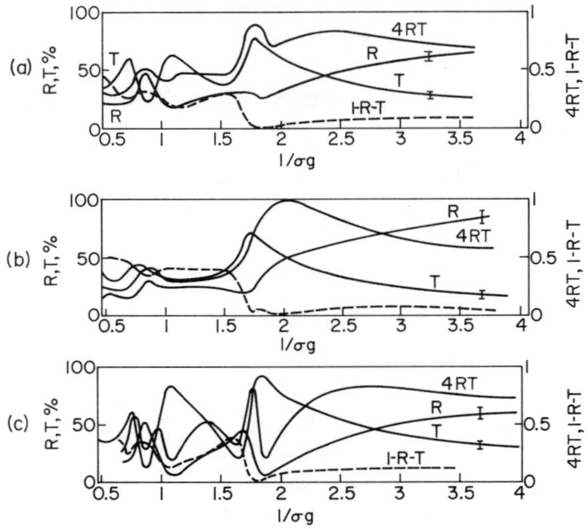


FIG. 9-18. Experimental reflectance $R$, transmittance $T$, relative efficiency $4RT$, and diffraction losses $1 - R - T$ for a metallic grid with sides of the square apertures in the plane of incidence. Curves (a) are for unpolarized radiation beams, curves (b) are for s-polarized radiation beams, and curves (c) are for p-polarized radiation beams. (After Vogel and Genzel [11].)

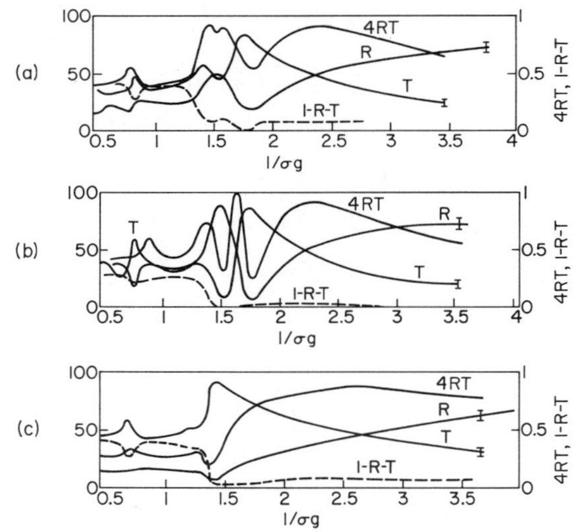


FIG. 9-19. Experimental reflectance $R$, transmittance $T$, relative efficiency $4RT$, and diffraction losses $1 - R - T$ for a metallic grid with sides of the square apertures at 45° to the plane of incidence. Curves (a) are for unpolarized radiation beams, curves (b) are for s-polarized radiation beams, and curves (c) are for p-polarized radiation beams. (After Vogel and Genzel [11].)

edges of the squares were at 45° to the plane of incidence. For each of the figures, curves (a) are for unpolarized radiation beams, curves (b) are for s polarization, and curves (c) are for p polarization. The relative efficiencies $4R_0T_0 \simeq 4RT$, the reflectances $R_0 \simeq R$, the transmittances $T_0 = T$, and the diffraction losses $1 - R_0 - T_0 \simeq 1 - R - T$ are given. For the wire-grid beamsplitters, there is no multiple internal reflection and the absorption is negligible. However, there are quite complicated phase shifts involved. From the figures, it is obvious experimentally that all of the beamsplitters have excellent relative efficiencies for wave numbers less than 0.5/g. Both polarizations have excellent efficiencies in excess of about 60% down to low wave numbers near $\sigma \simeq 0.25/g$. At the very low wave numbers, the reflectance approaches 100%, and the relative beamsplitter efficiency decreases. The properties of wire grids may be almost entirely described by the parameters $g$ and $a$. The thickness of the metal and the metallic material are of minor optical importance in the far-infrared. In the far-infrared region approaching $10\ \text{cm}^{-1}$, even woven wire cloth such as used in particle sieves may be used as beamsplitters. On the other hand, the electroformed grids are very fragile, and they may be made or held on substrates.

## CONCLUSIONS

In this chapter, we have discussed the major beamsplitters used in Michelson interferometers for Fourier transform spectroscopy. Theoretical relative efficiency curves were presented for Mylar and germanium dielectric beamsplitters and experimental data, as available, were given. Experimental data for the relative efficiencies of wire-grid beamsplitters were also presented.

The problems and uses of self-supporting dielectric sheets and of dielectric films on substrates for beamsplitters were treated separately. Conditions for optimum indices of refraction for the two separate cases were shown in Figs. 9-12 and 9-16. Polarization effects were also indicated, and it was shown that they can be quite large for Mylar beamsplitters with a relatively low index of refraction ($n \simeq 1.8$).

The efficiency of dielectric beamsplitters can be improved by using multilayer dielectric films. The multiple films can increase the peak efficiency and the wavenumber range over which the beamsplitter can be utilized. This technique was discussed by Tescher [12].

As mentioned before, the lamellar grating beamsplitter is described in Chapter 15. Two of the advantages of the lamellar grating beamsplitter over the Michelson interferometer beamsplitter are that it does not lose one-half or more of the flux, as shown in Fig. 9-3, and it does not polarize the radiation beam as much as Mylar beamsplitters do, as indicated in Table 9-1.

## *References*

*1.* J. E. Chamberlain, G. W. Chantry, P. D. Findlay, H. A. Gebbie, J. E. Gibbs, N. W. B. Stone, and A. J. Wright, The Spectral Transmission at Infra-Red Wavelengths of Michelson Interferometers with Dielectric Beam-Dividers, *Infrared Phys.* **6**, 195 (1966).

*2.* W. H. Steel, "Interferometry." Cambridge Univ. Press, London and New York, 1967.

*3.* G. Roland, Spectroscopie par transformation de Fourier. Thesis, Liege, Belgium (1965).

*4.* F. A. Jenkins and H. E. White, "Fundamentals of Optics," 3rd ed., p. 242. McGraw-Hill, New York, 1957.

*5.* H. A. Gebbie and N. W. Stone, A Michelson Interferometer for Far Infrared Spectroscopy of Gases, *Infrared Phys.* **4**, 85 (1964).

*6.* E. E. Bell, Measurement of the Far Infrared Optical Properties of Solids with a Michelson Interferometer Used in the Asymmetric Mode: Part 1, Mathematical Formulation, *Infrared Phys.* **6**, 57 (1966).

*7.* R. Cano and M. Mattioli, Far-Infrared Spectroscopy with a Michelson Interferometer and an InSb Detector, *Infrared Phys.* **7**, 25 (1967).

*8.* P. L. Richards, High Resolution Fourier Transform Spectroscopy in the Far Infrared, *J. Opt. Soc. Amer.* **54**, 1474 (1964).

*9.* M. Born and E. Wolf, "Principles of Optics," 2nd ed., p. 61. Pergamon Press, New York, 1964.

*10.* T. S. Moss, "Optical Properties of Semiconductors." Butterworth, London and Washington, D.C. and Academic Press, New York, 1959.

*11.* P. Vogel and L. Genzel, Transmission and Reflection of Metallic Mesh in the Far Infrared, *Infrared Phys.* **4**, 257 (1964).

*12.* A. G. Tescher, Beamsplitter Optimization for Fourier Spectroscopy, *Aspen Int. Conf. on Fourier Spectrosc., 1970*, p. 225. (G. A. Vanasse, A. T. Stair, Jr., and D. J. Baker, eds.). AFCRL-71-0019, 5 Jan. 1971, Spec. Rep. No. 114.

*Note added in proof*: Recently D. H. Martin and E. Puplett described a polarizing Michelson interferometer which uses a parallel wire beamsplitter [D. H. Martin and E. Puplett, Polarized Interferometric Spectroscopy for the Millimetre and Submillimetre Spectrum, *Infrared Phys.* **10**, 57 (1966).] The instrument has the advantages of 100% efficiency for an initially linearly polarized radiation beam, no channel spectra in the beamsplitter, the possibility of easily cooling the entire interferometer to liquid helium temperatures, separately recording the signals from *each* arm of the interferometer, and simplicity. Fabrication of the parallel wire beamsplitter is described by D. G. Vickers, E. I. Robson, and J. E. Beckman, Freestanding Polarizers for Submillimeter Interferometry, *Appl. Opt.* **10**, 682 (1971).

CHAPTER TEN

# *Spectral Filtering*

## INTRODUCTION

There must be a maximum wave number in the spectrum in any scan. As was pointed out in Chapter 7, if the maximum wave number is very large when the spectrum is computed, there will be aliasing. That is, there will be an unintelligible overlapping of spectra features, and one will not be able to unambiguously interpret the results. In fact, the maximum wave number $\sigma_{max}$ determines the path step size $\Delta\delta$ that may be used in a particular scan. With $\Delta\delta \leq 1/2\sigma_{max}$ one wants to keep $\sigma_{max}$ as small as possible so that $\Delta\delta$ may be as large as possible. Nonetheless, the spectral filtering problem for FTS is not at all as severe as it is for grating spectroscopy with high harmonic false radiation.

The spectral filtering may be accomplished in several ways. The most obvious way, to which most of this chapter is devoted, is to place long-pass spectral filters in the radiation beam. A wide variety of such filters are available, as can be seen in the figures in this chapter. Modulation filtering will be discussed for the far-infrared. Spectral filters which also are part of a chopper system and do not introduce radiant flux errors of their own are sometimes quite useful. Mathematical filtering is described in Chapter 5 along with the subject of apodization.

A very clever way to accomplish spectral filtering is to use electronic techniques which are described near the end of this chapter. Electronic filtering for fast, continuous scans depends on the most rapid fluctuations in the interferogram being caused by the very highest wavenumber contribution to the radiation beam. These high-frequency fluctuations in rapid scanning can be eliminated, along with high-frequency electronic noise, with electronic filters (Appendix D).

Spectral filtering may be accomplished with the beamsplitter material, using detector spectral sensitivity variations, with lens materials, etc. Some of the characteristics of these types of filters will be discussed.

For ordinary spectral filters dependent on the transmission and reflection characteristics of various materials, the book by Hadni [1] is recommended. For spectral filtering by electronic techniques in rapid scanning, the reader is referred to an article by Mertz [2] and to Appendix D.

We shall begin with spectral filters for long wavelengths and progress to filters for FTS in the visible and far-red. In each spectral region, the optical characteristics of several usable spectral filters will be either mentioned or shown. Finally, electronic spectral filtering will be briefly discussed.

Most filters which are discussed will only have room-temperature data presented. Sometimes, the low-temperature characteristics will be briefly mentioned. However, for many of the filters, low-temperature data are available in Dr. Hadni's book and in the open literature.

## SPECTRAL FILTERS FOR BELOW 400 $cm^{-1}$

In working in the far-infrared, one must eliminate the visible and near-infrared fluxes completely and use additional filters with cutoffs at lower wave numbers. In the far-infrared, Mylar beamsplitters are most often used; so, one should match the beamsplitter efficiency (Chapter 9) with the spectral filter characteristics. After choosing the wavenumber range, the beamsplitter thickness should be chosen so that one is first assured of good efficiency in the desired range. Second, the destructive interference in the beamsplitter can be used to assist in the spectral filtering. For the lamellar grating interferometers (Chapter 15), the filtering must be done entirely by transmission and reflection spectral filters.

In the early years of far-infrared spectroscopy, many good filters were developed. Some of them such as the isolation focusing technique, band-pass isolation with reflection from *reststrahlen* crystals, and short-wavelength intensity reduction with crystal chopper blades, are not used very often in FTS. The original methods are usually more expensive in terms of material or labor costs than the newer techniques, such as the Yoshinaga filters, which will be discussed.

One of the most common spectral filter materials, used almost universally in far-infrared spectroscopy, is black polyethylene. The material consists of very small (usually less than 500 Å diameter) particles of carbon black suspended at about a 3% concentration in soft polyethylene [3]. It is commercially available at almost any hardware or department store. However, its transmission characteristics vary considerably from manufacturer to manufacturer. The optical characteristics depend on the particle size, the concentration, and composition. In Fig. 10-1, the absorption coefficient versus wave number is shown for the range 3–4000 $cm^{-1}$ for black polyethylene from three manufacturers, together with some data at lower wave numbers for black polyethylene made by a fourth company.

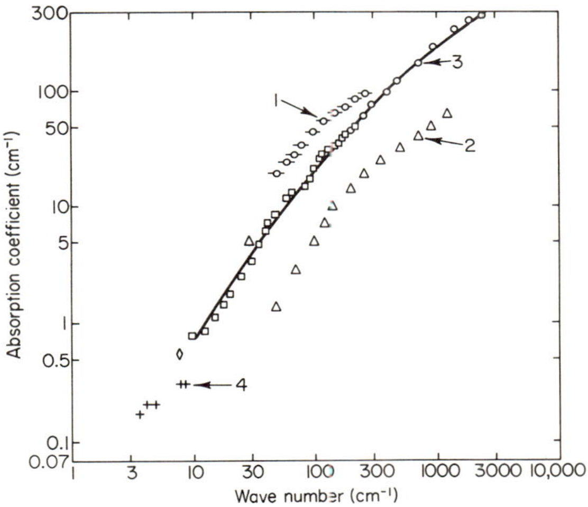


FIG. 10.1. Absorption coefficient of black polyethylene made by: (1) Monsanto, (2) USI Chemical, (3) Dow, (4) Imperial Chemical. The solid line is theory. (After J. M. Blea *et al.* [3].)

The low-temperature transmittance of black polyethylene is very nearly the same as that given by the high-temperature data.

Black polyethylene molds very well at about 90°C and has a nearly constant index of refraction of $1.50 \pm 0.03$ below 400 $cm^{-1}$. Therefore, it can be molded into a lens quite easily.

Black polyethylene is also used in other filters as a spacer or binder medium, and so it plays a dominant role in far-infrared spectroscopy. It is very useful for eliminating flux above about 1000 $cm^{-1}$ if the right thickness and type of black polyethylene are chosen.

The Yoshinaga filters are versatile filters for the far-infrared; they are made of absorbing materials suspended in a dielectric medium. One may finely grind (number 400 grit size or smaller) a crystal such as LiF, mix it with finely ground polyethylene, and at 90°C press the mixture into a slab. The

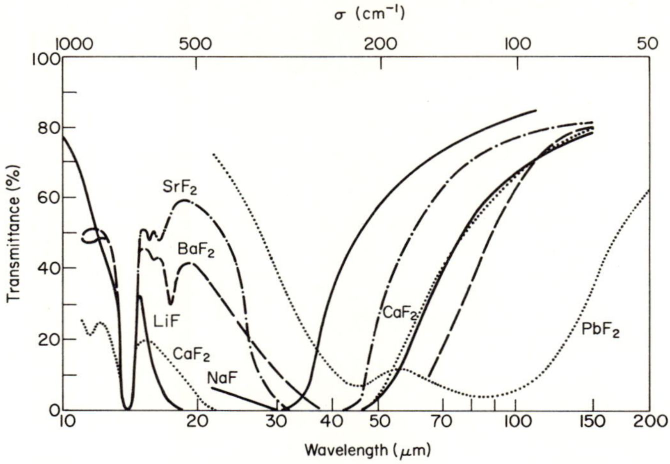


FIG. 10-2. Transmittance of polyethylene sheets containing LiF, NaF, $CaF_2$, $SrF_2$, $BaF_2$, and $PbF_2$ powders, respectively. Each material has been studied over a wide enough spectral range to exhibit transmission on both sides of the peak absorption near the *reststrahl* wave number. (After Yamada *et al.* [4].)

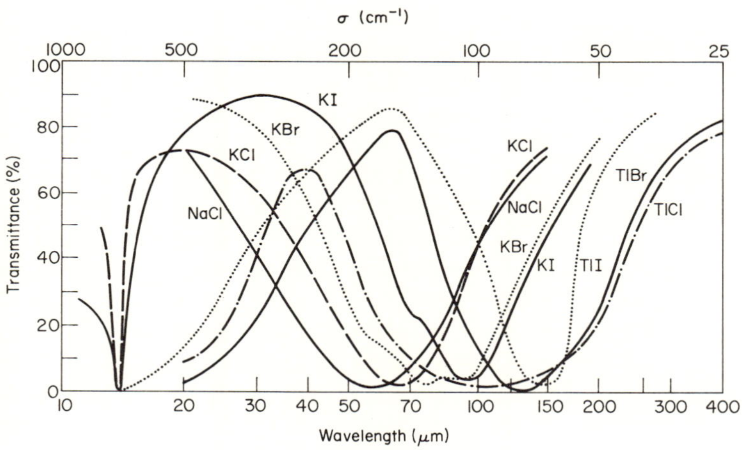


FIG. 10-3. Transmittance of polyethylene sheets containing NaCl, KCl, KBr, KI, TlCl, TlBr, and TlI powders, respectively. Each material has been studied over a wide enough spectral range to exhibit transmission on both sides of the peak absorption near the *reststrahl* wave number. (After Yamada *et al.* [4].)

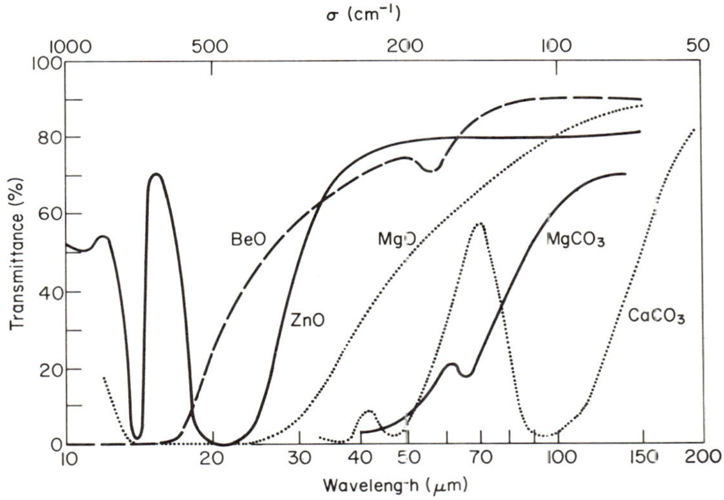


FIG. 10-4. Transmittance of polyethylene sheets containing BeO, ZnO, MgO, $MgCO_3$, and $CaCO_3$ powders, respectively. Most of the materials have been studied over a wide enough spectral range to exhibit transmission on both sides of the peak absorption near the *reststrahl* wave number. (After Yamada *et al.* [4].)

powdered alkali halide can almost completely remove the radiant flux in undesired wavenumber regions. The transmittances of several materials are shown in Figs. 10-2–10-4. Ordinary chemical-grade absorbing materials will suffice and only in quantities far less than 10 g (usually less than 1 g for a given filter). Several different absorbing materials can also be combined in one filter.

Sometimes, grating surfaces are pressed into the surface of Yoshinaga filters. With the grating constant given by $b$, then for $b\sigma$ greater than about 2.5 ($\lambda/b$ less than 0.4), the transmission grating is a good filter. By choosing the grating constant so that the Yoshinaga filter and the grating filter become transparent at nearly the same wave number, one can make a long-pass filter with a very rapidly rising transmittance at the cut-in wave number.

Transmission filters are easy to use because they do not have to be carefully aligned in the optical beam. However, they must be wedged if high-resolution data are to be obtained without channel spectra. In Fig. 10-5, the transmittances of several different filters are shown. All of these materials are readily available commercially. These filters were used to obtain the spectral data shown in Figs. 9-4–9-10 for the entire wavenumber range from 400 to 5 $cm^{-1}$. They, in the thicknesses given, are particularly well suited for work with

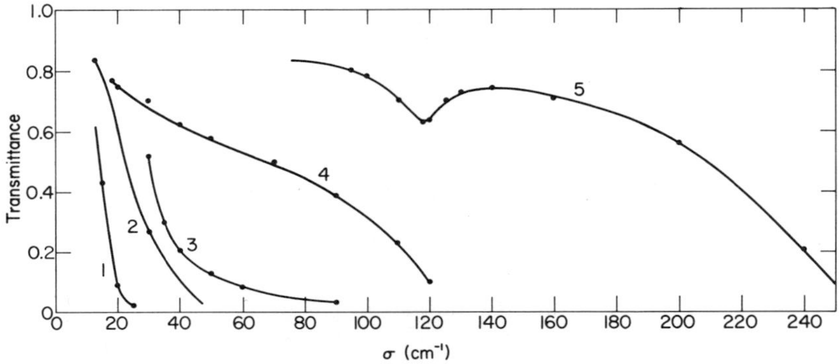


FIG. 10-5. Transmittances of several filters which are useful with black polyethylene for eliminating unwanted short-wavelength radiation flux: (1) 4.5 mm NaCl; (2) filter 7 (two parts black polyethylene and 0.4 parts each of TlI, TlC, KBr, NaCl, $BaF_2$, NaF, and LiF); (3) 1 mm LiF; (4) 0.56 mm fused quartz; (5) 1 mm crystal quartz.

Mylar beamsplitters. In more recent work, we have dispensed with the LiF because it is hygroscopic.

The following materials are well documented (in addition to Hadni's book [1]): Yoshinaga filters [4, 6], crystal quartz [7–10], fused quartz [9–11], LiF [10, 12] and NaCl [10, 13, 14].

There are three types of reflection filters presently in use (*reststrahlen* reflection filters are not used very often):

(1) A metal mesh (or wire cloth) stretched over a frame makes an excellent long-pass spectral filter. Figure 10-6 shows some typical reflectivity data. The high-wavenumber rejection (due to scattering and transmission) is greater than 95% and the low-wavenumber reflectance is flat and in excess

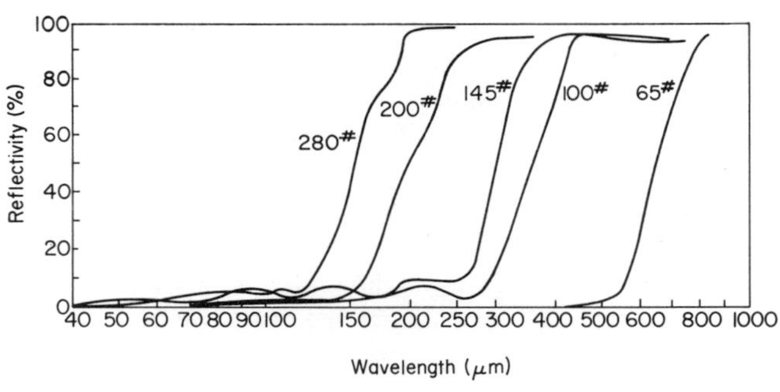


FIG. 10-6 Reflectivity of metal mesh filters. (Numbers in this figure show the mesh constant, i.e., the number of wires per inch.) (After Mitsuishi *et al.* [13].)

of 90%. The wire mesh filters have the sharpest "cut-in" region and are often used. One must be careful of scattered radiant flux from wire screens and provide absorbers. The reflection screens may be made by epoxying the mesh to a hollow metal frame.

(2) Another reflection filter that is used is the roughened surface on a metal or aluminized glass plate. A plane mirror may be easily roughened or replaced with one of these plates, and the visible and near-infrared fluxes will be scattered out of the radiant beam.

(3) Echellette gratings ruled on milling machines or shapers may be made and used as reflection filters in the zeroth order. These are described in detail in Hadni's book [1] and only a representative reflection curve is shown in Fig. 10-7. The Echelette gratings are used in pairs with the groves perpendicular to each other. This eliminates polarization effects and provides a sharper cut-in.

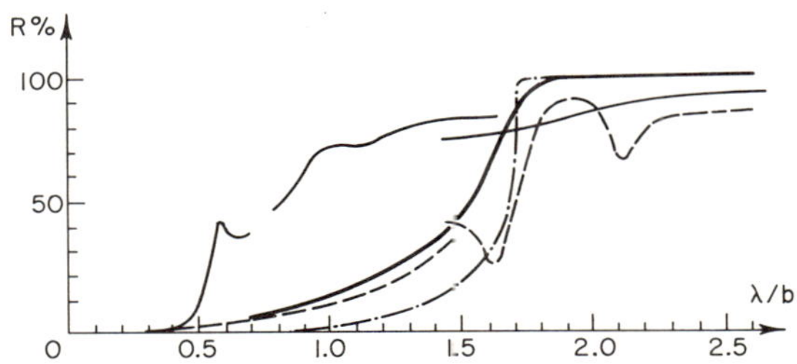


FIG. 10-7. Reflectivity of an echelette grating used with the grooves perpendicular to the plane of incidence. The thick line is theoretical, taking into account both series of faces of a groove. The broken line with dots is theoretical from the simple scalar theory. The thin solid line in experimental for the electric field parallel to the grooves. The dashed line is experimental for the electric field perpendicular to the grooves. The angle of incidence is 45°. (After A. Hadni [1].)

There are two more long-pass filters that deserve mentioning in far-infrared FTS. One is the multielement grid filters described by Ulrich [15] and demonstrated in Fig. 10-8. The components of these filters may be made in the laboratory (with much initial effort) or they may be purchased from Buckbee Mears (priced around $600 each at present). The multielement filters are superior long-pass filters because they so completely eliminate the short-wavelength flux and have a fast cut-in. Ulrich has coined the term "capacitive square" to describe a regular array of metallic square films supported by a plastic substrate (usually Mylar).

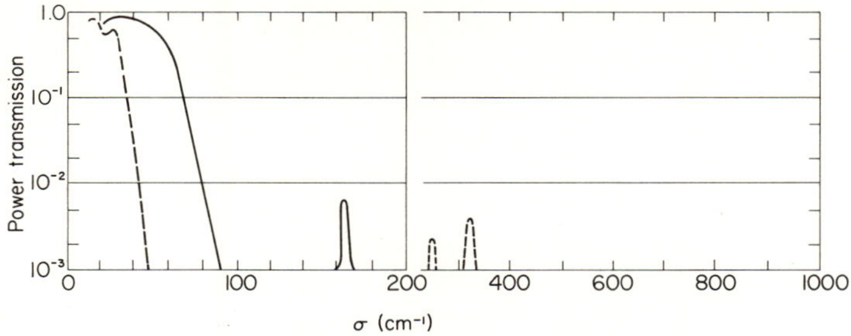


FIG. 10-8. Power transmission of two different long-pass filters. Each is made from four "capacitive square" grids with spacers between them. (After Ulrich [15].)

Thus, there are a wide variety of long-pass filters available in the far-infrared region for research in the wavenumber region below 400 $cm^{-1}$. For most purposes, the reader need only peruse Figs. 10–1-10-8 to find the type of spectral filter required.

## SPECTRAL FILTERS FOR BELOW 5000 $cm^{-1}$

In the near-infrared region, one may use a variety of spectral filters utilizing the transmission and the reflection characteristics of materials, the spectral sensitivity of photoconductive detectors, the properties of dielectric films, the transmission characteristics of beamsplitters material, etc. Electronic spectral filtering will be discussed in a later section.

Some useful filters for the complete elimination of ultraviolet, visible, and near-infrared radiant flux are semiconductors. Once the photon energy is large enough to excite electrons across the forbidden energy gap between the valence and conduction bands, the electromagnetic absorption becomes enormous [16, 17]. Samples of spectral filter materials which are thicker than 1 $\mu$m will completely eliminate radiant fluxes above about 7000 $cm^{-1}$.

Germanium and silicon [17] are also good beamsplitter materials and are also very useful filters (Chapter 9). Their transmittance is very flat except at the lower wave numbers, where lattice absorption becomes appreciable [18]. Near 330 $cm^{-1}$, germanium would be useless, but below 400 $cm^{-1}$, one usually uses Mylar beamsplitters anyway. Silicon presents problems at 1100 and 600 $cm^{-1}$ due to lattice absorption. The 1100-$cm^{-1}$ absorption in silicon can be greatly reduced by using oxygen-free silicon slabs which are commercially available. Dielectric antireflection coatings can be put on the semiconductors to reduce the large reflection losses.

Fortunately, long-pass and band-pass infrared filters are commercially available, from the Optical Coating Laboratory, Inc. of Santa Rosa, California. The company publishes periodically a list of the filters which it has in stock, and one can choose the ideal filter for a particular job. However, one should experimentally verify that the filter does not have spectral leakage at high wave numbers.

Some glasses make good band-pass filters in the infrared [19]. Manganal glass has an appreciable band pass between about 3300 and 10,000 $cm^{-1}$.

As black polyethylene lenses in the far-infrared make good spectral filters, so do other materials in the near-infrared. Arsenic trisulfide, germanium, and silicon lenses may be made.

Detector response for such detectors as PbSe above about 3800 $cm^{-1}$ is sufficiently low to eliminate the detection of high-wave number radiant fluxes. Various photoconductive detectors have the desired insensitivity at high wave numbers (see "detectors" in the bibliography).

## SPECTRAL FILTERS FOR BELOW 16,000 $cm^{-1}$

In research work below 16,000 $cm^{-1}$, there are many filters available. Not only is the Optical Coating Laboratory, Inc. able to make filters, but so also are such companies as Bausch and Lamb of Rochester, New York.

A beamsplitter which is usable in the wavenumber region between 16,000 and 2,800 $cm^{-1}$ is iron oxide ($Fe_2O_3$). This material is very strongly absorbing above about 30,000 $cm^{-1}$, as shown in Fig. 10-9. The exact cut-in point will vary and be strongly dependent on the film thickness, as shown by the various curves.

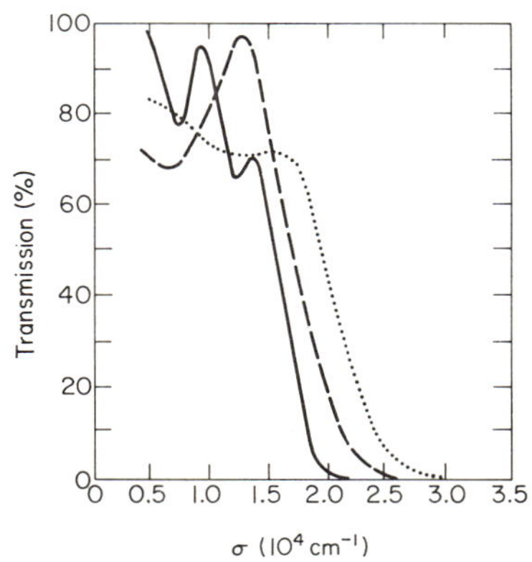


FIG. 10-9. Transmission versus wave number of iron oxide ($Fe_2O_3$). The average transmittance is about 75% below 15,000 $cm^{-1}$ due to reflection losses. The oscillations in the transmission are caused by multiple internal reflection. Film thicknesses are 3760, 1470, and 970 Å for the solid, dashed, and dotted curves, respectively. (After Lewis and Westwood [21].)

Another filter [20] is a solution of iodine in carbon tetrachloride, which is easily prepared and sealed in a glass container for use in a vacuum.

Detectors such as photomultipliers may be used in this range, and some are peaked for different spectral regions. They are, therefore, effectively spectral filters. Again, the photoconductive detectors are quite usable and provide spectral filtering (see "detectors" in the bibliography).

## SPECTRAL FILTERING WITH CHOPPERS

A reflection chopper [22, 23] might be a mirror which alternately reflects the radiation beam onto and beside a slit or aperture. If one roughens the reflecting surface, the flux at long wavelengths will be specularly reflected and will be moved on and off the aperture during the cycle of motion of the mirror. However, short-wavelength flux will be diffusely reflected, and its intensity on the aperture will not vary much during a cycle. Hence, the long-wavelength radiation signal will be chopped, and the short-wavelength flux will not be modulated. One of the additional advantages of this type of chopper is that it remains in the radiation beam path all the time, and there is no hot or cold chopper blade problem.

There are choppers based on frustrated total internal reflection that discriminate very strongly against short-wavelength flux. There are also crystal blade choppers with high rejection of the unwanted short-wavelength flux. Thin-film modulators are also under development.

## SPECTRAL FILTERING BY ELECTRONIC MEANS

If one scans very rapidly by moving the movable mirror quickly, the interferogram will show rapid oscillations as a function of time. The high-wavenumber oscillations will be at the highest oscillation frequency in the interferogram. The low-wavenumber contributions to the interferogram will oscillate at lower frequencies. Effectively, one may eliminate the high-wavenumber oscillations by using a long-pass *electronic* filter after the detector and before converting the output signal to a dc voltage. The filter will not pass oscillations in the interferogram above a certain frequency produced by the high-wavenumber flux. Such filtering also eliminates high-frequency noise.

The data can be digitized as a function of time (movable-mirror position) and the data analysis can be carried out in a normal fashion. The individual scans can be performed in less than 1 sec with the appropriate electronic filtering and then repeated to improve the signal-to-noise ratio at low wave numbers. The entire process can be controlled by minicomputers. Basically, this is the technique used in the Digilab equipment and is described in more detail in the article by Mertz [2]. The Digilab equipment is described in Chapter 19 and the rapid scan technique is outlined in more detail in Appendix D.

## CONCLUSIONS

Because of the aliasing problem described in Chapter 7, it is necessary to eliminate the radiant flux at high wave numbers. This can be done with spectral filters which are used in transmission or reflection, with beam-splitters, by electronic techniques, with discriminating choppers, etc.

Each wavenumber region has its own special filters; so, the chapter was broken into different spectral regions defined by an upper wave number stated at the beginning of the appropriate section. Many figures illustrating the optical characteristics of the various filters were given. In Fig. 10-5, the spectral characteristics of some very simple transmission filters for the far infrared are shown. The particular filters enable one to observe the entire submillimeter band. The remaining section was used to describe choppers which are also spectral filters. Finally, an electronic filtering technique was discussed for eliminating high-wavenumber flux from the interferograms in rapid-scanning FTS.

## *References*

*1.* A. Hadni, "Essentials of Modern Physics Applied to the Study of the Infrared," Chapters I-III. Pergamon Press, New York, 1967.

*2.* L. Mertz, Rapid Scanning Fourier Transform Spectroscopy, *J. Phys. (France)* **28**, 88 (1967).

*3.* J. M. Blea, W. F. Parks, P. A. R. Ade, and R. J. Bell, Optical Properties of Black Polyethylene from 3 to 4000 $cm^{-1}$, *J. Opt. Soc. Amer.* **60**, 603 (1970).

*4.* Y. Yamada, A. Mitsuishi, and H. Yoshinaga, Transmission Filters in the Far-Infrared Region, *J. Opt. Soc. Amer.* **52**, 17 (1962).

*5.* R. J. Bell, H. V. Romero, and J. M. Blea, Theory and Experiments for Multielement Grid Filters in a Dielectric, *Appl. Opt.* **9**, 2350 (1970).

*6.* K. D. Moller and R. V. McKnight, Far-Infrared Transmission Filter Gratings, *J. Opt. Soc. Amer.* **53**, 760 (1963).

*7.* D. W. Berreman, Low Pass Filters for Far Infrared Spectroscopy, *Rev. Sci. Inst.* **37**, 513 (1966).

*8.* D. E. McCarthy, The Reflection and Transmission of Infra-Red Materials: 1, Spectra from 2 to 50 Microns, *Appl. Opt.* **2**, 591 (1963).

*9.* R. Geick, Der Brechungsindex von Kristallinem und Geschmolzenem Quarz im Spektral bereich VM 100, *Z. Phys.* **161**, 116 (1961).

*10.* T. K. McCubbin and W. M. Sinton, Recent Investigations in the Far Infrared, *J. Opt. Soc. Amer.* **40**, 538 (1950).

*11.* H. Yoshinaga, Recent Techniques in Far Infrared Spectroscopy, *Jap. J. Appl. Phys. Suppl. I* **4**, 420 (1965).

*12.* R. C. Lord and T. K. McCubin, Infrared Spectroscopy from 5 to 200 Microns with a Small Grating Spectrometer, *J. Opt. Soc. Amer.* **47**, 689 (1952).

*13.* A. Mitsuishi, Y. Otsuka, S. Fujita, and H. Yoshinaga, Metal Mesh Filters in the Far Infrared Region, *Jap. J. Appl. Phys.* **2**, 574 (1963).

*14.* L. Genzel, H. Happ, and R. Weber, Dispersionsmessungen an NaCl, KCl, und KBr Zwischen 0.3 und 3mm Wellenlange, *Z. Phys.* **154**, 13 (1959).

*15.* R. Ulrich, Interference Filters for the Far Infrared, *Appl. Opt.* **7**, 1987 (1968).

*16.* H. Y. Fan, M. L. Shepherd, and W. Spitzer, "Photoconductivity Conference." Wiley, New York, 1956.

*17.* P. W. Kruse. L. D. McGlauchlin, and R. B. McQuistan, "Elements of Infrared Technology, Generation, Transmission, and Detection." Wiley, London, 1962.

*18.* M. Lax and E. Burstein, Infrared Lattice Absorption in Ionic and Homopolar Crystals, *Phys. Rev.* **97**, 39 (1955).

*19.* M. Hass and B. W. Hevis, Infrared Lattice Reflection Spectra of III-IV Compound Semiconductors, *J. Phys. Chem. Solids* **23**, 1099 (1962).

*20.* J. Lecomte, Spectroscopic dans l'Infrarouge, "Handbuch der Physik," Vol. 24, p. 283 1957.

*21.* D. C. Lewis and W. D. Westwood, The Spectra of Some Iron Oxide Films between 30,000 $cm^{-1}$ and 4000 $cm^{-1}$, *Can. J. Phys.* **42**, 2367 (1964).

*22.* R. J. Bell and T. E. Gilmer, Jr., A New Radiation Chopper Principle, *Appl. Opt.* **4**, 45 (1965).

*23.* S. I. Drasky and R. J. Bell, Radiation Chopper for the Submillimeter Wavelength Region, *Infrared Phys* **5**, 137 (1965).

*24.* N. J. Harrick, "Internal Reflection Spectroscopy." Wiley (Interscience), New York, 1967.

*Note added in proof*: The book by K. O. Möller and W. G. Rothschild, "Far-Infrared Spectroscopy," Wiley (Interscience), New York, 1971, is an excellent reference for Yoshinaga and multielement grid filters.

CHAPTER ELEVEN

# *Field of View*

## INTRODUCTION

If one has an extended source, the effect of the field of view on the interferogram and subsequently on the spectrum must be considered. So far, only point sources have been assumed; in this chapter, the modifications in the spectrum due to sources of finite extent will be treated. We shall find that the modifications to the interferogram which arise from the extended source can be generalized to cover other instrumental effects. The changes in the interferogram will be treated in detail for the extended source problem and also in general for later use (Chapter 15).

For the extended source, one should consider both rays on and rays oblique to the optical axis. The axial rays traverse the interferometer with an optical path difference of $2d$, but the oblique rays have a *smaller* path *difference* than $2d$. For a given relative mirror displacement, this yields an interferogram which is expanded because of the shorter path difference of the oblique rays. That is, the true optical path difference $\Delta$ is smaller than $\delta = 2d$. This expansion of the interferogram produces a measured spectrum which differs from the true spectrum in a manner which depends on both wave number and the solid angle subtended by the source. The modifications to the interferograms can be interpreted in terms of the fringes in the interferogram, as will be shown in the next to last section of this chapter.

The instrumental effect on the spectrum will be computed assuming that the source is monochromatic in frequency but extended in size. There are two resultant effects: (1) the computed spectrum is shifted to lower wave numbers by a small amount, and (2) there is a wavenumber spread introduced, i.e., the ILS function is no longer a sinc function as it was in Chapter 5. One might expect both of these effects upon consideration that a point source centered on the optical axis emits optical rays which have maximum possible optical path *differences* when considering the rays reflected from the two mirrors. Optical rays from points other than the center of the source traverse the interferometer along paths which produce smaller path differences. The shorter path differences for a given mirror displacement $d$ expand the oscillations in the interferogram towards larger $2d$ values. The same situation arises in the Fabry–Perot etalon with reflected optical rays at oblique incidence.) Since lower-frequency oscillations in the interferogram are associated with the lower wave numbers, one may reach the following conclusions. The computed spectrum from a monochromatic, extended source of frequency $\sigma_0$ will be spread out from $\sigma_0$ toward lower wave numbers. So one has a mean wavenumber shift and a resolution limitation with a finite source. We will derive the pertinent equations, show a computed instrumental spectral profile from a monochromatic, extended source, and numerically compute some typical wavenumber shifts and spreads.

Journal references for the subject of extended sources are Steel [1] and Loewenstein [2]. The references cited in Steel's article will give the subject some historical perspective.

## INTERFEROGRAM DUE TO AN EXTENDED SOURCE

If one has an extended source, there are optical rays from points off the optical axis which make angles $\theta$ with the optical axis. Consider Fig. 11-1 for an extended source, with a ring of source points on a circle indicated by the dashed lines. All source points such as $P$ emit radiation beams which can make an arbitrary angle $\theta$ with the optical axis as shown. In order to bring into consideration the entire extended source, we need only to sum the radiant flux from all angles between 0 and some maximum $\theta$ and weight the contributions according to the circumference of the concentric source ring (dashed). This is equivalent to integrating over the solid angle subtended by the source. To summarize this section: we will find the effective path difference through the interferometer for arbitrary oblique optical rays; we will use this effective path difference in the expression for the interferogram; and we can integrate the radiant intensity over the solid angle subtended by the source. This procedure will enable us to write the interferogram for an interferometer with an extended source.

In Fig. 11-1, the Michelson interferometer is shown with the pertinent

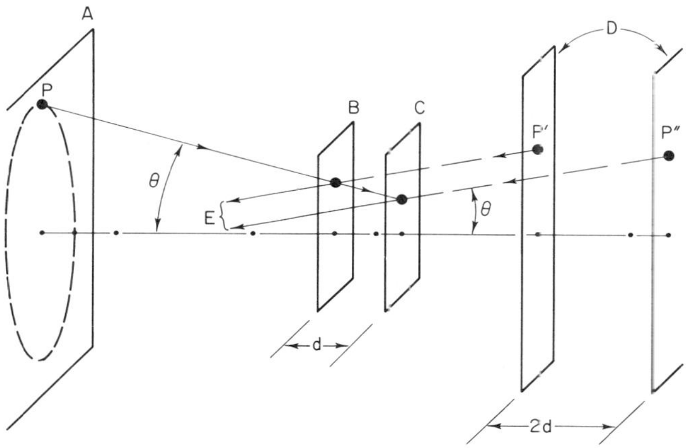


FIG. 11-1. An equivalent diagram of a Michelson interferometer. A: Extended source; B: movable mirror; C: image of fixed mirror; D: virtual image planes of the source; E: to detector.

optical elements in a linear array. The beamsplitter is omitted because it has no particular bearing on the extended source problem. The movable mirror is shown in relation to a parallel position with respect to the extended source. To linearize the optical diagram, the image of the fixed mirror is a distance $d$ behind the movable mirror. In this case, it happens that the movable mirror is a distance $d$ closer to the beamsplitter than the fixed mirror. The images of plane mirrors are behind them, and one usually locates the image with the source and image distance equidistant from the plane mirror. In the interferometer, the radiant beams have to go an optical path distance $2d$ to get to and from the fixed mirror relative to the movable mirror. Thus, the image for the fixed mirror $p''$ is a distance $2d$ behind the image for the movable mirror $p'$.

Figure 11-2 shows an optical ray incident on and reflected from the fixed and movable mirrors. The optical path difference for an oblique ray can be defined as $\Delta$. From Fig. 11-2, one has that

$$\Delta = (2d/\cos\theta) - 2d\tan\theta\sin\theta = 2d\cos\theta, \tag{11-1}$$

or in terms of the measured path difference for a ray on the optical axis, we have

$$\Delta = 2d\cos\theta = \delta\cos\theta. \tag{11-2}$$

Recall from Eq. (3-21) that the signal versus optical path difference (dropping multiplying constants) is given by

$$I_R(\Delta) = \int_0^\infty \mathscr{E}^2(\sigma)\,d\sigma + \int_0^\infty \mathscr{E}^2(\sigma)\cos(2\pi\sigma\Delta)\,d\sigma. \tag{11-3}$$

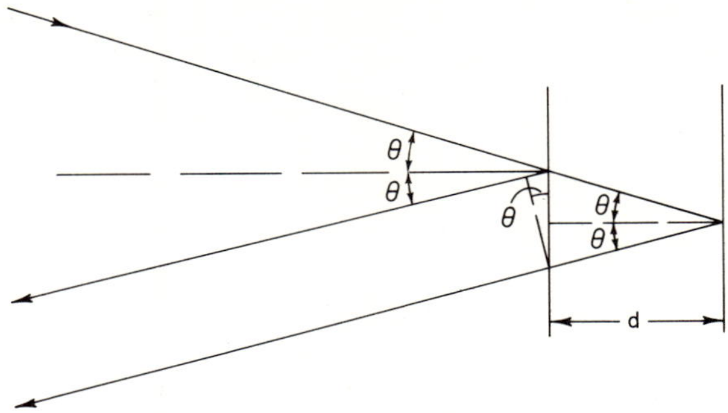


FIG. 11-2a. Optical diagram of the equivalent interferometer showing the movable mirror in front of the fixed mirror by a distance *d*.

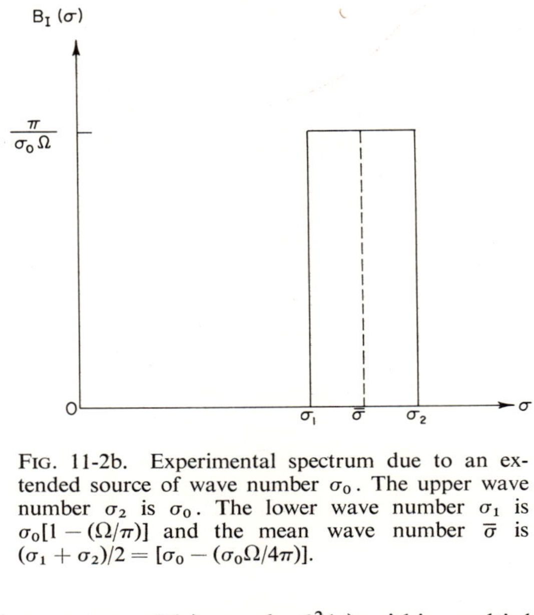


FIG. 11-2b. Experimental spectrum due to an extended source of wave number $\sigma_0$. The upper wave number $\sigma_2$ is $\sigma_0$. The lower wave number $\sigma_1$ is $\sigma_0[1-(\Omega/\pi)]$ and the mean wave number $\bar{\sigma}$ is $(\sigma_1+\sigma_2)/2 = [\sigma_0 - (\sigma_0\Omega/4\pi)]$.

Noting that the spectrum $B(\sigma)$ equals $\mathscr{E}^2(\sigma)$ within multiplying constants, substituting for the path of the oblique optical ray, and integrating over the source with the solid angle $\Omega$ subtended by the source, we have for the *normalized* signal versus optical path

$$I_R(\delta, \Omega) = (1/\Omega) \int_0^\infty B(\sigma) \int_0^\infty [1 + \cos(2\pi\sigma\delta/\cos\theta)]\, d\Omega\, d\sigma. \qquad (11\text{-}4)$$

But the solid angle is

$$\Omega = \int_0^{2\pi} d\varphi \int_0^{\theta} \sin\theta \, d\theta = 2\pi(1 - \cos\theta). \tag{11-5}$$

Using the expression for the solid angle, we have

$$I_R(\delta, \Omega) = (1/\Omega) \int_0^{\infty} B(\sigma) \left\{ \int_0^{\Omega} [1 + \cos(2\pi\sigma\delta - \sigma\delta\Omega)] \, d\Omega \right\} d\sigma. \tag{11-6}$$

Integrating and noting that for $\delta$ approaching infinity the sine terms average to zero, we have the interferogram

$$F(\delta, \Omega) \equiv [I_R(\delta, \Omega) - I_R(\infty, \Omega)] \tag{11-7}$$

given by

$$F(\delta, \Omega) = (1/\Omega) \int_0^{\infty} [B(\sigma)/\sigma\delta][\cos(2\pi\sigma\delta) \sin(\sigma\delta\Omega) - \sin(2\pi\sigma\delta) \cos(\sigma\delta\Omega) + \sin(2\pi\sigma\delta)] \, d\sigma. \tag{11-8}$$

Using trigonometric identities, we have the interferogram for some maximum solid angle $\Omega$ given by

$$F(\delta, \Omega) = \int_0^{\infty} \{B(\sigma) \operatorname{sinc}(\Omega\sigma\delta/2) \cos[2\pi\sigma\delta - (\Omega\sigma\delta/2)]\} \, d\sigma. \tag{11-9}$$

We could proceed in a straightforward manner (as we did in the first sections of Chapter 8) and compute the spectrum from the interferogram given in Eq. (11-9). However, the procedure can just as easily be generalized, and then Eq. (11-9) can be considered a special case. The generalization is especially useful for the treatment of interferometers without collimations, such as the spherical lamellar grating interferometer described in Chapter 15.

## GENERAL TREATMENT

If one has an interferogram, within a multiplicative constant, of the form

$$F(\delta) \equiv \int_0^{\infty} B(\sigma) s(\sigma, \delta) \cos[2\pi\sigma\delta + \varphi(\sigma, \delta)] \, d\sigma \tag{11-10}$$

the experimental or instrumental profile $B_{\exp}(\sigma)$ (spectrum produced from a monochromatic source of wave number $\sigma_0$) is given by

$$B_{\exp}(\sigma) = \tfrac{1}{2}\int_{-\infty}^{\infty} [s(\sigma_0, \delta)e^{i\varphi(\sigma_0,\delta)}]e^{i2\pi(\sigma_0-\sigma)\delta}\, d\delta, \tag{11-11}$$

where $s(\sigma_0, \delta)$ and $\varphi(\sigma_0, \delta)$ are defined by Eq. (11-10). It is the goal of this section to show that the conclusion reached in Eq. (11-11) follows for the general equation (11-10). One can see by comparing the equation for the interferogram of an extended source as given in Eq. (11-9) with the general equation as stated in Eq. (11-10)

$$s(\sigma, \delta) \equiv \text{sinc}(\Omega\sigma\delta/2) \tag{11-12}$$

and that

$$\varphi(\sigma, \delta) \equiv -\Omega\sigma\delta/2. \tag{11-13}$$

In order to proceed with the derivation of Eq. (11-11), we multiply both sides of Eq. (11-10) with $\exp[-2\pi\sigma'\delta]$, set up an integration over $\delta$ from $-\infty$ to $\infty$, and actually perform the first integration with $\sigma$ using a monochromatic source of wave number $\sigma_0$ represented by a Dirac delta function. We then have

$$\int_{-\infty}^{\infty} F(\delta)\exp(-i2\pi\sigma'\delta)\, d\delta$$
$$= \int_{-\infty}^{\infty} s(\sigma_0, \delta)\{\cos[2\pi\sigma_0\delta + \varphi(\sigma_0, \delta)]\}\exp(-i2\pi\sigma'\delta)\, d\delta. \tag{11-14}$$

We use Euler's formula, $e^{ix} = \cos x + i\sin x$, to separate the complex exponential into a cosine and a sine term, use trigonometric identities on the resultant products in Eq. (11-14), and then employ Euler's formula again to return to the complex exponential form. The results of the manipulation yield

$$\int_{-\infty}^{\infty} F(\delta)\exp(-i2\pi\sigma'\delta)\, d\delta$$
$$= \Big(\tfrac{1}{2}\int_{-\infty}^{\infty} \{s(\sigma_0, \delta)\exp[-i\varphi(\sigma_0, \delta)]\}\exp[-i2\pi(\sigma' + \sigma_0)\delta]\, d\delta$$
$$+ \tfrac{1}{2}\int_{-\infty}^{\infty} \{s(\sigma_0, \delta)\exp[i\varphi(\sigma_0, \delta)]\}\exp[i2\pi(\sigma_0 - \sigma')\,\delta]\, d\delta\Big). \tag{11-15}$$

For $s(\sigma_0, \delta)$ varying slowly enough, we can drop the first integral in Eq.

(11-15) because it has its main contribution at $\sigma' = -\sigma_0$ and has negligible values for positive $\sigma'$. For the experimental spectrum, we have

$$B_{\text{exp}}(\sigma') \equiv \int_{-\infty}^{\infty} F(\delta) \exp(-i2\pi\sigma' \, d\delta) \, d\delta; \tag{11-16}$$

so, on dropping the now unnecessary prime on $\sigma'$, we have

$$B_{\text{exp}}(\sigma) = \tfrac{1}{2} \int_{-\infty}^{\infty} [S(\sigma_0, \delta) \exp(i2\pi\sigma_0 \delta)] \exp(-i2\pi\sigma\delta) \, d\delta, \tag{11-17}$$

where

$$S(\sigma_0, \delta) \equiv s(\sigma_0, \delta) \exp[i\varphi(\sigma_0, \delta)]. \tag{11-18}$$

Equation (11-17) with Eq. (11-18) is identical to Eq. (11-11), so we have proved that the instrumental profile is the inverse Fourier transform of $S(\sigma_0, \delta) \exp(i2\pi\sigma_0 \delta)$. If one knows $S(\sigma_0, \delta)$ as found by expressing the interferogram in the form of Eq. (11-10), then Eq. (11-17) can be used to compute the line shape resulting from the interferogram.

For the extended source problem as for some other cases, if the phase $\varphi(\sigma_0, \delta)$ is linear in $\delta$ as shown in Eq. (11-13), one may easily combine all the complex exponential factors in Eq. (11-11) or, equivalently Eq. (11-17).

## APPLYING THE GENERAL TREATMENT TO THE EXTENDED SOURCE PROBLEM

We now want to combine the results of the first two sections. That is, we want to use the general treatment to handle the extended source problem. Essentially, all we have to do is substitute Eqs. (11-12) and (11-13) in Eq. (11-11) for the instrumental profile and evaluate the integral. The integral to be evaluated is

$$B_{\text{exp}}(\sigma) = \tfrac{1}{2} \int_{-\infty}^{\infty} \operatorname{sinc}(\Omega\sigma_0 \delta/2) \exp\{i2\pi\sigma_C[1 - (\Omega/4\pi)]\} \exp(-i2\pi\sigma\delta) \, d\delta. \tag{11-19}$$

The fact that the phase is linear in $\delta$ makes possible such a simple expression of the instrumental profile. One can evaluate the integral using contour integration, or one can use simple Fourier transforms since Eq. (11-19) is an inverse Fourier transform. Let us note that the Fourier transform of the rectangular function defined on the interval $\sigma_1$ to $\sigma_2$ ($\sigma_2 > \sigma_1$) is

$$\int_{-\infty}^{\infty} \operatorname{rect}(\sigma_1, \sigma_2) \exp(i2\pi\sigma\delta) \, d\sigma = \int_{\sigma_1}^{\sigma_2} \exp(i2\pi\sigma\delta) \, d\sigma,$$

or on completing the integral we have

$$\mathscr{F}\{\text{rect}(\sigma_1, \sigma_2)\}/(\sigma_2 - \sigma_1) = \text{sinc}[\pi(\sigma_2 - \sigma_1)\delta] \exp[i\pi(\sigma_1 + \sigma_2)\,\delta]. \quad (11\text{-}20)$$

Since the inverse Fourier transform of a Fourier transform of a function is the function, we have

$$\int_{-\infty}^{\infty} \{\text{sinc}[\pi(\sigma_2 - \sigma_1)\,\delta] \exp[i\pi(\sigma_1 + \sigma_2)\,\delta] \exp(-i2\pi\sigma\delta)\, d\delta$$
$$= [\text{rect}(\sigma_1, \sigma_2)]/(\sigma_2 - \sigma_1). \quad (11\text{-}21)$$

Comparing Eqs. (11-19) and (11-21), we have

$$\boxed{B_{\text{exp}}(\sigma) = [\pi/(\sigma_0\,\Omega)]\,\text{rect}(\sigma_1, \sigma_2),} \quad (11\text{-}22)$$

where

$$\boxed{\sigma_1 \equiv \sigma_0 - (\sigma_0\,\Omega/2\pi)} \quad (11\text{-}23)$$

and

$$\boxed{\sigma_2 \equiv \sigma_0\,.} \quad (11\text{-}24)$$

So, the general treatment has been used to set up the Fourier transform of $S(\sigma_0, \delta)$, and then the integration was performed to obtain the instrumental profile from an extended, monochromatic source.

## DISCUSSION OF THE INSTRUMENTAL PROFILE

Before getting involved with a detailed discussion of Eq. (11-22), let us note a difference between Eq. (11-22) and Steel's Eq. (3-2). Steel [1] obtained an instrumental or inherent profile given by $\pi/2\Omega\sigma_0$. The factor 2 appears because he used a one-sided interferogram, whereas we used a two-sided one.

In Fig. 11-3, the instrumental profile is plotted for the extended source which subtends a solid angle $\Omega$ and is of uniform brightness. As pointed out (in the introduction) for physical reasons, the spectrum is spread out toward lower wave numbers. The mean wave number $\bar{\sigma}$ is

$$\boxed{\bar{\sigma} = \sigma_0[1 - (\Omega/4\pi)]} \quad (11\text{-}25)$$

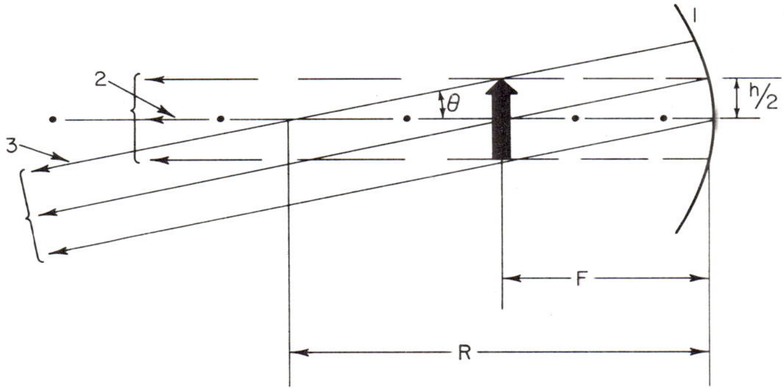


FIG. 11-3. Extended source in front of a collimating mirror for an interferometer. (1) Mirror; (2) from center; (3) from tip.

and the total wavenumber spread or resolution is

$$\boxed{\delta\sigma = \sigma_0 \, \Omega/2\pi} \tag{11-26}$$

and is entirely on the low-wavenumber side of the source frequency.

It is interesting to note that the throughput advantage or étendue is expressed in Eq. (11-26). Defining the resolving power as $R \equiv \sigma_0/\delta\sigma$ and using Eq. (11-26), one has

$$R\Omega = 2\pi \tag{11-27}$$

which is Eq. (2-8). Equation (2-8) is the basis of the throughput advantage of interferometers. Starting with Eq. (2-8) and using the physical reasoning in the introductory section of this chapter, all of the results shown in Fig. 11-2 could have been stated!

We want to establish the $\Omega$ is the solid angle subtended by the source and to express $\Omega$ as a function of easily measurable parameters. In Fig. 11-3, a circular source of diameter $h$ is in the focal plane of a collimating mirror. As seen from the diagram, optical rays from the center of the source and the edge of the source cross at an angle $\theta$. This is the maximum angle at which the optical rays can intersect the optical axis. Therefore, the extreme rays define the solid angle which is given in Eq. (11-5). Noting from the figure that

$$\tan\theta = h/2F, \tag{11-28}$$

and using the small-angle approximations in Eqs. (11-5) and (11-28), we have

$$\Omega = \pi h^2/4F^2. \tag{11-29}$$

But, from Fig. 11-3, $\pi h^2/4F^2$ is exactly the solid angle subtended by the source with the center of the collimator mirror at the vertex. So, on substituting Eq. (11-29) in Eq. (11-27), the resolving power is

$$R \equiv \sigma/\delta\sigma \simeq 8F^2/h^2. \tag{11-30}$$

To furnish numerical information about the resolving power limitations caused by finite sources, we consider a typical Michelson interferometer with various source diameters. If the focal length of the collimator is $F = 115$ mm and we use source diameters of 3, 5, and 10 mm, we would have resolving power limitations of $1.2 \times 10^4$, $4.1 \times 10^3$, and $9.3 \times 10^2$, respectively. Therefore, in the laboratory with ordinary collimation optics, the source size cannot be unlimited if superior resolution is required. Nonetheless, the bounds on the source size are usually no problem in simple experiments.

For use of the interferometer in astronomy, the source subtends such a small solid angle that Eq. (11-27) allows essentially unlimited resolutions. Extremely high resolution in astrophysical data has been obtained with long scan times and large arm displacements.

## INTERFERENCE FRINGES AND AN EXTENDED SOURCE

What is happening optically when the interferogram is modulated by $\text{sinc}(\Omega\sigma\delta/2)$ as in Eq. (11-9)? The sinc function is decreasing the energy in the spectrum with increasing solid angle $\Omega$. For $\Omega = 2\pi/\sigma\delta$, the sinc function reaches its first zero, and for $2\pi/\sigma\delta < \Omega < 4\pi/\sigma\delta$, the sinc function is negative. The negative values of the sinc function express the fact that the interference fringes have reversed phase and allow destructive interference. Thus, if $\Omega > 2\pi/\sigma\delta$, energy is effectively lost from the spectrum. One should keep in mind that the smallest critical solid angle value $\Omega_c$ occurs for $\delta = L \equiv$ maximum optical path displacement, and $\sigma = \sigma_{\text{max}}$. Thus, we have for the first zero of the sinc function that

$$\Omega_c = 2\pi/L\sigma_{\text{max}}. \tag{11-31}$$

We would like to know if the zero in the interferogram reached at $\Omega_c$ strongly affects the central spot in the interference pattern. The central pattern is controlled by the term $\cos(2\pi\sigma\Delta)$ in Eq. (11-3) with $\Delta$ defined in Eq. (11-2). Using the small-angle approximation for $\cos\theta$, one has

$$\cos(2\pi\sigma\Delta) \simeq \cos(2\pi\sigma\delta - \pi\sigma\delta\theta^2). \tag{11-32}$$

The central ray is described for $\theta = 0$ or for a phase of $2\pi\sigma\delta$. As $\theta$ is increased, the intensity in the central spot will decrease to a minimum at a phase of $2\pi\sigma\delta \pm \pi$. Thus, the interference fringes will have the first minimum for

$$\theta_{\mathrm{m}}{}^2 = 1/\sigma\delta. \tag{11-33}$$

Since, for small angles, $\Omega \simeq \pi\theta^2$ [see Eq. (11-5)], one has the smallest solid angle at which the interference pattern reaches a minimum in intensity given by

$$\Omega_{\mathrm{m}} = \pi/L\sigma_{\mathrm{max}}. \tag{11-34}$$

Let us compare Eq. (11-31), which gives the solid angle at which the fringes will reverse phase and subtract energy from the spectrum, and Eq. (11-34), which gives the solid angle at which the central spot in the interference fringe reaches its first intensity minimum. It is obvious that

$$\Omega_{\mathrm{c}} = 2\Omega_{\mathrm{m}}, \tag{11-35}$$

so the central spot of the interference pattern reaches a minimum at $\frac{1}{2}$ the solid angle, at which the sinc function predicts phase reversal. Therefore, as far as the central spot is concerned, the sinc-function modulation only represents a minor distortion. For the outer fringes, the sinc function does represent phase reversal and some energy is lost. When the minimum at the edge of the central spot has been reached, $\Omega = \pi/\sigma\delta$ and the amplitude of the sinc function has dropped to $\mathrm{sinc}(\pi/2) = 0.64$, which is not a serious loss in energy.

## CONCLUSIONS

If one uses an extended source in an interferometer, the *computed* spectrum [using $\delta = 2d$ in Eqs. (11-10) and (11-11)] should be shifted down in wave number by a mean value given by Eq. (11-25). Also, there is a wavenumber spread given in Eq. (11-26) caused by an extended source. The effects can be interpreted physically.

A generalized equation describing optical effects was introduced in Eq. (11-10). Equation (11-11) or Eq. (11-17) provides a general method of obtaining the final effect of the instrument on the spectrum. The general statement is useful in Chapter 15.

## *References*

*1.* W. H. Steel, Interferometers without Collimation for Fourier Spectroscopy, *J. Opt. Soc. Amer.* **54**, 151 (1964).

*2.* E. V. Loewenstein, Fourier Spectroscopy: An Introduction, *Aspen Int. Conf. on Fourier Spectrosc., 1970*, p. 3. (G. A. Vanasse, A. T. Stair, Jr., and D. J. Baker, eds.). AFCRL-71-0019, 5 Jan. 1971, Spec. Rep. No. 114.

CHAPTER TWELVE

# Phase Error and Sampling Problems

## INTRODUCTION

If, in sampling the interferogram as described in Chapter 7, a measurement or sample is not taken at the true origin, serious difficulties can appear in the computed spectrum. An extreme example has already been seen in the case of the asymmetric Fourier interferometer. For this case, the true origin is displaced from the measured origin by an amount $D \simeq (\bar{n} - 1)l$ as was shown in Chapter 8. If $n(\sigma) \neq \bar{n}$ for all $\sigma$ in the spectrum, then the true origin is a function of $\sigma$ also. The displacement of the true origin from the measured origin is indeed just $[n(\sigma) - 1]l$. Consider for a moment a sample with no absorption, so that $\tau_0(\sigma) = 1$; then, if we consider transmission measurements where multiple internal reflections are ignored, Eq. (7-46) shows that the computed spectrum $B_c(\sigma)$ is

$$B_c(\sigma) = B_0(\sigma) \exp\{i2\pi\sigma[n(\sigma) - 1]l\}. \tag{12-1}$$

Hence, we use the term "phase error." Missing the origin has led to an error that can be expressed as $e^{i\varphi(\sigma)}$. Even if our interferometer is symmetric, such an error can arise because a sampled interferogram does not contain a sample for $\delta_T = 0$. Instead, the origin is displaced by an amount

$$\beta \equiv \delta_T - \delta_M. \tag{12-2}$$

where $\delta_M$ is the origin assumed for the interferogram, i.e., $\delta_M$ corresponds to the maximum in the measured, sampled interferogram, and $\delta_T$ is the true optical path difference.

Poor optical alignment of an interferometer can also lead to asymmetric interferograms which yield phase errors, even nonlinear phase errors.

In the following, we first consider the case of having missed the true origin in a sampled, one-sided interferogram. This is referred to as a sampling phase error. A Lorentzian profile will be considered for the spectrum, and the resultant distortion illustrated. We then show that phase errors may be completely eliminated from a double-sided interferogram. Next, we consider an approximate general treatment for phase errors. The important approximation is that the phase error be a slowly varying function of wave number. With this assumption, one can use a low-resolution, two-sided interferogram to determine the phase error to correct a high-resolution, one-sided interferogram. All of the phase-error-correction techniques can be carried out on computers without manual steps.

There are three papers which are of particular use in the study of phase errors. The most general is one by Sakai *et al.* [1]. This paper discusses four classes of phase errors and examines the subject in a broader scope than is done in this chapter. Two other noteworthy papers are by Mertz [2] and Forman *et al.* [3]. The results reached in the last two papers are equivalent to the treatment by Sakai *et al.* The initial portion of the section, General Phase Error will follow some of the work of Sakai *et al.* quite closely.

## SAMPLING PHASE ERRORS

Suppose one has recorded a one-sided interferogram in order to achieve maximum resolving power, to conserve scan time, or to reduce computation time. However, let us assume that the origin was missed by an optical path difference of $\beta$ as given in Eq. (12-2).

We now want to inquire about the effect on the computed spectrum due to missing the origin when a one-sided interferogram is used. For the sake of discussion, assume that the source is emitting radiation with an intensity distribution $B_S(\sigma)$ given by

$$B_S(\sigma) = A\varepsilon/[(\sigma - \sigma_0)^2 + \varepsilon^2]. \tag{12-3}$$

Equation (12-3) describes a Lorentzian line shape that is often encountered in atomic, molecular, and solid-state absorption and emission spectra. The peak intensity at $\sigma = \sigma_0$ is $A/\varepsilon$ and the full-width at one-half peak intensity (fwhp) is $2\varepsilon$.

First, we must determine what the interferogram is for the Lorentzian profile in Eq. (12-3) with no sampling error. We can compute the interferogram or the Fourier transform of $B_S(\sigma)$ most easily by using the shift rule to move the peak of the Lorentzian to the origin. We wish to compute the interferogram from

$$I_R(\sigma) - I_R(\infty) = A\varepsilon \int_{-\infty}^{\infty} \{[\exp(i2\pi\sigma\delta)]/[(\sigma - \sigma_0)^2 + \varepsilon^2]\}\, d\sigma; \qquad (12\text{-}4)$$

so, the form of the shift rule that we want is

$$\int_{-\infty}^{\infty} f(\sigma - \sigma_0) \exp(i2\pi\sigma\delta)\, d\sigma = \exp(i2\pi\sigma_0\delta) \int_{-\infty}^{\infty} f(\sigma) \exp(i2\pi\sigma\delta)\, d\sigma. \qquad (12\text{-}5)$$

So, we compute the Fourier transform of the shifted Lorentzian profile by contour integration, and we have

$$A\varepsilon \int_{-\infty}^{\infty} \{[\exp(i2\pi\sigma\delta)]/[\sigma^2 + \varepsilon^2]\}\, d\sigma = A\pi \exp(-2\pi\varepsilon|\delta|). \qquad (12\text{-}6)$$

We now note that if one has an interferogram and computes the spectrum, he obtains spectra for both the positive and negative wave numbers. If energy is to be conserved and one wishes to compute an interferogram from a spectrum, he must enter the spectrum and its mirror image about the wave-number origin. These problems were encountered with the Dirac delta function in Chapter 5 and with aliasing in Chapter 7. So, one must add a spectrum centered about $-\sigma_0$. Mathematically, in Eq. (12-5), one may change $\sigma_0$ to $-\sigma_0$ and apply the shift rule. The interferogram for the Lorentzian profile on adding the two contributions becomes [combining Eqs. (12-3), (12-5), and (12-6)]

$$I_R(\delta) - I_R(\infty) = [A\pi \exp(-2\pi\varepsilon|\delta|)][\exp(i2\pi\sigma_0\delta) + \exp(-i2\pi\sigma_0\delta)] \quad (12\text{-}7)$$

or

$$I_R(\delta) - I_R(\infty) = [2\pi \mathrm{A} \exp(-2\pi\varepsilon|\delta|)] \cos(2\pi\sigma_0\delta). \qquad (12\text{-}8)$$

So, the Lorentzian source produces a damped cosine interferogram, in contrast to the cosine interferogram for a monochromatic source. Now, suppose one has the correct interferogram given by Eq. (12-8), but in the computation of the spectrum, used the wrong origin (maybe due to simply not knowing the location of the true origin.) Then, one would compute a spectrum $B_c(\sigma)$ given by [using Eq. (12-8) for the interferogram of a Lorentzian profile]

$$B_c(\sigma) = 2(2\pi A) \int_0^{\infty} [\exp(-2\pi\varepsilon\delta_M)] \cos(2\pi\sigma_0\delta_M) \cos(2\pi\sigma\delta_T)\, d\delta_T, \quad (12\text{-}9)$$

where the extra factor 2 in front of the integral is introduced because we are now considering a one-sided instead of a two-sided interferogram. Using Eq. (12-2) for the true path difference, we have

$$B_c(\sigma) = [4\pi A \exp(2\pi\varepsilon\beta)] \int_0^\infty [\exp(-2\pi\varepsilon\delta_T)] \cos[2\pi\sigma_0(\delta_T - \beta)] \cos(2\pi\sigma\delta_T)\, d\delta_T. \tag{12-10}$$

On using trigonometric identities, we can break Eq. (12-10) into a sum of simple integrals, and upon integration, we have

$$B_c(\sigma) = B_S(\sigma)[\exp(2\pi\varepsilon\beta)]\{\cos(2\pi\sigma_0\beta) + [(\sigma_0 - \sigma)/\varepsilon] \sin(2\pi\sigma_0\beta)\}, \tag{12-11}$$

where the Lorentzian profile $B_S(\sigma)$ is given by Eq. (12-3). For $\beta = 0$, one retrieves the Lorentzian profile, as he should. However, one can see that the departures from the Lorentzian profile are a strong function of $\sigma_0\beta$.

In Fig. 12-1, the computed spectrum for various values of $\beta\sigma_0$ are plotted. One should consider that for any experiment, the central wave number of the Lorentzian resonance is less than the maximum wave number $\sigma_{max}$ in the spectrum. For most experiments in which the interferogram is recorded in steps, one will have scannings steps $\Delta\delta \simeq 1/2\sigma_{max}$ (Chapter 7). Then, with $\beta \leq \Delta\delta/2$ for the worst error one will have

$$\beta\sigma_0 \lesssim \tfrac{1}{4}. \tag{12-12}$$

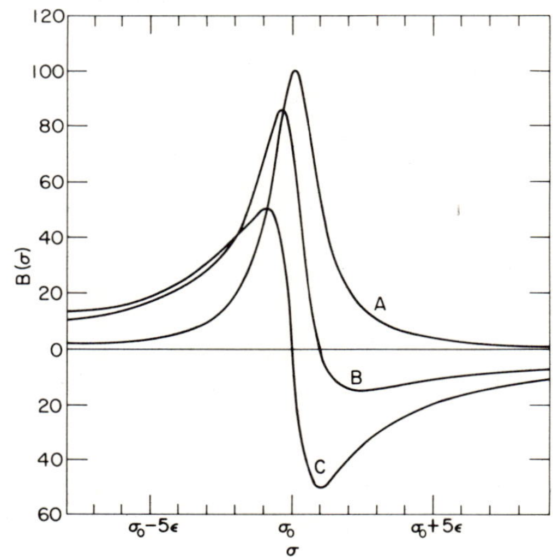


FIG. 12.1. Appearance of a Lorentzian line when an error has been made in the location of the origin; curve $A$, $\beta\sigma_0 = 0$; curve $B$, $\beta\sigma_0 = \frac{1}{8}$; and curve $C$, $\beta\sigma_0 = \frac{1}{4}$.

So, the values of $\frac{1}{8}$ and $\frac{1}{4}$ for $\beta\sigma_0$ used in Fig. 12-1 are certainly magnitudes of the error in $\beta$ that can be encountered in experiments. If $\sigma_0 \simeq \sigma_{max}$, the value $\beta\sigma_0 = \frac{1}{4}$ would correspond to actually having the origin midway between recorded steps in the interferogram. With these possible errors in locating the origin, one can grossly distort the computed spectrum. The resonance wave number is shifted, the full-width at one-half peak intensity is increased, and the spectrum can have negative values.

The phase error in missing the origin cannot be ignored and there are other phase errors due to alignment problems. So, in the next sections, we present ways to overcome various phase errors. We can take two-sided interferograms and eliminate the phase error from the computed spectrum; we can measure the phase error and correct the computed spectrum; and, if the phase error is due solely to missing the origin, we can fit the data at the interferogram peak with a parabolic function and accurately calculate where the origin should be.

## SAMPLING PHASE ERRORS AND TWO-SIDED INTERFEROGRAMS

In this section, we will show that if a two-sided interferogram is used in computing any spectrum, the origin does not have to be located. It is only necessary to have the origin near the center of the recorded interferogram for scans from $-L$ to $+L$.

Recall from Eq. (8-19) that the spectrum for a two-sided interferogram is given by

$$B(\sigma) = \int_{-\infty}^{\infty} \{[I_R(\delta) - I_R(\infty)] \exp(-i2\pi\sigma\delta)\}\, d\sigma, \tag{12-13}$$

where we have assumed that the reflectivities of the fixed and movable mirrors are equal. If one missed the origin by a distance $\beta$ as in Eq. (12-2), then the computed spectrum would be

$$B_c(\sigma) = \int_{-\infty}^{\infty} \{[I_R(\delta - \beta) - I_R(\infty)] \exp(-i2\pi\sigma\delta)\}\, d\delta. \tag{12-14}$$

Applying the shift rule and substituting the true spectrum of Eq. (12-13), we have the computed spectrum $B_c(\sigma)$ given by

$$B_c(\sigma) = B(\sigma) \exp(-i2\pi\sigma\beta). \tag{12-15}$$

If we separate the computed inverse Fourier transform in Eq. (12-14) into its real and imaginary parts $P_c(\sigma)$ and $-iQ_c(\sigma)$, respectively, then we have

$$|B_c(\sigma)| = [B_c(\sigma)B_c^*(\sigma)]^{1/2} = [P_c^2(\sigma) + Q_c^2(\sigma)]^{1/2}. \tag{12-16}$$

Also, taking the square root of the product of $B(\sigma)\exp(-i2\pi\sigma\beta)$ with its complex conjugate in Eq. (12-15), we have

$$|B_c(\sigma)| = [P^2(\sigma) + Q^2(\sigma)]^{1/2} = |B(\sigma)|. \tag{12-17}$$

On comparing Eqs. (12-16) and (12-17), we can find the actual spectrum $B(\sigma)$ from the square root of the sum of the squares of the real and imaginary parts of the computed spectrum $B_c(\sigma)$, and the results are *independent of* $\beta$!

As long as a two-sided interferogram is used to compute the spectrum, with Eq. (12-17), one has the correct spectrum without any knowledge of the location of the origin. The results can be generalized to any linear phase error. If there is a phase error which is linear in wave number, the argument of the interferogram can be written as ($\delta \pm$ constant). Then, the shift rule can be applied and Eqs. (12-15) and (12-17) result. So, any linear phase error in wave number can be exactly eliminated by computing the spectrum from two-sided interferograms.

In the section on general phase errors, we will show that as long as the phase error is a slowly varying function of wave number, we can eliminate the phase error from one-sided interferograms with computations based on a short, two-sided interferogram. The phase error does not have to be small or even linear as long as it is smoothly varying.

There are drawbacks to using Eq. (12-17). The most obvious disadvantage is that two-sided interferograms take twice as long to record, take twice the computer memory capacity, and require about twice as long to compute the spectrum as one-sided interferograms. Also, the process of obtaining the absolute value by the square root of the sum of two squares is nonlinear, and the noise in the final spectrum is no longer constant but increases as the signal decreases and is always positive. If $P_c(\sigma)$ and $Q_c(\sigma)$ have deviations of $\pm\Delta P_c(\sigma)$ and $\pm\Delta Q_c(\sigma)$, the final spectrum will be given by

$$B_c(\sigma) = \{[P_c(\sigma) \pm \Delta P_c(\sigma)]^2 + [Q_c(\sigma) \pm \Delta Q_c(\sigma)]^2\}^{1/2}, \tag{12-18}$$

where there is random correlation betwen the $\pm$ signs. For very low spectral signals compared to the deviations $\Delta P_c(\sigma)$ and $\Delta Q_c(\sigma)$, the noise in the final spectrum is given by

$$\Delta B_c(\sigma) \simeq \{[\Delta P_c(\sigma)]^2 + [\Delta Q_c(\sigma)]^2\}^{1/2}, \tag{12-19}$$

which is always positive. If the spectral signals $P_c(\sigma)$ and $Q_c(\sigma)$ are large compared to the noise, the noise in the computed spectrum is given approximately by

$$\Delta B_c(\sigma) \simeq [\pm P_c(\sigma)\, \Delta P_c(\sigma) \pm Q_c(\sigma)\, \Delta Q_c(\sigma)]/B_c(\sigma), \tag{12-20}$$

where there is random correlation between the pairs of $\pm$ signs. If the signal-to-noise ratio in the interferogram is high, the signal-to-noise ratio in $P_c(\sigma)$ and $Q_c(\sigma)$ will be about the same, and from the expression above, one can see that the signal-to-noise ratio will also be about the same in the spectrum computed from Eq. (12-17).

We now summarize the signal-to-noise situation when the spectrum is computed from a two-sided interferogram: At high signal-to-noise ratios, the spectrum has essentially an accurate noise spectrum. As the signal decreases, the true spectral level progressively drops farther below the center of the noise spectrum. When the signal has fallen to zero, the noise is entirely positive.

## GENERAL PHASE ERRORS

In this section, we shall discuss how one can use one-sided interferograms and correct the spectrum for the phase error in general situations. This section treats only one class of phase error (those for which only the interferogram is asymmetric), and the reader is referred to the paper by Sakai *et al.* [1] for the other cases. The asymmetric interferogram introduces the most important type of phase error, and the other types are rarely encountered.

We assume that the phase error is a slowly varying function of the wave number. This enables us to expand the phase error in a Taylor series and to retain terms only through the first derivative of the phase with respect to the wave number. The following derivation will show that this assumption is sufficient to allow correction of the computed spectrum from a one-sided interferogram. The spectral results are amenable to a simple corrective procedure which can be performed entirely on the computer.

The one-sided interferogram (which actually has a short portion recorded on the other side of the origin) has several advantages. The scan and computing times are halved; the computer memory bank requirements are reduced; the highest possible resolving power for a given mirror drive system is attained; and the noise is unaffected, in contrast to the way it was affected in the computation of the spectrum from a two-sided interferogram.

If the origin of the interferogram has been mislocated or the instrument is not properly aligned, phase errors $\varphi(\sigma)$ which are entirely a function of wave number may be introduced. Phase errors which are a function of the optical path $\delta$ and wave number $\sigma$ are discussed in Chapters 11 and 15. In general, we can describe the phase error $\varphi(\sigma)$ in the equation for the modulus of the spectrum $B_0(\sigma)$ with the following equation for the two-sided interferogram:

$$B_0(\sigma) \equiv \int_{-\infty}^{\infty} F(\delta) \exp\{-i[2\pi\sigma\delta + \varphi(\sigma)]\} \, d\delta. \tag{12-21}$$

In this equation, $\varphi(\sigma)$ expresses any errors in the measured values of $\delta$ due to wavenumber-dependent effects. For example, if one had missed the origin by an optical path of $\pm\beta$, then $\varphi(\sigma)$ would equal $\pm 2\pi\sigma\beta$ and the argument of the exponential function in Eq. (12-22) would become $2\pi\sigma(\delta \pm \beta)$. This factored form of the argument is an expression of an error in locating the origin as it should be.

Since $\varphi(\sigma)$ is not a function of $\delta$, it may be factored from Eq. (12-21), and we have

$$B(\sigma) \equiv B_0(\sigma) \exp[i\varphi(\sigma)] \equiv \int_{-\infty}^{\infty} F(\delta) \exp(-i2\pi\delta) \, d\delta. \tag{12-22}$$

In Eq. (12-22), $B(\sigma)$ is the spectrum one would compute from the interferogram $F(\delta)$ if a phase error $\varphi(\sigma)$ were present. Our problem is to eliminate the phase error and compute $B_0(\sigma)$, and this is complicated by having to take finite scans of the interferogram and utilizing apodization. We may obtain the interferogram from the spectrum by taking the Fourier transform of $B(\sigma)$, and we have

$$F(\delta) = \int_{-\infty}^{\infty} B_0(\sigma) \exp\{i[2\pi\sigma\delta + \varphi(\sigma)]\} \, d\sigma. \tag{12-23}$$

If $F(\delta)$ is to be real, then $B_0(\sigma) = B_0(-\sigma)$ and $\varphi(\sigma) = -\varphi(-\sigma)$, i.e., they are complex symmetric. [$B_0(\sigma)$ is said to be Hermitian, as shown in the footnote on page 35 of Chapter 3.] Thus, for the one-sided interferogram, we would have

$$F(\delta) = 2 \int_0^{\infty} B_0(\sigma) \cos[2\pi\sigma\delta + \varphi(\sigma)] \, d\sigma. \tag{12-24}$$

Since $\varphi(\sigma) = -\varphi(-\sigma)$ and $F(\delta) = F(-\delta)$, we may also express the modulus of the spectrum for a one-sided interferogram as [using Eq. (12.21)]

$$B_0(\sigma) = 2 \int_0^{\infty} F(\delta) \cos[2\pi\sigma\delta + \varphi(\sigma)] \, d\delta, \tag{12-25}$$

which is a real function.

As discussed in Chapter 5, one is forced to make scans over finite ranges of $\delta < L$, and this introduces sidelobes in the computed spectrum which are corrected by apodizing functions. The instrumental spectrum $B_I(\sigma)$ is computed from the experimental interferogram, which implicitly contains phase errors. If the finite range of $\delta < L$ is expressed by the rectangular function and if the interferogram is apodized, the instrumental spectrum becomes

$$B_I(\sigma) = \int_{-\infty}^{\infty} [\text{rect}(\delta) A(\delta, L) F(\delta) \exp(-i2\pi\sigma\delta)]\, d\delta. \tag{12-26}$$

This is the best spectrum one can compute unless phase errors are corrected.

Let us define

$$C(\delta) \equiv \text{rect}(\delta) A(\delta, L) \tag{12-27}$$

and its inverse Fourier transform (called the "scanning function")

$$D(\sigma) \equiv \mathscr{F}^{-1}\{C(\delta)\}. \tag{12-28}$$

Using the convolution theorem on the inverse Fourier transform of the product $C(\delta)F(\delta)$ expressed in Eq. (12-26), we have

$$B_I(\sigma) = B(\sigma) \star D(\sigma) \tag{12-29}$$

where $B(\sigma)$ is the inverse Fourier transform of the interferogram as shown in Eq. (12-22).

With the phase error $\varphi(\sigma)$ defined in Eq. (12-22), we now want to simplify the convolution in Eq. (12-29) so that the phase error will be determinable by measurement and available for correction.

The phase error may be expressed as a Taylor's series about some arbitrary wave number $\sigma_0$ as

$$\varphi(\sigma) = \varphi(\sigma_0) + \left.\frac{d\varphi(\sigma)}{d\sigma}\right|_{\sigma=\sigma_0} (\sigma - \sigma_0) + \frac{1}{2!}\left.\frac{d^2\varphi(\sigma)}{d\sigma^2}\right|_{\sigma=\sigma_0} (\sigma - \sigma_0)^2 + \cdots. \tag{12-30}$$

For a phase error varying slowly with wave number, the high derivatives of $\varphi(\sigma)$ will be vanishingly small, and we approximate Eq. (12-30) as

$$\varphi(\sigma) \simeq \varphi(\sigma_0) + \left.\frac{d\varphi(\sigma)}{d\sigma}\right|_{\sigma=\sigma_0} (\sigma - \sigma_0). \tag{12-31}$$

The justification for assuming $\varphi(\sigma)$ is slowly varying is provided by experiment. We can measure $\varphi(\sigma)$ using a short, two-sided interferogram scanned from $-L_1$ to $+L_1$. As will be shown in Eq. (12-48), we can compute the phase error from

$$\varphi(\sigma) = \arctan\left\{\frac{\text{Im}\left[\int_{-\infty}^{\infty} \text{rect}(\delta) A(\delta, L) F(\delta) \exp(-i2\pi\sigma\delta)\, d\delta\right]}{\text{Re}\left[\int_{-\infty}^{\infty} \text{rect}(\delta) A(\delta, L) F(\delta) \exp(-i2\pi\sigma\delta)\, d\delta\right]}\right\}. \tag{12-32}$$

One can always obtain the phase of any complex quantity such as $B_I(\sigma)$ as expressed in Eq. (12–26). However, the point of the derivation leading to Eq. (12-48) lies in proving that the phase of $B_I(\sigma)$ and the phase error $\varphi(\sigma)$ are equal [under the assumptions that $C(\delta)$ is symmetric and that $\varphi(\sigma)$ is slowly varying]. In general, the two phases may not be equal. So, it will be shown that Eq. (12-32) is valid if $\varphi(\sigma)$ is a slowly varying function of $\sigma$ and $C(\delta) = C(-\delta)$. No restriction is placed on $F(\delta)$.

So we can assume $\varphi(\sigma)$ needs only to be approximated through the first derivative as in Eq. (12-31). Now, we proceed to develop a way of utilizing the knowledge of the phase for use with one-sided interferograms and justifying Eq. (12-32).

We can express the instrumental spectrum as a convolution with the phase approximation Eq. (12-31), and we have [using Eq. (12-29)] that

$$B_I(\sigma) \simeq \int_{-\infty}^{\infty} B_0(\sigma')\left(\exp\left\{i\left[\varphi(\sigma_0) + d\varphi(\sigma)/d\sigma\bigg|_{\sigma=\sigma_0}(\sigma-\sigma_0)\right]\right\}\right)D(\sigma-\sigma')\,d\sigma', \tag{12-33}$$

Since $\sigma_0$ was arbitrary, we can set it equal to the arbitrary wave number $\sigma$ in Eq. (12-33), and we can rearrange the terms to write

$$B_I(\sigma) = \exp[i\varphi(\sigma)] \int_{-\infty}^{\infty} B_0(\sigma')D(\sigma-\sigma')\exp\left\{-i(\sigma-\sigma')\,d\varphi(\sigma)/d\sigma\bigg|_{\sigma}\right\}d\sigma'. \tag{12-34}$$

The term $\exp[i\varphi(\sigma)]$ can be extracted from the integral because $\varphi(\sigma)$ is independent of $\sigma'$. Now, $D(\sigma - \sigma')$ only has appreciable values near $\sigma = \sigma'$, and so the integral is effectively determined by its value near $\sigma = \sigma'$. We can prove this by evaluating $D(\sigma - \sigma')$ for a typical case.

The value of $D(\sigma - \sigma')$ can be determined by assuming a form for the apodizing function $A(\delta, L)$ and evaluating the inverse Fourier transform of $C(\delta)$ using the definitions in Eqs. (12-27) and (12-28). Assuming triangular apodization as introduced in Chapter 5 and integrating Eq. (12-28), we have

$$D(\sigma - \sigma') = (1/L)\,\mathrm{sinc}^2[\pi(\sigma - \sigma')], \tag{12-35}$$

where we used the apodizing function $A(\delta, L)$ given by

$$A(\delta, L) = (L - |\delta|)/L, \qquad -L < \delta < L, \tag{12-36}$$

on the same range as the rectangular function.

From Eq. (12-35), we see that almost all of the contributions to the integral come from its values near $\sigma \simeq \sigma'$. Note that since $\varphi(\sigma)$ is slowly varying, $d\varphi(\sigma)/d\sigma$ is slowly varying. Therefore, we need the value of $d\varphi(\sigma)/d\sigma$ over such a small range of wave numbers that we can assume it is a *constant*.

That is, for a particular wave number $\sigma$ in the convolution expressed by Eq. (12-34), we have

$$\left.\frac{d\varphi(\sigma)}{d\sigma}\right|_{\sigma} \simeq \text{const} \tag{12-37}$$

for $\sigma \simeq \sigma'$. So, we now have an integrand of an integral from $-\infty$ to $\infty$ which is in proper form to be expressed as a convolution.

So, using the definition of a convolution, we have

$$B_{\mathrm{I}}(\sigma) = \{\exp[i\varphi(\sigma)]\}(B_0(\sigma) \star \{D(\sigma) \exp[-i\sigma\, d\varphi(\sigma)/d\sigma]\}). \tag{12-38}$$

We will now show that the term in curly brackets may be approximated with just $D(\sigma)$. Noting that $\sigma\, d\sigma(\sigma)/d\sigma$ is a function of $\sigma$ alone, we may write

$$D(\sigma) \exp[-i\sigma\, d\varphi(\sigma)/d\sigma] = \int_{-\infty}^{\infty} C(\delta) \exp\{-i2\pi\sigma[\delta + (1/2\pi)\, d\varphi(\sigma)/d\sigma]\}\, d\delta. \tag{12-39}$$

Changing variables in the integral to $[\delta + (1/2\pi)\, d\varphi(\sigma)/d\sigma]$, we have

$$D(\sigma) \exp[-i\sigma\, d\varphi(\sigma)/d\sigma] = \int_{-\infty}^{\infty} \{C[\delta - (1/2\pi)\, d\varphi(\sigma)/d\sigma] \exp(-i2\pi\sigma\delta)\}\, d\delta. \tag{12-40}$$

Since $C[\delta - (1/2\pi)\, d\varphi(\sigma)/d\sigma]$ is given by

$$C[\delta - (1/2\pi)\, d\varphi(\sigma)/d\sigma] = \text{rect}[\delta - (1/2\pi)\, d\varphi(\sigma)/d\sigma]\, A[\delta - (1/2\pi) d\varphi/d\sigma, L], \tag{12-41}$$

we note that the range of the rectangular and apodizing functions is shifted from being $-L$ to $L$ to being $[-L - (1/2\pi)\, d\varphi/d\sigma]$ to $[L - (1/2\pi)\, d\varphi/d\sigma]$. If

$$L \gg (1/2\pi)\, d\varphi(\sigma)/d\sigma, \tag{12-42}$$

the shift for the rectangular and apodizing functions is negligible. [It is worth noting, from Chapter 6, that the resolution $\delta\sigma$ is given approximately by $1/L$. Then, the condition $d\varphi/d\sigma \ll 2\pi/\delta\sigma$ is equivalent to Eq. (12-42).] So we may state, to a good approximation, that

$$C[\delta - (1/2\pi)\, d\varphi(\sigma)/d\sigma] \simeq C(\delta), \tag{12-43}$$

if the condition of Eq. (12-42) is maintained. Then we have from Eqs. (12-39) and (12-28) that

$$D(\sigma) \exp\{-i\sigma\, d\varphi(\sigma)/d\sigma\} \simeq D(\sigma), \tag{12-44}$$

and Eq. (12-38) becomes

$$B_{\mathrm{I}}(\sigma) \simeq \{\exp[i\varphi(\sigma)]\}\{B_0(\sigma) \star D(\sigma)\}. \tag{12-45}$$

If we now make the assumption that the apodizing and rectangular functions are symmetric, i.e., $C(\delta) = C(-\delta)$, we can write $D(\sigma)$ in Eq. (12-28) as a real function for the one-sided interferogram (noting that the sine transform becomes zero)

$$D(\sigma) = 2\int_0^\infty C(\delta)\cos(2\pi\sigma\delta)\,d\delta. \tag{12-46}$$

Recall that $B_0(\sigma)$ is also real and can be found for a one-sided interferogram by Eq. (12-25). Thus, the convolution $B_0(\sigma) \star D(\sigma)$ is a real function. Therefore, we have

$$B_{\mathrm{I}}(\sigma) \simeq (\text{real number})\exp[i\varphi(\sigma)]; \tag{12-47}$$

so we may calculate $\varphi(\sigma)$ from Eq. (12-47). If $B_{\mathrm{I}}(\sigma)$ were written in complex polar form, one would have an equation identical to the form of Eq. (12-47), where the modulus would be the real number and the phase would be $\varphi(r)$. Thus, the phase error $\varphi(\sigma)$ is, under the assumptions, also the phase of $B_{\mathrm{I}}(\sigma)$. Explicitly, we would compute the phase from the following equation:

$$\varphi(\sigma) = \arctan\{[\mathrm{Im}\,B_{\mathrm{I}}(\sigma)]/[\mathrm{Re}\,B_{\mathrm{I}}(\sigma)]\}. \tag{12-48}$$

Thus, we have the modulus of the true spectrum convolved with the product of the rectangular and apodizing function given by

$$B_{\mathrm{c}}(\sigma) \equiv B_0(\sigma) \star D(\sigma) = B_{\mathrm{I}}(\sigma)\exp[-i\varphi(\sigma)]. \tag{12-49}$$

The convolution $[B_0(\sigma) \star D(\sigma)]$ is the instrumental spectrum that one desires because it represents the phase-corrected, true spectrum computed from a interferogram taken over a finite range of $\delta$ which has been apodized. [So $B_{\mathrm{c}}(\sigma)$ is the computed, phase-corrected spectrum.]

The procedure for using Eq. (12-49) is as follows. One starts an interferogram at some $\delta = -L_1$ which is a short distance to the left of the origin and records the interferogram past the origin to a positive $\delta = L\,(L \gg L_1)$. The value of $L$ is chosen by the desired resolution $\delta\sigma$ of the spectrum, since from Eq. (6-9), one has $\delta\sigma \simeq 1/L$. We use the two-sided portion of the interferogram from $\delta = -L_1$ to $\delta = L_1$ in Eq. (12-32) to compute the phase $\varphi(\sigma)$. In Eq. (12-32), the rectangular and apodizing functions are both adjusted to the range $-L_1 < \delta < L_1$. Since we assumed that $\varphi(\sigma)$ is a slowly varying function of $\sigma$, we do not need to know its values with excellent resolution. Therefore, $L_1$ does not have to be any larger than dictated by the resolution required for $\varphi(\sigma)$, and $L$ may be much greater than $L_1$.

With $\varphi(\sigma)$ known, we may compute $B_c(\sigma)$ using Eq. (12-49) with the *one-sided* interferogram used in

$$B_I(\sigma) = 2 \int_0^\infty \text{rect}(\delta) A(\delta, L) F(\delta) \exp(-i2\pi\sigma\delta)\, d\sigma. \qquad (12\text{-}50)$$

That is, the recording of the interferogram from $\delta = 0$ to $L$ is all that is needed once the slowly varying phase error $\varphi(\sigma)$ is known. We do not need to know $F(\delta)$ on the range of $\delta$ from $-L$ to 0 because the knowledge of $F(-\delta)$ is supplied by the factor $\exp[-i\varphi(\sigma)]$.

If all the assumptions of $\varphi(\sigma)$ slowly varying and $C(\delta)$ symmetric are met, the product $B_I(\sigma)\exp[-i\varphi(\sigma)]$ should be real since $B_0(\sigma) \star D(\sigma)$ is real. If $B_c(\sigma)$ is not a real function, the phase error $\varphi(\sigma)$ was not estimated accurately enough.

There are several ways to compute the terms $\varphi(\sigma)$ and $B_I(\sigma)$ for calculating $B_c(\sigma)$. One way to calculate the terms is by using the straightforward calculation of the inverse Fourier transform and picking the *same* $\sigma$ for the calculation of both $\varphi(\sigma)$ and $B_I(\varphi)$. This method does not involve curve-fitting the phase-error data to mate values of $\varphi(\sigma)$ and $B_I(\sigma)$ at the chosen wave number. Computation of straightforward inverse Fourier transforms is described in Chapter 16. If the single-sided portion of the interferogram has many points, it may be advantageous to use the Cooley–Tukey algorithm to compute the inverse Fourier transform. Since the Cooley–Tukey algorithm computes the spectrum at only fixed wavenumber increments, one has the problem of mating the values of $\varphi(\sigma)$ with those of $B_I(\sigma)$ *at the same wave number*. The experimentalist has several options:

A. He may compute $\varphi(\sigma)$ either by using the straightforward inverse Fourier transforms or using the Cooley–Tukey algorithm and fitting the $\varphi(\sigma)$ data over an acceptable range of $\sigma$ values with an equation. Since $\varphi(\sigma)$ is a slowly varying function, a second-order polynomial equation of $\sigma$ will suffice. As the Cooley–Tukey Fourier-transform-algorithm values of $B_I(\sigma)$ are determined at each wave number, $\varphi(\sigma)$ is computed from the polynomical equation. This can be done entirely on the computer.

B. He may compute $B_I(\sigma)$ using the Cooley–Tukey algorithm and then determinine those values of $\sigma$ for which $\varphi(\sigma)$ is desired. If one uses the Cooley-Tukey algorithm, we show in Chapter 17 that the spectral data are at wave numbers given by

$$1/N(\Delta\delta), \quad 1/(N-1)(\Delta\delta), \quad 1/(N-z)(\Delta\delta), \ldots, \quad 1/\Delta\delta,$$

where $N$ is the number of data points in the interferogram between $\delta = 0$ and $\delta = L$ and $\Delta\delta$ is the *optical*-path step size used in the sampling of the interferogram. Then, $\varphi(\sigma)$ may be computed at just those wave numbers listed above using a straightforward calculation of the inverse Fourier transforms for Eq. (12-32).

C. Another way to mate the values of $\varphi(\sigma)$ and $B_I(\sigma)$ is to use interpolation. Both $B_I(\sigma)$ and $\varphi(\sigma)$ may be computed using the Cooley–Tukey algorithm. Since the number of data points in the one-sided interferogram far exceed the number of data points in the two-sided interferogram used to determine $\varphi(\sigma)$, the values of $B_I(\sigma)$ are much more closely spaced than those for $\varphi(\sigma)$. Interpolation can be used to compute the intermediate points in the $\varphi(\sigma)$ spectrum. One can linearly interpolate, or, more accurately, use the interpolation procedure outlined by Connes [4].

None of the procedures requires one to locate the origin accurately. If the origin for computation is taken as the point in the interferogram for which one has maximum signal, the origin will be missed by less than one-half a step. The phase error can be computed from the short, two-sided part of the interferogram using Eq. (12-32) and the uncorrected spectrum can be computed from Eq. (12-50). The phase-corrected spectrum can be calculated from Eq. (12-49).

Though missing the origin is a large contribution to the phase error, there are other causes of phase errors. As long as those phase errors are slowly varying, the procedures outlined in this section are usable.

It was shown by Sakai *et al.* [1] in the last section of their article that this general technique of phase determination is used for the convolution technique by Forman *et al.* [3] and for the multiplicative technique by Mertz [2].

## ORIGIN SHIFTS CORRECTED BY CURVE FITTING

If one records the interferogram and later plots the data versus $\delta$ in the region of the origin, he can locate the origin very accurately. With the origin precisely fixed, the measured value $\delta_m$ may be shifted to the true value $\delta_T$ by appropriately adding or subtracting the displacement of the maximal sampled data point from the origin. It has been found that a parabolic curve fitted to the region of the peak of the interferogram is satisfactory to locate the actual peak. All of the computations can be performed on the computer (with no plotting), and one needs only to record the sampled interferogram, specify the step size, and take care of the origin shift in the computer (see Chapter 17).

This procedure is entirely satisfactory for eliminating not only origin location problems, but *all linear phase errors*. All linear phase errors are equivalent to missing the origin and vice versa, as can be seen by manipulation of Eq. (12-10) with $\beta$ redefined as the linear phase error. So as long as the main phase error is linear, one can use the parabolic fit of the interferogram peak to eliminate phase errors. This procedure is much simpler than the phase-error-correction methods previously mentioned. Because of its simplicity, the method is suitable for real-time analysis.

In Fig. 12-2, the reflectance of $Mg_2Ge$ as computed from both single-sided and double-sided interferograms is shown. The one-sided interferogram was phase-corrected by fitting a parabola to the three highest points in the interferogram. In the middle of the spectral range, the two methods of computation yield results which are in excellent agreement. At the ends of the spectrum, the correct values can be found by overlap experiments using different beamsplitters and spectral filters.

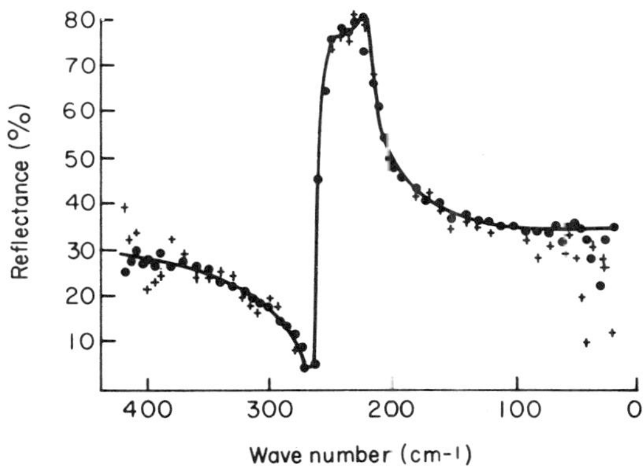


FIG. 12-2. Reflectivity of $Mg_2Ge$ computed from single-sided (crosses) and double-sided (dots) interferograms (after Perry *et al.* [5]).

## CONCLUSIONS

In this chapter, we have shown that mislocating the origin in an interferogram can grossly distort the computed spectrum. For a Lorentzian emission profile, this was graphically illustrated in Fig. 12-1. In the figure, $\beta$ was the optical path distance by which the origin was missed, and $\sigma_0$ was the central wave number of the Lorentzian profile.

Missing the origin is only one type of phase error, as other phase errors are introduced by poor optical alignment. These phase errors may be eliminated by several methods. One very popular method is to compute the spectrum from a two-sided interferogram, with results that are then independent of phase error. However, the two-sided interferogram has several drawbacks in signal-to-noise, computer, experimental scan-time considerations, and ultimate resolution.

In the section entitled General Phase Errors, it was shown that if the phase error is a slowly varying function of wave number, there are simple ways of correcting one-sided interferograms. All that was further required was that the rectangular function and the apodizing function be symmetric. The methods of correcting phase errors were such that the entire correction could be performed on a computer.

Finally, if phase errors linear in $\sigma$ were the main source of phase error, a curve-fitting procedure on the computer could be used to correct for the error. This procedure would correct for more than just missing the origin, but would not correct dispersive phase errors.

In addition to being needed to shorten overall scan time, etc., one-sided interferograms are extemely useful for real-time spectral computation. If one starts an interferogram scan slightly to the left (or right) of the origin, an on-line computer can quickly compute the phase correction and proceed to calculate the entire spectrum *as* the positive (or negative) path difference gets larger. This real-time analysis has to be performed with a straightforward inverse Fourier transform as outlined in Chapter 18. (The Cooley–Tukey algorithm, Chapters 16 and 17, is only usable for numbers of data points that are powers of two, and for long scans, the points at which the computation can be performed are too far apart.) As the resolution improves with increasing optical path displacement, one can watch the phase-corrected (or uncorrected, if desired) spectrum develop and make critical judgments about the experiment. The visual display can be realized with an oscilloscope connected to the output of an interface between the computer and the oscilloscope. If the realtime analysis of the one-sided interferogram does not include too many samples, mimicomputers can be used as described in Chapter 18.

## *References*

*1.* H. Sakai, G. A. Vanasse, and M. L. Forman, Spectral Recovery in Fourier Spectroscopy, *J. Opt. Soc. Amer.* **58**, 84 (1968).

*2.* L. Mertz, Auxilliary Computation for Fourier Spectroscopy, *Infrared Phys.* **7**, 17 (1967).

*3.* M. L. Forman, W. H. Steel, and G. A. Vanasse, Correction of Asymmetric Interferograms Obtained in Fourier Spectroscopy, *J. Opt. Soc. Amer.* **56**, 59 (1966).

*4.* J. Connes, Computing Problems in Fourier Spectroscopy, *Aspen Int. Conf. on Fourier Spectrosc., 1970*, p. 83. (G. A. Vanasse, A. T. Stair, Jr., and D. J. Baker, eds.). AFCRL-71-0019, 5 Jan. 1971, Spec. Rep. No. 114.

*5.* C. H. Perry, R. Geick, and E. F. Young, Solid State Studies by Means of Fourier Transform Spectroscopy, *Appl. Opt.* **5**, 1171 (1966).

CHAPTER THIRTEEN

# *Procedures for Choosing Experimental Parameters*

## INTRODUCTION

We have covered the major theoretical considerations of symmetric and asymmetric Michelson interferometers. (In Chapter 15, we will discuss lamellar grating interferometers from both theoretical and experimental viewpoints.) Now, it seems appropriate to illustrate how the theory is put into practice. The experimentalist needs to know how to choose the parameters for a given scan. So, in this chapter, we explain how the experimental conditions are chosen after the wavenumber region and resolution are decided (with an assumed signal-to-noise ratio).

In this chapter, we will use the symmetric Michelson interferometer as an example. The basic theory for the lamellar grating interferometer is quite similar to that for the Michelson interferometer, and so the basic experimental considerations are about the same. Therefore, no special treatment for the lamellar grating is given in this chapter and the reader is referred to Chapter 15 for special details. As far as source size, resolution, sampling step size, maximum scan displacement, spectral filters, etc., are concerned, the symmetric and asymmetric Michelson interferometers are essentially the same. Thus, the considerations in this chapter apply to both instruments. However, the sample containment and preparation procedures are very different, and the reader should see Chapter 8 for those details concerning the asymmetric interferometers. Also, the methods of computing the spectra are different.

Some mention will be made of data acquisition systems and minicomputers. Basic questions as to whether to use a one-sided or two-sided interferogram will be involved. More detailed discussions of these subjects are in Chapters 16–18 and Appendix B.

Sample interferograms and their spectra are shown in Chapter 14, where several experimental problems and solutions are presented.

## EXPERIMENTAL PARAMETERS

There are several prime criteria for an experiment. The chemist, physicist, ceramist, etc. will decide at what temperature the sample should be and whether transmittance or reflectance studies will be made, he will have to accept some phase for the sample; he may not have any choice concerning the sample size; etc. After all of these conditions are decided, the experimentalist then asks, "How are the pertinent parameters for an interferometer scan chosen?"

We will assume as an example that the experimental situation dictates that the transmittance be measured of a sample between 40 and 250 $cm^{-1}$ with resolution of at least 5 $cm^{-1}$.

The first decision to be made is whether or not to record a two-sided or a one-sided interferogram. The criteria on which this decision is based are described in Chapter 12 in general terms. The advantages of a two-sided interferogram are that:

1. It is unnecessary to locate the origin accurately, and so no elaborate set-up procedure is required.
2. All phase errors linear in wave numbers are eliminated in the calculation of the spectrum.
3. In the interferogram, if there is thermal or electronic drift which is essentially linear in the optical path difference $\delta$, the two-sided interferogram eliminates the effect from the computed spectrum.
4. Catastrophic noise in the interferogram due to electrical pick-up from phenomena not associated with the experiment can be tolerated if it is not too frequent. Since the interferogram is essentially reproduced on both sides of the origin, one side can be folded onto the other side at the point(s) of the catastrophe. The unjustified noise is edited out.
5. If the interferometer has lost its alignment, either slowly or quickly, the trouble may show up in the lack of symmetry in a two-sided interferogram and be corrected.

The disadvantages of the two-sided interferogram are that:

1. It takes about twice as long to record the two-sided interferogram as it does for one-sided interferograms.

2. The computer memory capacity for the two-sided interferogram has to be about twice as large for the same resolution as it does for the one-sided interferogram.

3. The computation time on the computer is about twice as long for double versus single-sided interferograms.

4. The maximum optical path difference for a given movable-mirror drive system is half as long as it is for the one-sided interferogram. With the considerations in Eq. (6-9), the best obtainable resolution with the two-sided interferogram is one-half that for the one-sided interferogram.

5. As discussed in Chapter 12, the spectral noise is worse for low signal levels computed from two-sided interferograms than from one-sided interferograms. For one-sided interferograms, if phase error corrections (as described in Chapter 12) are made, the noise in the computed spectrum is pretty much the same as encountered in the interferogram.

6. The spectrum cannot be computed from a large, two-sided interferogram in real-time analysis. Short, two-sided scans can be performed quickly after an experiment, but not during an experiment.

The listed disadvantages of the two-sided interferogram are also the main advantages of the one-sided interferogram. However, the one-sided interferogram has a major disadvantage which must not be treated lightly—phase errors.

Since the one-sided interferogram does not automatically correct for phase errors which are linear in the wave number (such as missing the origin with a sampling step), the experimentalist must use one of several procedures to correct for the phase error. The phase-error corrections are described in detail in Chapter 12. In the section, General Phase Errors in Chapter 12, it is explained how one may correct for *any* slowly varying (with wave number) phase error. One extends the one-sided interferogram past the origin so that he also (in the same scan) has a short, two-sided interferogram. The length of the two-sided portion of the interferogram is determined by the resolution required in the phase error $\varphi(\sigma)$. From Eq. (6-5), the resolution $\delta\sigma$ is given by

$$\delta\sigma \simeq 1/L_1, \tag{13-1}$$

where $L_1$ is the optical *path* difference from the origin to the maximum value of the short, two-sided interferogram. That is, the two-sided interferogram is scanned a total optical path distance of $2L_1$. With the phase error computed from the short, two-sided portion of the interferogram using Eq. (12-32), we insert $\varphi(\sigma)$ in Eq. (12-49) to obtain the final, computed, phase-corrected spectrum from the one-sided interferogram. The uncorrected spectrum computed from the one-sided interferogram is calculated from Eq. (12-50). In these equations, there are several functions which need defining: $F(\delta)$ is equal to $[I_R(\delta) - I_R(\infty)]$ and is the interferogram; rect($\delta$) is the rectangular function defined as being equal to unity on the $\delta$ interval from $-L_1$ to $+L_1$ and zero elsewhere; and $A(\delta, L_1)$ is the apodizing function, which is described in Chapter 5. We use $A(\delta, L)$ to correct for the fact that we can only scan over finite optical path differences (as expressed by the rectangular function). Several different apodizing functions are listed in Table 5-1, and the choice of which one to use is not very critical. Computational details are described in Chapters 16–18 and computer programs are given in Appendix B.

Also in Chapter 12, in the section, Origin Shifts Corrected by Curve Fitting, a very simple procedure is described for correcting *any* phase error which is linear in the wave number. The procedure only requires that one record a few points (10 or less) in the two-sided interferogram.

All of the phase-error-correction procedures can be done entirely on the computer. The experimentalist does not need to locate the origin with high precision. The several techniques are described in Chapter 12. If the experimentalist wishes to avoid the use of the computer to eliminate the most common phase error—missing the origin (and any other linear phase error)—he can take extreme care in locating the origin. He can do this by stepping the movable mirror back and forth over the origin and manually shifting the movable mirror until a step falls exactly on the peak. Then, the long, one-sided interferogram is initiated at the true, experimental peak, and linear phase errors are eliminated. This can be a bothersome procedure requiring a few minutes or worse, it may require special adaptation of equipment in some instruments. Also, the experimentalist must be aware of the fact that the origin can drift from day to day with thermal provocation.

So, the preceding considerations determine whether or not one will make a one- or a two-sided recording of the interferogram. Suppose that a two-sided interferogram has been agreed upon; we now use the spectral requirements in deciding the experimental parameters.

Let us choose the beamsplitter and spectral filters first. To isolate the 40–200-$cm^{-1}$ range, we refer to Chapters 9 and 10. From Figs. 9-4 and 9-5, one can see that the 25- and 50-gauge Mylar beamsplitters are to be chosen. The 50-gauge beamsplitter will be used for good signal-to-noise ratio at the lower wave numbers. In Figs. 9-4 and 9-5, the solid lines are actual normalized power spectra found by experiments using these beamsplitters with the filters listed in the captions. The dashed lines are the theoretical beamsplitter efficiencies with no consideration of source intensity, spectral filters, or detector sensitivity. For the 25-gauge beamsplitter, the major filter is the black polyethylene lens. When the 50-gauge beamsplitter is used a 0.6-mm-thick piece of wedged crystal quartz is added to the spectral filters. From the references for the transmittance of crystalline quartz listed in Chapter 10, one can see that it removes all measurable radiant flux between about 250 and 2000 $cm^{-1}$.

Since a maximum resolution of $\delta\sigma = 5\ \text{cm}^{-1}$ was asked for, recall from Eq. (6-5) that

$$L \simeq 1/\delta\sigma, \tag{13-2}$$

where $L$ is the maximum optical path distance from the origin. [In Eq. (13-1), on the other hand, $L_1$ was the maximum optical path for the short, two-sided interferogram used to determine phase errors alone.] So, for the one-sided interferogram, we must use an optical path distance from the origin of 0.2 cm. The maximum distance from the origin that the movable mirror must be driven is $0.2/2 = 0.1$ cm. So, one wishes to allow the mirror carriage to traverse from $-0.1$ cm through the origin to $+0.1$ cm in order to have resolution at least as good as $5\ \text{cm}^{-1}$. The maximum scan distance is influenced by the use of the Cooley–Tukey algorithm in that the total number of sample points must be two raised to some power. We will therefore return to the maximum scan distance.

Recall from Eq. (7-21) that the upper limit of the optical path sampling interval is given by

$$\Delta\delta \leq 1/2\sigma_{max}. \tag{13-3}$$

Figure 9-4 shows that the experimental upper wave number $\sigma_{max}$ is about $480\ \text{cm}^{-1}$ for the 25-gauge beamsplitter. Therefore, the maximum optical path sampling interval should be less than 10.4 $\mu$m. The next *smallest* sampling step (on the Beckman FS-720) is 10 $\mu$m, so we can use that step size and be very efficient in that we can use about as large a step as is allowed. For the 50-gauge beamsplitter, we used a crystal quartz spectral filter and $\sigma_{max} = 250\ \text{cm}^{-1}$. Thus, one should use a maximum sampling step of 20 $\mu$m with the crystal quartz filter.

One must be very careful never to violate the inequality in Eq. (13-3). However, for the sake of experimental efficiency, the upper limit should be closely approached. If one does not respect the inequality, the computed spectrum can have false information (aliased) as described in Chapter 7.

We must have a maximum optical path scan from the origin equal to at least 0.2 cm to achieve the $5\text{-cm}^{-1}$ resolution, and we want to see what effect this has on the number of data points we need in the interferogram. Let us assume that the Cooley–Tukey algorithm is going to be used to compute the spectrum. Since the Cooley–Tukey algorithm requires $2^N$ data points, we must provide either experimental values up to the full $2^N$ points or add zeros to the *interferogram* until we have that number of points for computation. If we record the interferogram and have the number of samples of the interferogram that are less than two to some power, we can add zeros and not affect the results in the spectrum. However, if we do not wish to be bothered with putting the zeros symmetrically about the origin in the data decks (or, if we would be just as content to have slightly better resolution), we need to calculate the number of data points required for a total number of points equal to $2^N$.

For the 25-gauge beamsplitter, we could take samples at a path interval of 10 $\mu$m, so on counting data points on both sides of the interferogram, we would have a total of $(0.2 \times 2)/(10 \times 10^{-4}) = 400$ interferogram samples. But $2^8 = 256$ and $2^9 = 512$; so, we must (if we are going to avoid adding $512 - 400 = 112$ zeros) extend the scan to include 256 samples on each side of the origin. So we maintain the 10-$\mu$m path steps and increase $L$ to $256 \times (10 \times 10^{-4}) = 0.256$ cm. Now, we will have a better resolution of $1/0.256 = 3.9$ cm$^{-1}$. More recording time is involved, but the measurement procedure has been made simple. For the 50-gauge beamsplitter and 20-$\mu$m sampling path step, we would only have 256 samples in the total two-sided interferogram instead of 512 as we have for the 25-gauge beamsplitter. For both beamsplitters, if the maximum path scan distance $L$ is fixed at 0.256 cm, the resolution will be 3.9 cm$^{-1}$. In order to provide a little extra margin in the experiment, one might change $L$ to 0.260 cm and have a few points at the extremes of the interferogram which are not used in the computation. The extra points make locating the origin less critical, so that the requisite number of data points are available on both sides of the interferogram. Also, the extra time after initiation of the scan when electronic transients are produced allows one to prune (on the computer) any of the start-up, short-term electrical pulses.

In deciding upon the size of the source, we use the material presented in Chapter 11. From Eq. (11-30), we have that the resolving power $R$ is limited because of source size to

$$R \equiv \sigma/\delta\sigma = 8F^2/h^2, \tag{13-4}$$

where $F$ is the focal length of the collimating mirror (or lens) and $h$ is the diameter of the source. With $\delta\sigma \simeq 3.9$ cm$^{-1}$ and the largest wave number 480 cm$^{-1}$ (for the 25-gauge beamsplitter), one wants $R$ from extended source considerations to be at least as large as $(480/3.9) = 123$. For the FS-720 Beckman Michelson interferometer with a focal length of the collimator $F = 11.5$ cm and the largest source diameter of 1 cm, one would have the extended source limit the resolving power to 930 as calculated by Eq. (13-4). Thus, as far as resolution is concerned, the source size is unimportant. However, the source size is important as an intensity control. Before leaving the question of the source size, we should consider another matter which can be pertinent for spectra of excellent resolution. As shown in Chapter 11, the size of the source also affects the wave number that one would have in the computed spectrum. If one ignored the size of the source and computed the spectrum from the interferogram, he would have wave numbers which are too high. Using Eq. (11-25), the values of the wave number for any spectra feature must be lowered on the average by $\sigma\Omega/4\pi$. That is, the wave numbers are too high by a factor which is linearly dependent on the wave number. The solid angle subtended by the source is given approximately in Eq. (11-29) by $\Omega \simeq \pi h^2/4F^2$. Thus, the computed wave numbers must be reduced by

$$\Delta\sigma \simeq (h^2/16F^2)\sigma. \tag{13-5}$$

For the example just given, $h = 1$ cm and $F = 11.5$ cm, so $\Delta\sigma \simeq 4.8 \times 10^{-4}\sigma$. The maximum reduction in wave numbers for the 25-gauge beamsplitter would be about $4.8 \times 10^{-4} \times 480 \simeq 0.23$ cm$^{-1}$. Since the resolution is 3.9 cm$^{-1}$, only in the most careful work would we bother making the corrections in wave number due to having an extended source.

For use of the interferometer in astronomy, the source subtends such a small solid angle that Eqs. (13-4) and (13-5) often place meaningless restrictions on the resolution and wavenumber corrections.

The next experimental parameter to determine is the signal-to-noise ratio that is permissible. Usually, the spectral features sought will dictate what spectral signal-to-noise ratio is minimal. Without a great deal of elaboration, the signal-to-noise ratio is about the same in the interferogram and its computed spectrum. Qualifying remarks concerning this are made in Chapters 7 and 12, and there is ample literature listed in the subject bibliography.

A simple criterion for establishing the signal-to-noise ratio $S/N$ in the interferogram is to use the value of $I_R(\infty)$ for the signal and to observe the noise at some particular optical path displacement away from the origin. Let us assume that a $S/N$ ratio of 20 is satisfactory for the experiment. Assume that we learned, by observation of the signal-to-noise ratio in the interferogram, that a 1-sec electronic time constant $\tau_e$ for the detector signal amplifier is sufficient for $S/N \gtrsim 20$. So, as we see in Chapter 14, we need a sampling period $\tau_s$ that is about 30% or more longer. Thus, we should have $\tau_s \gtrsim 1.3$ sec. For the FS-720 interferometer, the closest equal or larger sampling period available is 2.13 sec. With $\tau_s \gtrsim 1.3\tau_s$, one does not distort the spectral features at large wave numbers because of false depletion, as pointed out in Chapter 7 and illustrated in Chapter 14. For the 25-gauge beamsplitter with a total of about 512 samples in the interferogram, the scan time will be 512(2.13) = 1090 sec = 18 min. For the 50-gauge beamsplitter (assuming one can use the same electronic amplifier time constant and step period) with only 256 samples required for the same resolution, the scan time would be about 9 min.

Thus, if one uses a two-sided interferogram for procurement of spectra between 40 and 250 cm$^{-1}$ with at least 5 cm$^{-1}$ resolution, he may pragmatically settle for:

1. The use of the 25- and the 50-gauge Mylar beamsplitters to cover the entirewave number range.
2. A black polyethylene lens and crystal quartz as spectral filters.
3. A scan from $-0.256$ cm to $+0.256$ cm in optical path or $-0.128$ to $+0.128$ cm in movable-mirror displacement.
4. A resolution of 3.9 $cm^{-1}$ (see note later in the chapter concerning the effects of apodization).
5. A total number of interferogram points of 512 for the 25-gauge beamsplitter and 256 points for the 50-gauge beamsplitter.
6. An optical path sampling interval of 10 $\mu$m for the 25-gauge beamsplitter and 20-$\mu$m for the 50-gauge beamsplitter.
7. A signal-to-noise ratio of about 20, a 1-sec amplifier time constant, and a 2.13-sec sampling period.
8. A total scan time of 18 min for the 25-gauge beamsplitter and 9 min for the 50-gauge beamsplitter.

Obviously, one would like to reduce the scan time, and there are two basic ways to do this. In the example, it was assumed that one was using an Unicam Golay cell with a noise equivalent power NEP of about $10^{-10}$–$10^{-11}$ W. Much more sensitive low-temperature detectors are commercially available having an NEP as low as $10^{-14}$ W. So, it is possible to use more sensitive detectors which will allow shorter recording times and higher $S/N$ ratios. Another way to halve the interferometer scan time is to use one-sided interferograms as outlined earlier in this chapter and in Chapter 12. The one-sided interferogram has every advantage except with regard to phase errors. If one has programmed phase-error corrections as outlined in Chapter 12, much more efficient experimental operation can be performed. (The rapid-scan technique described in Appendix D is by far the largest time-saver.)

In passing, we should mention the requirements of the data acquisition system. The data are usually so voluminous and being received so rapidly that *solely* recording the information on strip-chart recorders is out of the question. It is wise to have the data on strip charts, but one also needs magnetic or punched paper tape recordings unless a computer is being time-shared. The data acquisition system only needs to have one channel for recording the interferograms; the other channels can be used for anything else. The one channel is sufficient because the interferogram samples are recorded in sequence, are integrated for the same time period (see Chapter 7 for a special technique), and are equally spaced in terms of optical path difference. However, if laser movable-mirror controls are employed, one could easily demand at least another channel to record the fringe count from the laser beam. With punched-paper-tape data retrieval, one must often convert the tapes to punched cards for the computer. Even with automatic tape-to-card converters, much time can be wasted with such an operation. It is far better to have a tape reader on the computer or use magnetic tape.

As an alternative procedure, one can directly couple the interferometer to a minicomputer situated in the laboratory. Real-time analysis can be performed for modestly long (less than 1024 samples) one-sided interferograms (with phase correction). Also, the minicomputer can be used as a storage and coupling unit or remote terminal for a larger central computer. For the real-time analysis, one can display the computed spectrum within seconds and minutes after a particular datum point in the interferogram has been stored. The continually developing display can be on an oscilloscope screen, and the final record can be obtained via a printout and/or plot by an *x–y* recorder or even a photograph.

Often, the data acquisition system is as expensive as the interferometer. Good data acquisition systems start in at about \$7000 or \$8000 and easily cost as much as \$15,000 for adequate data handling. There are not many economic savings to be realized from buying a single-channel unit versus a multichannel unit. The magnetic tape units are several thousand dollars more expensive than the punched-paper-tape systems. With most punched-paper-tape units, it is usually difficult to record any faster than once every 200 or 300 msec because of the slow mechanical components. Magnetic tape systems do not suffer this drawback. Also, most large computers can much more easily digest data from magnetic rather than paper tape. On the other hand, the minicomputer usually can receive information more easily from punched paper tape. The complete minicomputer systems with at least 8K memory units and 16-bit words will be priced in the neighborhood of \$20,000 to \$10,000 and necessary peripheral and interface equipment will at least double the price.

We have outlined how one can make a single scan and record the data. Since we assumed that a transmittance measurement was to be made, we will now discuss a few pertinent guidelines. Most of the interferometers are single-beam instruments—a sample-in and a sample-out measurement must be made separately. The transmittance can be computed by dividing the power spectrum for the sample-in scan by the power spectrum for the sample-out scan (on the computer). For the experimentalist, it is important to maintain the same conditions for both the sample-in and the sample-out scans. Unless special software programs are written, the scans must be taken with the same sampling step size, the same maximum scan distance $L$, the same spectral filters, and the same source size. It is not necessary to use the same electronic time constant $\tau_e$ or sampling period $\tau_s$ since the sample-out or background radiant signal is usually much stronger than the sample-in signal. So the same signal-to-noise ratio can be maintained in each scan with different time constants. If one wishes to control the source size to vary radiant intensity levels, this can be done easily with commercially available irises. One only needs to ratio the spectral signals for the different openings at which the iris is set and use the ratio in the computation of the transmittance. A rough estimate of the spectral ratio for a well-aligned interferometer can be obtained by taking the ratio of the large optical displacement signals $I_R(\infty)$ for the different iris diameters.

Now that one has the interferogram recorded, he must compute the spectrum. He will use the apodizing function as described in Chapter 5 and carry out the actual computation as detailed in Chapters 16–18 with the computer software in Appendix B. The apodizing function $A(\delta, L)$ eliminates false, major sidelobes in the computed spectrum produced by making scans over finite optical path differences. The interferogram is multiplied by the apodizing function point by point at each optical path difference $\delta$, and then the inverse Fourier analysis is calculated on the computer.

If one used a triangular apodizing function as pointed out in Table 5-1, the resolution of the computed spectrum will be degraded by 48%. Thus, for the example, instead of having a resolution of 3.9 $cm^{-1}$ one would have a value of $1.48 \times 3.9 = 6.8$ $cm^{-1}$ in the final, computed spectrum.

We shall not go into the computer techniques or software in this chapter. Suffice it to say that there are many alternatives—all of which are easy to use once one has his original computer programs. Several useful computer programs are given in Appendix B and real-time analysis is treated in Chapter 18.

## CONCLUSIONS

We have shown by example and reference to other chapters how one chooses the proper beamsplitter, spectral filters, sampling step size, maximum scan distance, electronic time constants, and source size for an experiment. The experimental conditions were dictated by a scan over a particular wavenumber range with at least a specified resolution (and desired signal-to-noise ratio in the spectrum). The question of whether one should record a two-sided or one-sided interferogram was discussed. As far as resolution and experimental scan time are concerned, the one-sided interferogram is favored.

Data acquisition systems, computers, and minicomputers were also mentioned. It was pointed out that in these items one is dealing with questions that possibly need more scrutiny in selection than is sometimes given.

No detailed description of commercial equipment was given, since this is discussed in Chapter 19. The Digilab, Inc. Fourier transform spectrophotometers are so different from what we have considered—they are automatic—that the experimentalist can be relatively uninformed about Fourier transform spectroscopy.

## *References*

*1.* R. J. Bell and H. V. Romero, A Study of an Array of Square Openings, *Appl. Opt.* **9**, 2341 (1970).

*2.* R. Ulrich, Interference Filters for the Far Infrared, *Appl. Opt.* **7**, 1987 (1967).

*3.* R. J. Bell, H. V. Romero, and J. M. Blea, Theory and Experiments for Multielement Grid Filters in a Dielectric, *Appl. Opt.* **9**, 2350 (1970).

*4.* R. A. Hanel, B. Schlachman, F. D. Clark, C. H. Prokesh, J. B. Taylor, W. M. Wilson, and L. Chaney, The Nimbus III Michelson Interferometer, *Aspen Int. Conf. on Fourier Spectrosc., 1970*, p. 231. (G. A. Vanasse A. T. Stair, Jr., and D. J. Baker, eds.). AFCRL-71-0019, 5 Jan. 1971, Spec. Rep. No. 114.

CHAPTER FOURTEEN

# *Sample Interferograms and Spectra*

## INTRODUCTION

In this chapter, we wish to collect interferograms and spectra that will illustrate some of the capabilities and problems of Fourier transform spectroscopy. An attempt has been made to exhibit data which is of interest to chemists, physicists, astronomers, etc. As can be seen from the Table of Contents, the sections are aggregated according to function and not field of study. It is hoped that the eight different categories (sections) will give the reader a feeling for some of the successful applications involving Fourier transform spectroscopy.

Some data from asymmetric interferometers are presented in this chapter, but the reader should consult Chapter 8 for more. For both the plane and spherical lamellar grating interferometers, the material in Chapter 15 is pertinent for further examples.

We have presented data taken with equipment from several different manufacturers. We have tried to make the point with this presentation that there are several manufacturers with excellent interferometers on the market. To help the reader in mating his needs, his budget, and his options in a given commercial field, we refer him to Chapter 19.

Most sample preparation techniques are the same as for grating spectroscopy except that the advantages in Fourier transform spectroscopy make the experiments easier. In FTS, the radiant beam is not in the shape of a long image which is difficult to fit onto samples; the throughput advantage makes it possible to use more highly absorbing samples; the lack of false radiation makes it possible to study transmittances as low as 0.1%; the high resolution yields resolved spectral features which the experimentalist could not separate before; etc.

The data also extend all the way from the ultraviolet to below 10 $cm^{-1}$, illustrating the wide spectral range over which Fourier transform spectroscopy has been used. The range from 17,000 to 10 $cm^{-1}$ is readily available in commercial equipment. However, above 17,000 $cm^{-1}$, the instruments are still developmental, although, by the time this has appeared, they probably will be commercially available.

Several special advantages are pointed out in this chapter. The small size and weight of interferometers were used to good advantage for satellite studies as described in the section Planetary Atmospheres and Astronomy. Throughout the chapter, the throughput and multiplex advantages are mentioned in context. Finally, for many experimentalists, just being free of a long, narrow, slit image is a big advantage. The excellent resolution obtainable with Fourier transform spectroscopy is often evident.

## REPRODUCIBILITY OF SCANS AND SIGNAL AVERAGING

One way to increase the signal-to-noise ratio in a spectrum is to repeat the scan several times. Of course, this involves averaging the spectra from all of the scans, but this is particularly easy in Fourier transform spectroscopy. With the interferogram recorded for computer analysis, one may average the spectra very easily on the computer. In Fig. 14-1, spectra of a polystyrene film which has been averaged from 256 scans is shown. In the sequence as explained in the figure caption, the signal-to-noise ratio drastically improves with the number of spectra averaged.

Let us consider an experimental illustration of the results of the signal averaging technique outlined in the Experimental Comments section of Chapter 7. Figure 14-2 shows spectra obtained under identical optical conditions: one spectrum was computed from an interferogram which used *RC* time constant control of the signal to noise ratio (solid line) and another was computed from an interferogram which used signal averaging (dashed line). For the *RC* time constant control, an electronic time constant $\tau_e = 3$ sec was used to control the signal-to-noise ratio and a longer stepping period $\tau_s = 4.27$ sec was mated with it so that $\tau_e/\tau_s = 0.70$. The spectrum computed from the interferogram which used the signal averaging technique had the same sampling period $\tau_s = 4.27$ sec. In the signal averaging process, the 4.27 sampling seconds were utilized in the following way: With the amplifier *RC* time constant set at 0.1 sec (about equal to the response time of the Golay

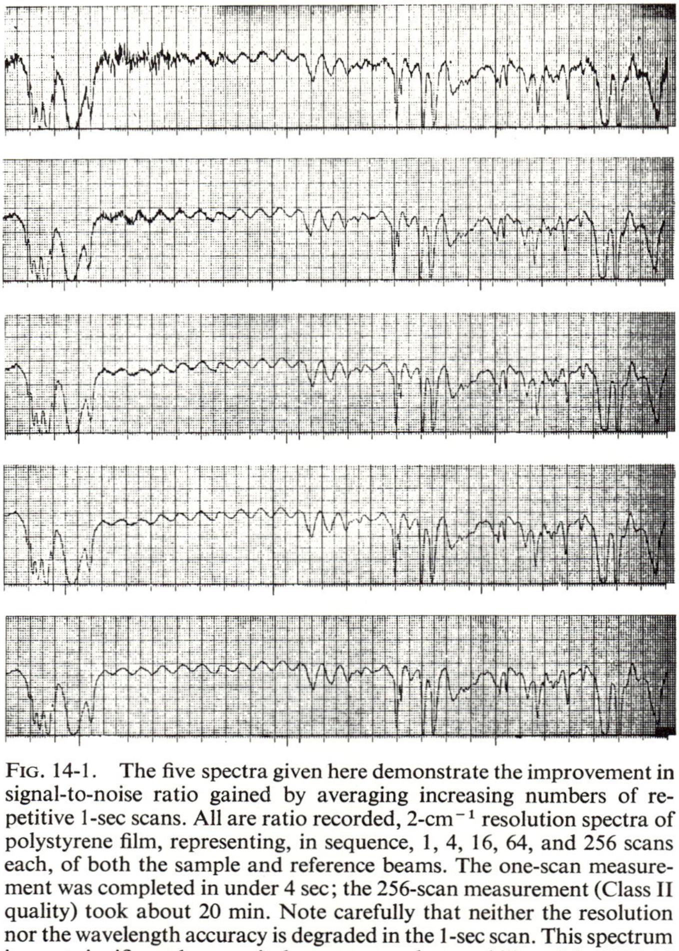

FIG. 14-1. The five spectra given here demonstrate the improvement in signal-to-noise ratio gained by averaging increasing numbers of repetitive 1-sec scans. All are ratio recorded, 2-$cm^{-1}$ resolution spectra of polystyrene film, representing, in sequence, 1, 4, 16, 64, and 256 scans each, of both the sample and reference beams. The one-scan measurement was completed in under 4 sec; the 256-scan measurement (Class II quality) took about 20 min. Note carefully that neither the resolution nor the wavelength accuracy is degraded in the 1-sec scan. This spectrum is most significant because it demonstrates the capability of the FTS-14 to measure qualitative spectra of short-lived species or gas chromatograph effluent peaks "on the fly." (After Digilab, Inc., Cambridge, Massachusetts.)

cell detector), a mirror step was taken and for 1 sec after the step, no information was averaged. During the 1-sec interval (equal to 10 times the *RC* time constant), the signal was coming to its equilibrium value. At the equilibrium level, the signal is very noisy because the *RC* time constant is quite short for the detector signals encountered. After the first second, the amplified and rectified detector signal, in the data acquisition system, is converted to a frequency proportional to the signal voltage and cycles are counted for a

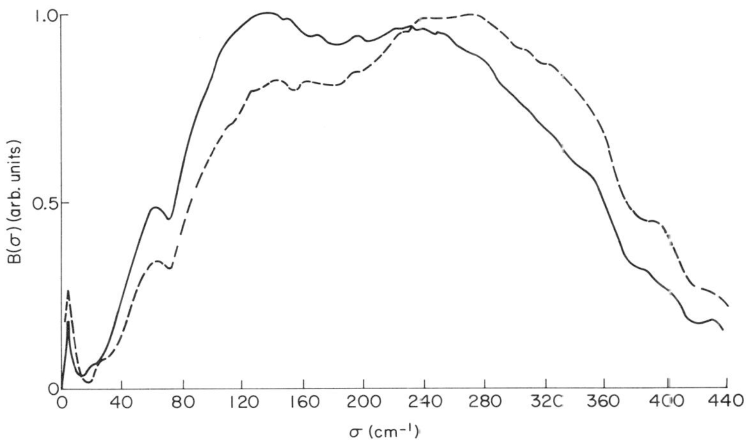


FIG. 14-2. Normalized spectra recorded under identical optical conditions and with the same sampling period. The spectrum from the interferograms recorded using the *RC* time control of the signal-to-noise ratio is represented by the solid line. The spectrum from the interferogram using the signal averaging technique is represented by the dashed line.

3-sec period. The total number of cycles is proportional to the true signal average during the 3-sec period. The true signal average is attained because the signal was not averaged until at least 10 *RC* time constants after the mirror was moved or stepped. Therefore, there was no signal lag problem. For 0.1 sec after the signal was averaged, the data acquisition system punched the paper tape to record the data, and the remaining 0.17 sec were discarded. So, the first second was used to allow equilibrium to be reached, the next three seconds were used to signal average, the following 0.1 sec were needed to record the data, and the last 0.17 sec of the 4.27-sec sampling period were not used.

## READING INTERFEROGRAMS

One of the former chief objections to Fourier transform spectroscopy was that the experimentalist could not quickly discern the spectra features by "reading" the interferogram. In this section, we want to illustrate what the practiced eye and a slide rule can do to read interferograms.

The first step in reading interferograms is to attempt mathematically to compute the Fourier transforms of some of the anticipated spectra—or vice versa. In Table 14-1, we present some types of spectra and their Fourier

TABLE 14-1[a]

| Spectrum (centered at $\sigma_0$) | Interferogram[b] |
|---|---|
| Monochromatic, $\delta'(\sigma_0)$ | $\cos(2\pi\sigma_0\delta)$ |
| Lorenztian, $A\varepsilon/[(\sigma-\sigma_0)^2+\varepsilon^2]$ | $\cos(2\pi\sigma_0\delta)\exp(-2\pi\varepsilon\lvert\delta\rvert)$ |
| Gaussian, $\{\exp[-(\sigma-\sigma_0)/2\sigma']\}/(2\pi)^{1/2}\sigma'$ | $\cos(2\pi\sigma_0\delta)\exp[-(1/2)(2\pi\sigma'\delta)^2]$ |
| Chi-squared,[c] $K(\sigma/\sigma_0)^{3/2}\exp-(\sigma/2\sigma_0)$ | $[\cos(10\pi\sigma_0\delta)][1+(4\pi\sigma_0\delta)^2]^{-5/4}$ |

$\delta'(\sigma_0)\equiv$ Dirac delta function; $\sigma'\equiv$ standard deviation; $A$ and $K$ are constants defined in the text.

[b] Multiplying constants have been dropped.

[c] Five degrees of freedom.

transforms, i.e., interferograms. In this table, we are trying to call attention to near-monochromatic systems. All of the spectra have a cosine modulation and as the source spectra broadens, the dampening increases. The monochromatic source was discussed in Chapter 5; the Lorenztian profile was examined in Chapter 12; and the Gaussian resonance can be treated similarly to the problem of the two close Gaussians of Chapter 4; the chi-squared spectrum can be easily shown.

The constants in the functions in Table 14-1 can be quickly obtained from the interferogram and rough estimates of the spectra are thereby made.

Channel spectra are quite common in far-infrared studies of plane samples with parallel faces. Sometimes, channel spectra are quite troublesome, especially in filters, source envelopes, detectors, and detector windows. However, channel spectra can be used to determine indices of refraction very accurately. Channel spectra also produce distinct characteristics in interferograms. In Fig. 14-3. one has an interferogram which shows strong, equally spaced signatures at $\delta=\pm 0.35$, $\pm 0.70$ cm. These signatures are caused by channel spectra which are apparent in the spectrum in Fig. 14-4. It was shown by C. M. Randall and R. D. Rawcliffe [3] that the signatures are at

$$\delta = 2snl, \tag{14-1}$$

where $s$ is the set of integers, $l$ is the sample thickness, and $n$ is the index of refraction of the sample. With signatures at $\pm 0.35$ and $\pm 0.70$ cm and $l\simeq 0.51$ mm, we would estimate an index of refraction of about 3.4 from the interferogram in Fig. 14-3. The value of $n$ as obtained from the equal separation of the channel peaks in Fig. 14-4 is 3.46. So, channel spectra are indicated by distinct, isolated, equally spaced signatures which can be used to roughly estimate the index of refraction.

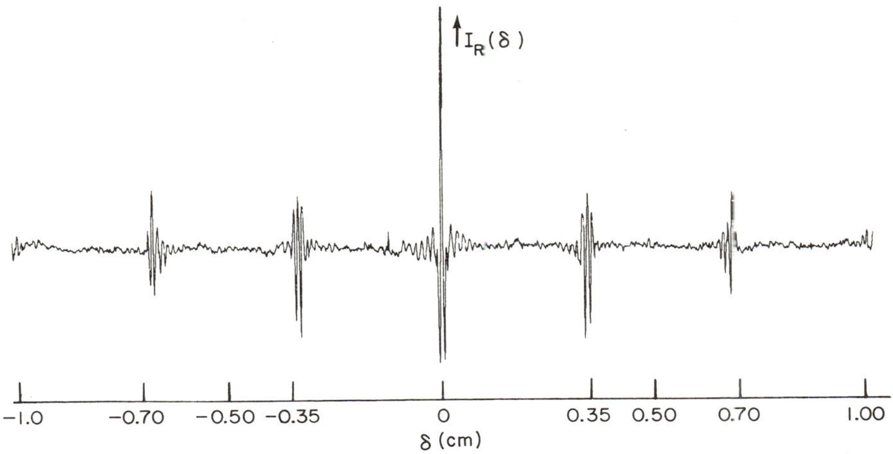


FIG. 14-3. Interferogram exhibiting the effect of channel spectra by signatures at $\pm 0.35$, $\pm 0.70$, and $\pm 1.05$ cm optical path displacements for eight silicon slabs. For each of the slabs, $n = 3.46$ and $l = 0.508$ mm. The signatures are at $\delta = 2snl$, where $s$ is the set of all integers.

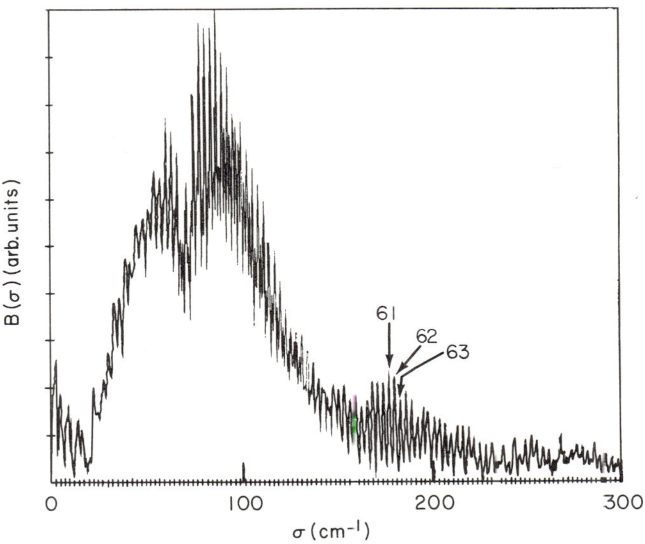


FIG. 14-4. Power spectrum using a 25-gauge Mylar beamsplitter exhibiting the channel spectra produced in eight silicon slabs. For the silicon slabs, $n = 3.46$ and $l = 0.508$ mm. The 61st, 62nd, and 63rd orders of interference are marked on the curve for reference.

The index of refraction can also be determined with greater accuracy from the spectrum. If the sample thickness $l$ is known and one has several orders of interference $m$ in the spectra, the index of refraction $n$ can be determined from (for normal incidence)

$$\sigma_m = m/2nl, \tag{14-2}$$

where $\sigma_m$ is the wave number at the $m$th constructive interference peak. Assume the index is not varying too rapidly between, say the $m$th and the $(m + 1)$th constructive interference peaks at $\sigma_1$ and $\sigma_2$ respectively. Upon subtracting the two wave numbers and using Eq. (14-2), we have the average index of refraction $\bar{n}$ given by

$$\bar{n} \simeq 1/2l(\sigma_2 - \sigma_1). \tag{14-3}$$

This value of $\bar{n}$ should agree fairly closely with the value read from the interferogram. For a particular channel spectral peak, we can substitute $\bar{n}$ into Eq. (14-3) and obtained the order number from

$$m \simeq \sigma_m/(\sigma_2 - \sigma_1) \tag{14-4}$$

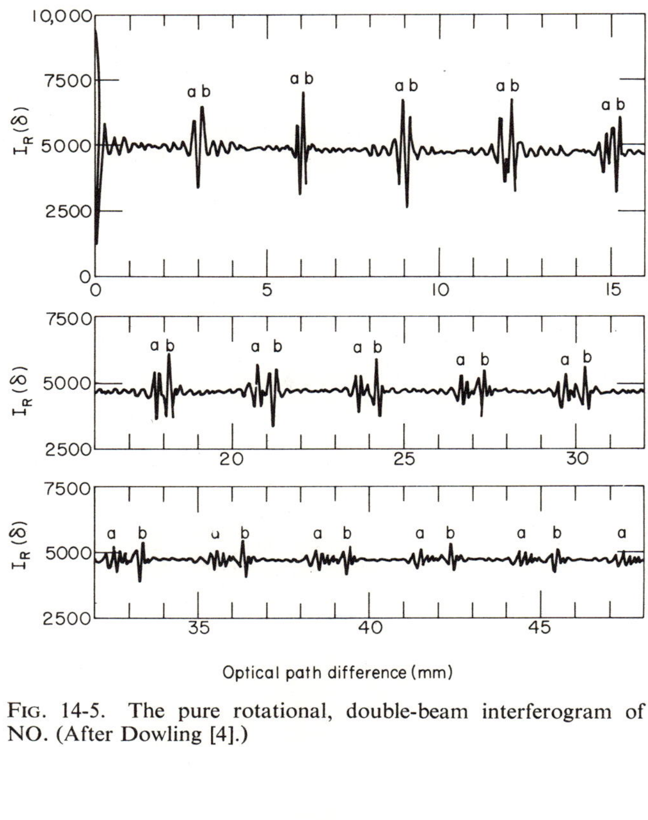


FIG. 14-5. The pure rotational, double-beam interferogram of NO. (After Dowling [4].)

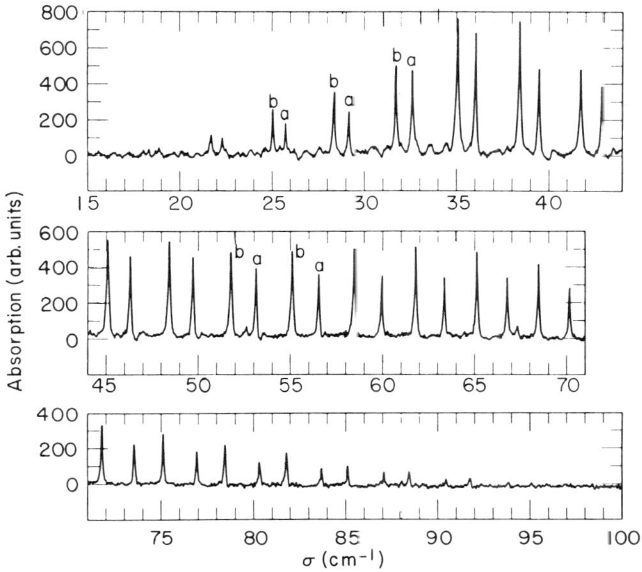


FIG. 14-6. The pure rotational spectrum of NO. The ground state splits into $^2\Pi_{1/2}$ and $^2\Pi_{3/2}$ components by spin–orbit interaction. (After Dowling [4].)

Of course, the order is an integer, so the value computed in Eq. (14-4) is rounded to the nearest integer. Once one of the orders of constructive interference is known, all of the others can be sequentially written down. Now, with $m$, $\sigma_m$, and $l$ accurately known, the index of refraction can be calculated quite accurately from Eq. (14-2). If the values of $n$ are desired at closer spacing than given by Eq. (14-2), the experimentalist can make use of the destructive interference wave numbers, too. This method can be used to obtain $n(\sigma)$ values from only a single scan with the sample in. It is not necessary to make a background scan.

Another class of repetitive structural features that can be noted in the interferogram are those due to the rotational spectra of gases. The spacing between absorption lines is uniform for rotational spectra and produces signatures very much like that caused by channel spectra. In Fig. 14-5, a long interferogram obtained from the gas NO is shown. The rotational spectra are shown in Fig. 14-6. The spacing between a set of rotational absorption lines $\Delta\sigma$ is given by

$$\Delta\sigma = 1/\Delta\delta, \tag{14-5}$$

where $\Delta\delta$ is the spacing in the interferogram between the signatures. In the

NO case, we actually have two sets of rotational lines due to the ground state splitting into two components. For a given set of signatures, say set *a*, one sees from Fig. 14-5 that the signatures are about 3.0 mm apart. Therefore, the set of rotational lines *a* should be about 3.3 $cm^{-1}$ apart. A check of Fig. 14-6 shows that this is the case. Likewise, the set of rotational components *b* are also found to be about 3.3 $cm^{-1}$ apart.

Any spectral feature which is equally spaced is going to produce these signatures, so vibrational, rotational, and channel spectra should be easily interpretable. Also, the transmittance of a Fabry–Perot (channel spectra) or any other such filter should give the equally spaced characteristic signatures.

## TRANSMISSION STUDIES OF SOLIDS

Polarization studies are quite simple using wire-grid polarizers, which are good components up to about 4000 $cm^{-1}$. They may be purchased for about \$750 from the Perkin-Elmer Corporation, Norwalk, Connecticut and are about 95% efficient. The wire-grid polarizers are approximately 22 mm in diameter and work as well in uncollimated radiation beams as in collimated ones. Other polarizers in the infrared are stacked sheets of soft polyethylene, AgCl, selenium, etc. In Fig. 14-7, some transmission spectra using wire-grid polarizers are presented.

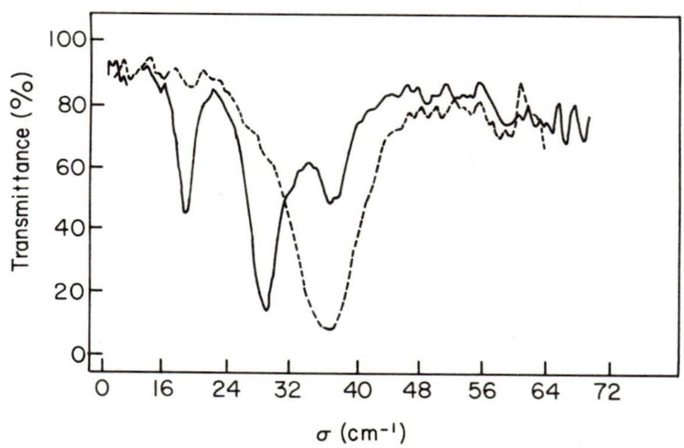


FIG. 14-7. Polarized transmission spectra of a 2-mm-thick single crystal of iododurene, recorded with the beam electric vector **E** parallel to either of two major crystallographic axes: (——)**E** ∥ *a*, (- - -)**E** ∥ *b*. Wire-grid polarizers of 10 μm period (1GP 221) were used for these measurements. Scan time: 20 min per interferogram; time constant: 1 sec. (After Beckman Instruments, Inc., Fullerton, California.)

Powered sample studies are quite common in chemical studies, and in Fig. 14-8, a powdered sample transmission spectrum is presented. The procedures that are used for prism or grating instrument work are used as well for Fourier transform spectroscopy—with greater ease. With no false radiation problem in Fourier transform spectroscopy, lower radiant signal levels can be tolerated.

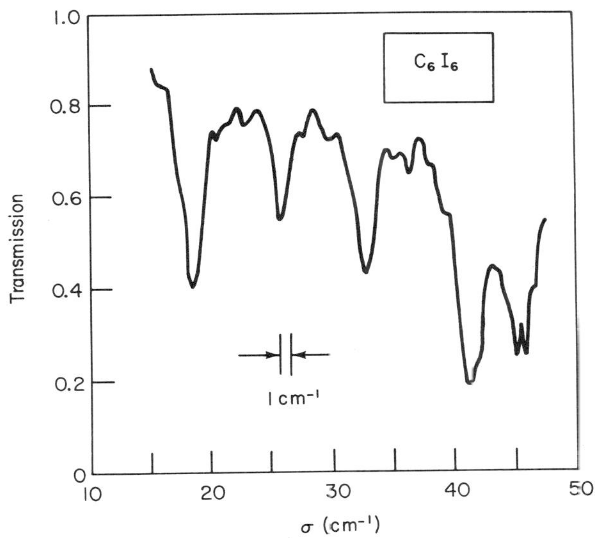


FIG. 14-8. The low-frequency transmission spectrum of a 2-mm powdered disk of hexaiodobenzene measured with the lamellar grating interferometer. The absorption lines at 18.5, 25.5, 33, 41, and 45 $cm^{-1}$ are attributed to internal molecular vibrational modes. The "sample" and "background" interferograms of $\pm 1$ cm optical path difference were each scanned in 20 min using an electronic time constant of 2 sec. (Beckman Instruments, Inc. equipment, after Milward [1].)

## TRANSMISSION STUDIES OF LIQUIDS

It is very simple to perform transmission studies of liquids because the interferometer has such a large throughput advantage. By using wedge liquid cells in which the windows and the sample compartment are all wedged, channel spectra can be avoided. In Fig. 14-9, the transmission spectra of several liquid solutions are presented. In curve *A*, the spectra of water is

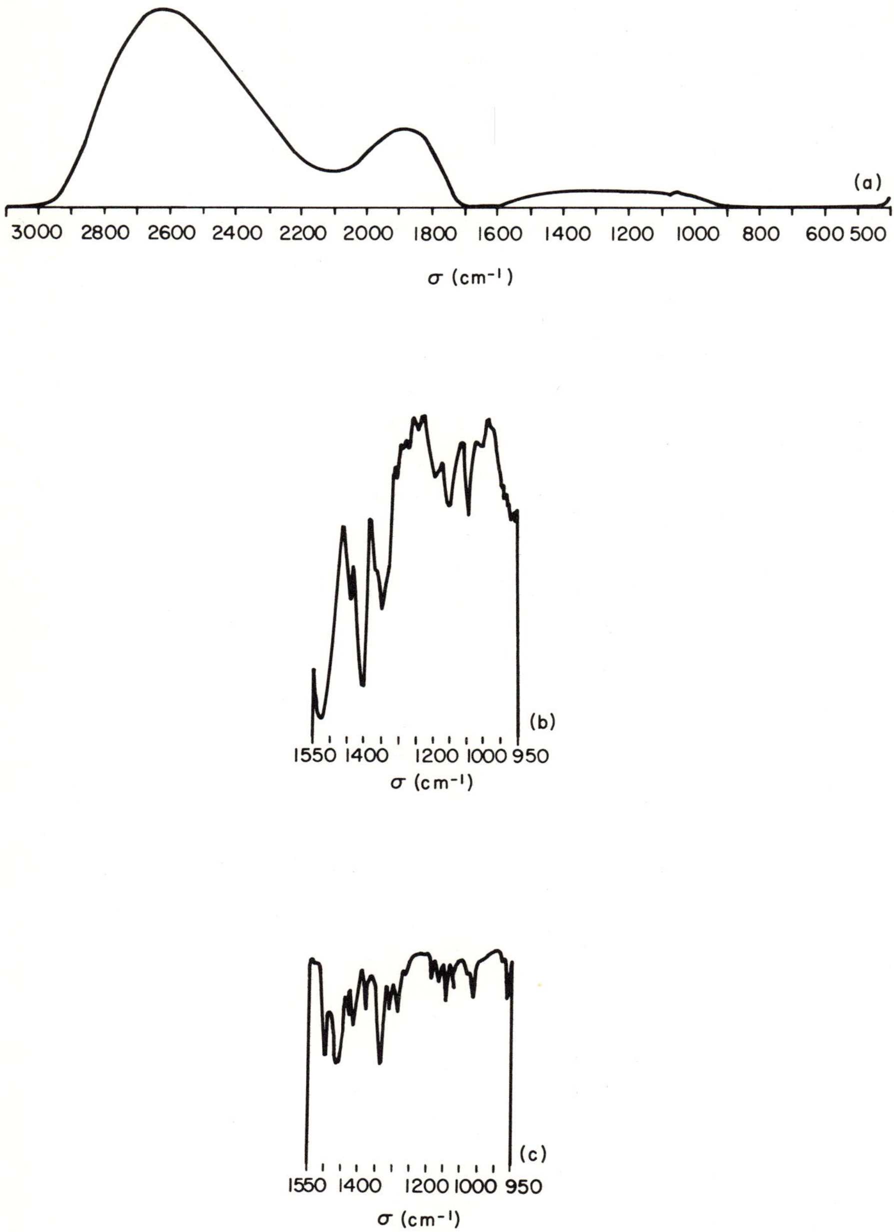


FIG. 14-9. Infrared spectra of liquid solutions. (*A*) Spectrum of water for a sample which is 0.05 mm thick. (*B*) Spectrum of histidine hydrochloride in water 0.05 mm thick. (*C*) Spectrum of histidine hydrochloride in a Nujol mull. (After Digilab, Inc., Cambridge, Massachusetts.)

shown. In curves *B* and *C*, one has the transmittance of histidine hydrochloride in water and Nujol, respectively. These spectra illustrate another type of spectrum which can be obtained by standard, simple techniques even at short wavelengths.

## TRANSMISSION STUDIES OF GASES

Gases can be studied in interferometers with almost any type of gas cell one desires. In Fig. 14-10, the vibrational–rotational spectrum of ammonia is well resolved. From this information, one can deduce the molecular spectral constants.

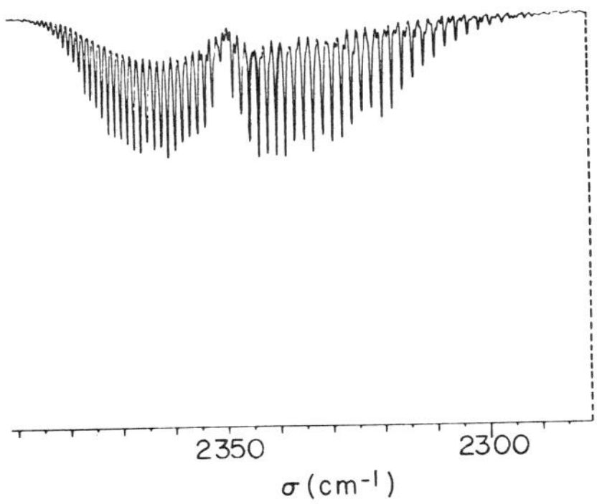


FIG. 14-10. Rotational spectrum of ammonia exhibiting the vibrational–rotational transitions. (After Digilab, Inc., Cambridge, Massachusetts.)

Figure 14-11 shows an example of isotope splitting of the pure rotational spectrum of HCl due to the $^{35}$Cl and $^{37}$Cl isotopes. Fourier transform spectroscopy can easily resolve the lines. Figure 14-12, from double-beam interferometry, shows the $J = 16 \rightarrow 17$ transition well defined in an absorption spectrum. The absorption resonance is very symmetric, as represented by the open circles. In order to locate the resonance wave number very

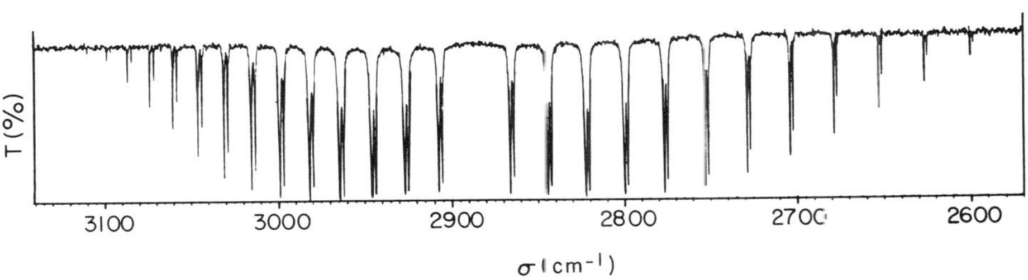


FIG. 14.11. Pure rotational spectrum of HCl exhibiting the isotope splitting due to $H^{35}Cl$ and $H^{37}Cl$. The spectrum was taken with the Digilab, Inc. FTS-14 system. (After Low [8].)

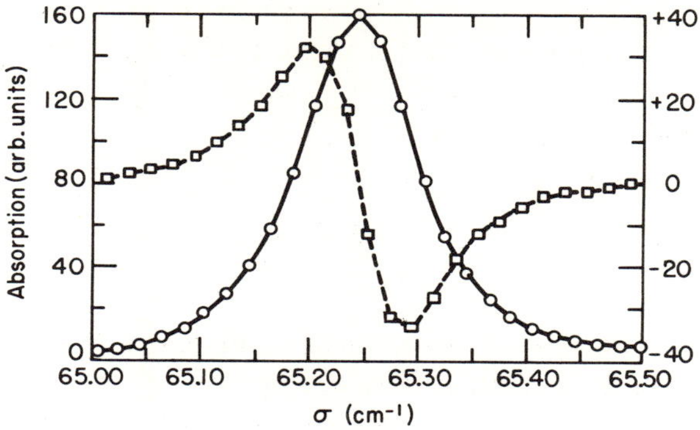


FIG. 14-12. The $J = 16 \rightarrow 17$ transition of CO. The circles are intensity points computed from the interferogram. The squares are the first derivative points computed from the intensity points. (After Dowling [4].) Aerospace Corporation, Los Angeles, California.)

accurately, the derivative curve has been plotted using squares. The derivative may be taken on the computer right after the spectrum has been calculated from the interferogram. So, with relative ease, resonance wave numbers can be located to better than 0.01-$cm^{-1}$ accuracy.

## REFLECTION STUDIES

Reflection studies may be performed in interferometers with almost any one of the many types of reflection attachments that have been designed. Most of the manufacturers have reflection attachments readily available, and, if not, they can be easily made. Again, having a circular radiation beam geometry is an advantage in design.

In Fig. 14-13, reflection measurements over a wide temperature range are shown. This type of measurement can be made with Dewars and specially designed reflection attachments. The radiation beam approaching the detector from most interferometers is usually such that one can easily incorporate reflection attachments (with vacuum seals which connect the Dewar and interferometer).

## EMISSION STUDIES

Emission spectra of many different types have been obtained with Fourier transform spectrometers. Because of their small size and weight, interferometers are suitable for field studies. A chemiluminescence experiment in the

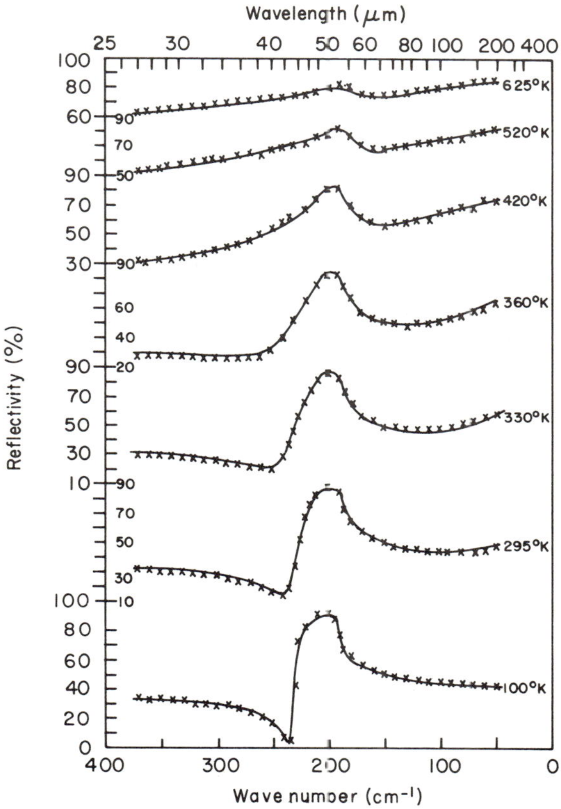


FIG. 14-13. Reflectivity of $Mg_2Sn$ at various temperatures. Curves: experimental data; crosses: calculated by means of classical dispersion formula. (After Perry *et al.* [5].)

visible spectral region is shown in Fig. 14-14. Because of the throughput advantage of interferometers, very weak luminescent sources may be studied. Air pollution studies have been made by observing the emission from stack effluents. In the data in Fig. 14-15, one can see that $CO_2$ and $SO_2$ have been identified in the plume.

It should be noted that Fourier transform spectrometers could also be used in Raman spectroscopy. In Raman spectroscopy, one always needs a spectrometer with very high resolving power to study the scattered flux, and a interferometer is ideal because of its large throughput advantage. With a throughput advantage over grating instruments of the order of 200 (Chapter 2), one can study weaker sources. With virtually no restrictions on the source

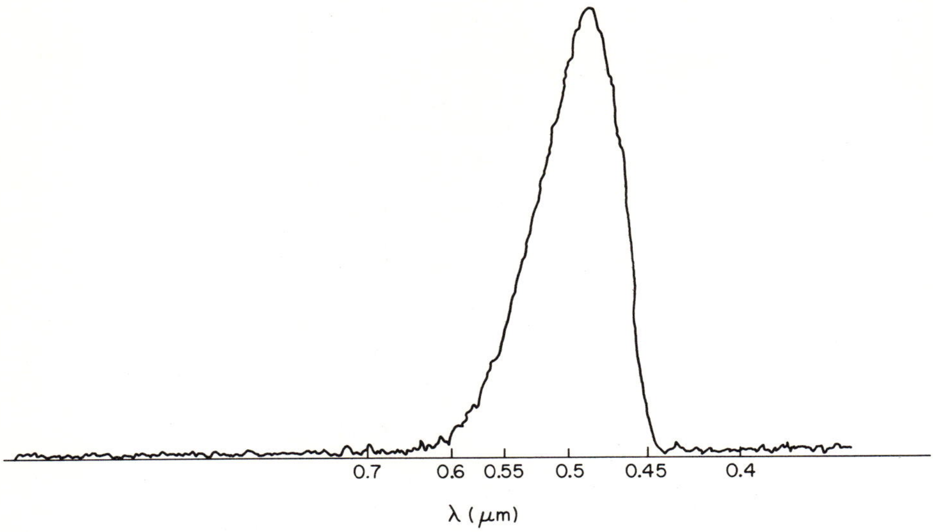


FIG. 14-14. Chemiluminescence emitted by the thermal decomposition of the transannular peroxide of 1.4-dimethoxy 9,10-diphenylanthracene in solution in toluene at 55°C. (After Digilab, Inc., Cambridge, Massachusetts.)

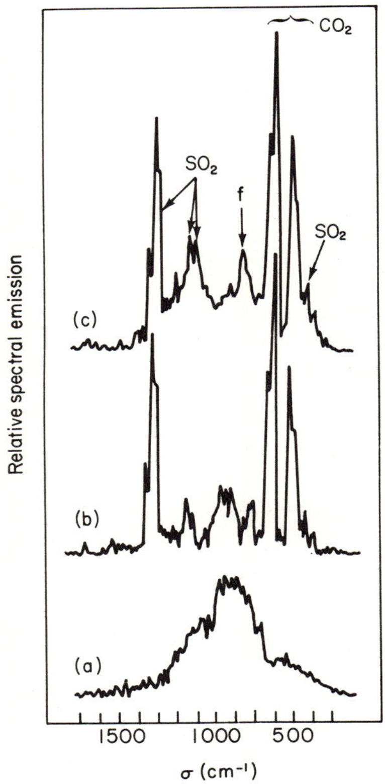


FIG. 14-15. Infrared emission spectra of stack effluents; (*A*) Sky near plume; (*B*) plume plus sky; and (*C*) plume only (*A* subtracted from *B*). (After Digilab Inc., Cambridge, Massachusetts.)

shape, the interferometer can be coupled easily to the radiation beam from the sample. With the resolution dependent essentially only on the movable-mirror displacement, one can attain the desired resolution without using double, grating spectrometers. If pulsed lasers are used, the Raman spectrum can be obtained by flashing the laser at least once for each movable-mirror position. Information about the entire Raman spectrum is obtained at each movable-mirror position and the desired resolution determines how many steps or pulses are necessary.

## PLANETARY ATMOSPHERES AND ASTRONOMY

Many of the problems involved in using Fourier transform spectroscopy in astronomy have been solved by P. Connes of France, L. Mertz of the Smithsonian Observatory of the U.S., and R. Hanel of the Goddard Space Flight Center of the U.S. and their colleagues.

We can begin the presentation with studies of the earth's atmosphere. In Fig. 14-16, a transmission spectrum of a one-mile optical path in the atmosphere is presented. These data exhibit water vapor and carbon dioxide lines

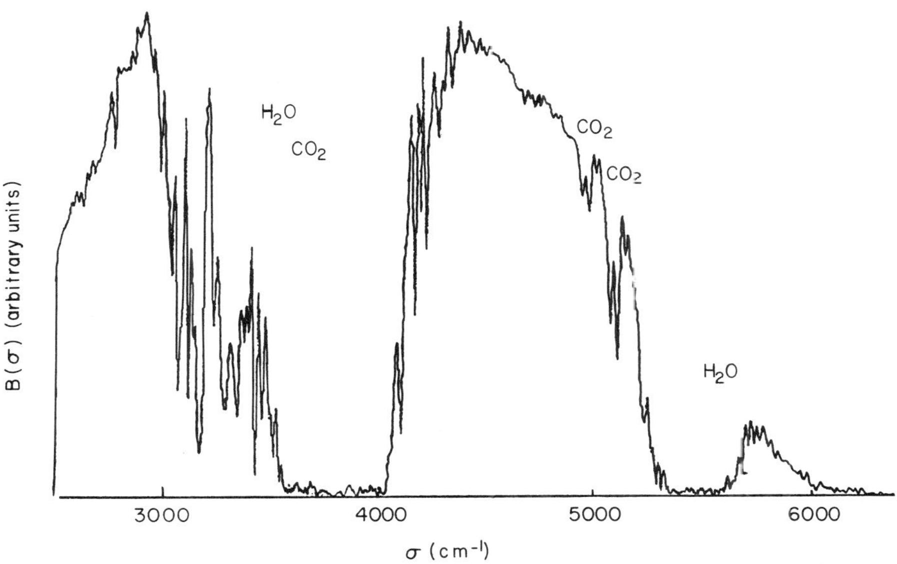


FIG. 14-16. Atmosphere absorption spectrum, one-mile path, uncorrected for instrument response. (After Digilab, Inc., Cambridge, Massachusetts.)

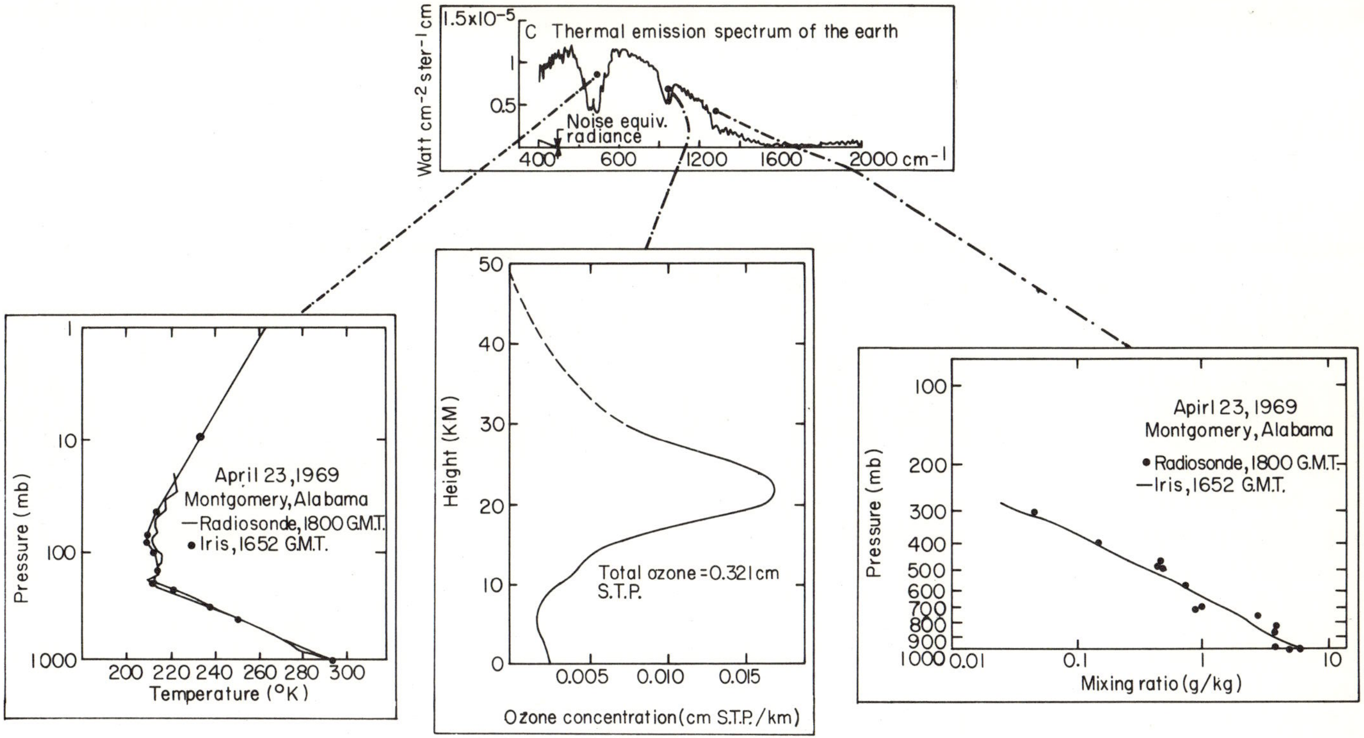


FIG. 14-17. Infrared interferometer atmospheric emission experiments on Nimbus III. (After Laboratory for Atmospheric and Biological Sciences, Goddard Space Flight Center, Greenbelt, Maryland.)

with good resolution. In Fig. 14-17, an emission observed with the Nimbus III satellite is shown. From the scan, pressure versus temperature profiles may be determined, as can ozone concentrations as a function of altitude. Mixing ratios are also available. These thermal emission spectra of the earth's atmosphere are obtained in just 13 sec with a 3° field of view.

Figure 14-18 shows an absorption spectrum of Venus. The spectrum of $^{12}C^{16}O^{18}O$ is very well resolved, and is compared with a solar spectrum. Such superior resolution was obtained with a 60-in. telescope by P. Connes. (The references are given in an article by Hunten [11].)

In Fig. 14-19, the spectrum of the star Betelgeuse as viewed through the

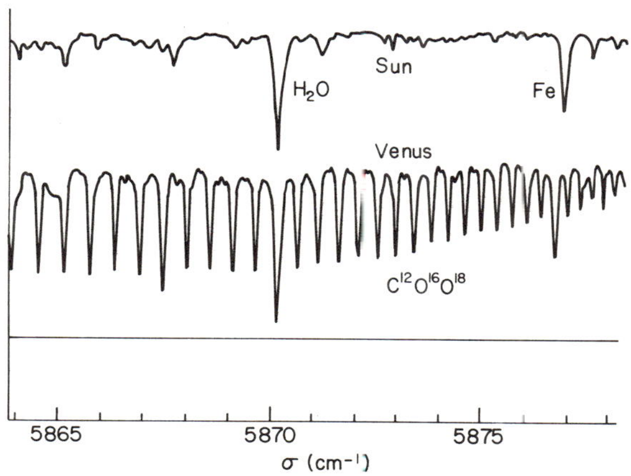


FIG. 14.18. Very small part of one of the Connes's spectra of Venus, along with a solar spectrum displaced upward. (After Hunten [11].)

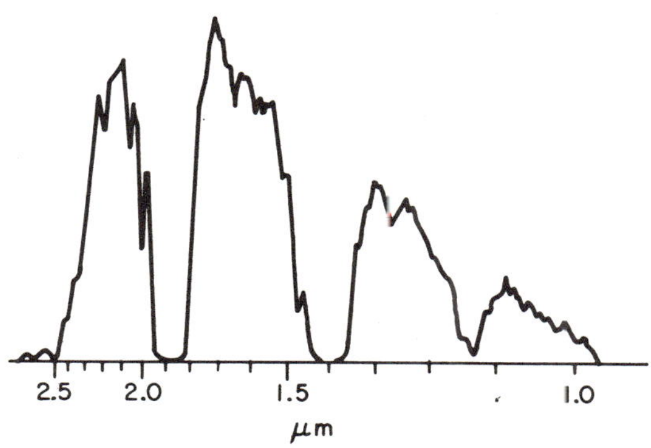


FIG. 14.19. Scans of Betelgeuse using rapid scanning FTS and a 25 cm telescope; 3 III 65; $Z = 40°$; 274 scans at 2 per second. (After Mertz [12].)

earth's atmosphere is presented. Its spectrum, along with several other interesting ones, are shown in the article by Mertz [12].

Finally, a great deal of planetary data taken by Fourier transform spectroscopy is available in an atlas by Connes *et al.* [13]. Never before has such high-quality, superior resolution data been obtained in astronomy in the infrared.

## CONCLUSIONS

In this chapter, we have shown by example, that Fourier transform spectroscopy is quite versatile and useful to the chemist, physicist, astronomer, environmental scientist, etc. We have shown experimental capabilities and problems, and we have indicated how one can go about reading interferograms. The transmission, reflection, and emission studies of a wide variety of samples were presented. In the emission section, we showed how Fourier transform spectroscopy could be of great benefit to the Raman spectroscopist. Finally, we showed some of the data from telescopes. In the hands of the pioneering FTS astronomers, the spectra of Saturn's rings have been resolved, the water vapor content of the atmosphere of Mars has been determined, the "Sinton lines" have been resolved, etc. Fourier transform spectroscopy has just begun to return excellent data from satellites and space probes.

Throughout the chapter, we pointed out some of the major advantages of interferometric studies. Primary are the throughput (Jacquinot) advantage and the multiplex (Fellgett) advantage.

## *References*

*1.* R. C. Milward, A Complete Fourier Spectrophotometer System for the Far-Infra-Red, "Molecular Spectroscopy," p. 81. *Institute of Petroleum*, Adlard and Son, Dorking, England, 1968.

*2.* P. G. Hoel, "Introduction to Mathematical Statistics," 2nd ed., p. 166. Wiley, New York, 1954.

*3.* R. D. Rawcliffe, Refractive Indices of Germanium Silicon, and Fused Quartz in the Far Infrared, *Appl. Opt.* **6**, 1889 (1967).

*4.* J. Dowling, Aerospace Corp., Los Angeles, California.

*5.* C. H. Perry, R. Geick, and E. F. Young, Solid State Studies by Means of Fourier Transform Spectroscopy, *Appl. Opt.* **5**, 1171 (1966).

*6.* R. C. Milward, A Small Lamellar Grating Interferometer for the Very Far-Infrared, *Infrared Phys.* **9**, 59 (1969).

*7.* M. J. D. Low, Fourier Transform Spectrometers, Part Three, *J. Chem. Ed.* **47**, A349 (1970).

*8.* M. J. D. Low. Fourier Transform Spectrometers, Part Two, *J. Chem. Ed.* **47**, A255 (1970).

*9.* R. B. Sanderson, Measurement of Rotational Line Strengths in HCl by Asymmetric Fourier Transform Techniques, *Appl. Opt.* **6**, 1527 (1967).

*10.* R. B. Sanderson and H. E. Scott, A High Resolution Far Infrared Interferometer, *Appl. Opt.* **10**, 1097 (1971).

*11.* D. M. Hunten, Fourier Spectroscopy of Planets. *Science* **162**, 313 (1968).

*12.* L. Mertz, Rapid Scanning Fourier Transform Spectroscopy, *J. Phys.* **28**, C2-87 (1967).

*13.* J. Connes, P. Connes, and J. P. Maillard, "Atlas des Spectres dans le Proche Infrarouge de Venus, Mars, Jupiter, et Saturne," Editions du Centre National de la Recherche Scientifique, Paris, 1969.

CHAPTER FIFTEEN

# *Lamellar Grating Interferometers*

## INTRODUCTION

The lamellar grating interferometer was invented by Professor John Strong. He designed the plane, lamellar grating in the 1950's, with Dr. G. A. Vanasse's help, while he was at the Johns Hopkins University. He developed the spherical, lamellar grating in the late 1960's, with Dr. N. P. Hansen's assistance, when he was at the University of Massachusetts. Both of these instruments will be described in some detail in this chapter. Dr. Strong's instruments are essentially single-beam spectrometers. However, a double-beam, plane, lamellar grating interferometer has been made by Dr. J. Dowling at the Aerospace Corporation. The reader is referred to Dr. Dowling's publications for information about the double-beam, plane, lamellar grating interferometer.

The first plane, lamellar grating interferometer was described in articles by Strong and Vanasse [1, 2] and in Strong's book [3], "Concepts of Classical Optics."

The lamellar grating interferometer is more efficient than the Michelson interferometer at wave numbers below about 100 $cm^{-1}$. This has been experimentally determined by Richards [4]. Dr. Richards built a plane, lamellar grating interferometer, a grating spectrometer, and a Michelson interferometer, operated them all in the far-infrared region, and proved for wave numbers below 100 $cm^{-1}$ that the lamellar grating interferometer was the most efficient. The reasons for the greater efficiency will be explained in this chapter.

The plane, lamellar grating will be described first, and some of its merits mentioned. Then, it will be shown that the basic equation for Fourier transform spectroscopy applies to the lamellar grating interferometer as well as the Michelson interferometer. The subject of diffraction by the lamellar grating as it could affect the operation of the interferometer is treated. It will be shown that there is a maximum grating constant for a given source size, radius of curvature of the collimating mirror, and maximum wave number. However, it is shown that, due to waveguide cavity-type effects, there is also a minimum grating constant below which large errors in the optical path difference can occur for long wavelengths. This error becomes more important as the ratio of grating constant to wavelength decreases. In the cavity effect, the group velocity of a wave polarized parallel to the grooves is lower than for waves with perpendicular polarization. Then, the true optical path difference is not the same for both waves. Consequently, the spectrum is affected.

Finally, the spherical, lamellar grating interferometer is described. It is a simpler instrument than the plane, lamellar grating interferometer. However, the complicated subject of noncollimated optical beams and interferometry makes it necessary to study the instrumental line-shape function in depth. With the spherical, lamellar grating interferometer, we shall also describe Michelson interferometers with noncollimated optics.

## PLANE, LAMELLAR GRATING INTERFEROMETERS AND EFFICIENCY OF THE BEAMSPLITTER

The Michelson interferometer splits wave amplitudes at the beam splitter, but the lamellar grating interferometer divides the wavefront. The reader should observe Fig. 15-1, which shows the plane, lamellar grating interferometer which was made by Dr. John Strong at Johns Hopkins University. The chopped radiation from the source is focused in the focal plane of the collimating mirror. From the collimating mirror, the wavefront proceeds to the lamellar grating. At the lamellar grating, the wavefront is divided such that one-half of the beam is reflected from the front facets or strip mirrors and one-half goes farther to be back facets before being reflected. If the back facets are a distance $d$ behind the front facets, an optical path difference of $2d$ is introduced between the two parts of the wave. The collimated beams return to the second half of the large spherical mirror (called the Ebert mirror) and are focussed on the detector. The lamellar grating operates in the zeroth order, as will be proved later in this chapter. The disassembled lamellar grating components are shown in Fig. 15-2. A good photograph of

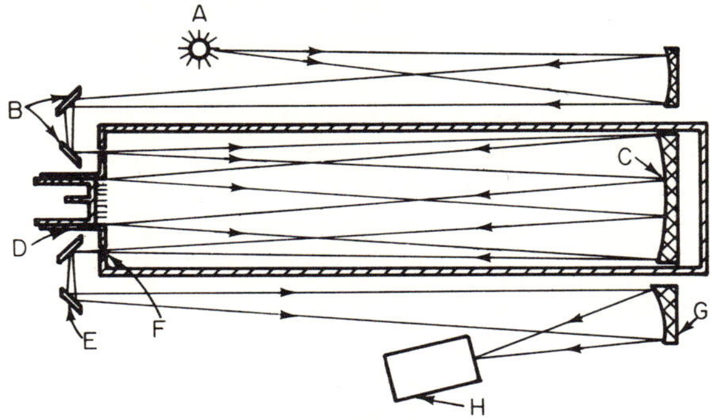


FIG. 15-1. The Johns Hopkins University plane, lamellar grating interferometer made by Professor John Strong's group. A: High-intensity mercury lamp. B: Entrance aperture, residual ray, crystals. C: Ebert mirror. D: Interferometric modulator. E: Mirrors or crystals. F: Exit aperture. G: Off-axis ellipse. H: Detector. (After Strong [3].)

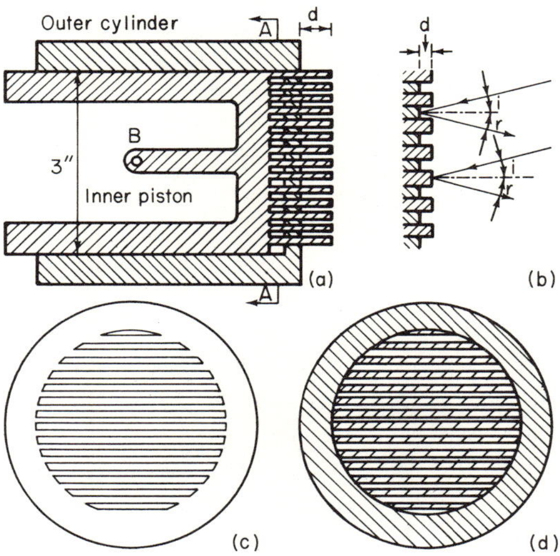


FIG. 15-2. The lamellar grating modulator made by the group of Professor John D. Strong at Johns Hopkins University. (a) The assembled plane, lamellar grating. (b) The grating showing the incident and reflected components off front and back strip mirrors. Front views of the disassembled plane lamellar grating parts: (c) slotted cylinder with piston removed; (d) piston in cylinder section A–A. (After Strong [3].)

a plane, lamellar grating may be found in Connes's article [5], "How Light is Analyzed."

One set of facets or strip mirrors is half fixed in position and the other set of facets, which are on the piston, are moved in and out so that the optical path difference can be varied. Usually, the lamellar grating interferometer is used for the production of only one-sided interferograms. That is, the piston is moved from the vicinity of the origin (including the origin position) to some large displacement *d*.

One can see that the entire wavefront is used in the lamellar grating interferometer. On the other hand, the Michelson interferometer, even if the beamsplitter were perfectly efficient, loses one-half of the total flux. Figure 15-3 shows a crude drawing of a Michelson interferometer. The wavy lines indicate the reflected and transmitted waves that return to the source and never reach the detector. Also as was shown in Chapter 9 on beam splitters, the Mylar beamsplitter is only about 60% as efficient as an ideal beamsplitter, which has reflectance and transmittance each equal to 50%. Since the dielectric beamsplitter is at an angle of 45° to the incident beam, there are polarization effects. The wave for which the component of the electric vector is

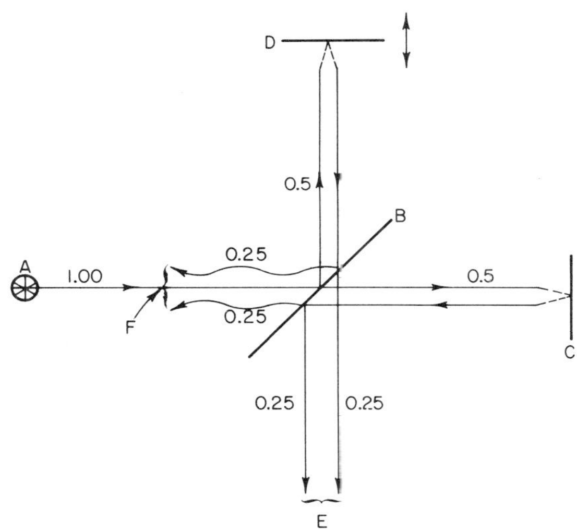


FIG. 15-3. Crude drawing of a Michelson interferometer showing that two components of the exit beams (F) return toward the source and never reach the detector. A: Source; B: beamsplitter; C: fixed mirror; D: movable mirror; E: to detector. If the incident flux magnitude is 1.00, only 0.5 of the incident flux reaches the detector. The displacements of the beams are only for illustrative purposes.

perpendicular to the plane of incidence is reflected more strongly than the component parallel to the plane of incidence. As a result of these limitations, the Michelson interferometer with a Mylar beamsplitter can have a maximum efficiency of $(0.50 \times 0.60) \times 100 \simeq 30\%$ (*For the latest information concerning the efficiency of interferometers, see the added note about polarizing interferometers at the end of Chapter 9.*)

It is recognized that the lamellar grating efficiency is less than 100%. The efficiency of the lamellar grating interferometer drops at high frequencies, when diffracted waves cause cancellation in the exit slit. As will be presently shown, the efficiency begins to drop at $\sigma_c = F/as$, where $F$ is the focal length of the collimator, $a$ is the grating constant, and $s$ is the aperture diameter. At very long wavelengths as $\sigma$ approaches $1/0.3a$, the cavity effect begins to decrease the modulation of the wave for which the electric vector is parallel to the sides of the cavity. The group velocity for the wave with the electric vector parallel to the grooves is decreased and there is an error in the effective optical path difference which one would use in the computation. The error in the optical path difference cannot be computed accurately because no one has solved the cavity problem that the lamellar grating interferometer presents for finite displacements $d$. These limitations of the lamellar grating interferometer's efficiency will be treated in more detail later in this chapter. Typical beamsplitter efficiency estimates are shown in Fig. 15-4. The lamellar grating efficiency is by far the largest. The wire-grid beamsplitters described in Chapter 9 are second best, but are usually not employed because they are rather delicate and more expensive than the Mylar beamsplitters. Wire-grid beamsplitters are available as electroformed grids from the Buckbee-Mears Company. It is also obvious that the Mylar beamsplitter exhibits strong channel spectral features, as shown in its oscillations in Fig. 15-4. In order to operate at peak efficiency, the thickness of the dielectric beamsplitter must be changed for different spectral regions. Very thick beamsplitters may be used at very long wavelengths. Thus, on the basis of the beamsplitter, the lamellar grating interferometer is much more efficient than the Michelson interferometer at low wave numbers. The data used for Fig. 15-4 are taken from an article by Richards [4]. Some consider the Michelson interferometer easier to build. The alignment and movement in the lamellar interferometer are quite critical and all the facets have to be flat and parallel. Except for very long wavelengths, usually longer than 100 $\mu$m or below 100 $cm^{-1}$, the tolerances are too high for most machine shops. The facets can be ground together and out of a wide variety of materials, the piston movement can be guided by commercially available high-precision screws and rods, etc., but even with these techniques it is still difficult to use the lamellar grating interferometer above 100 $cm^{-1}$.

The Michelson interferometer is easy to operate with both the positive

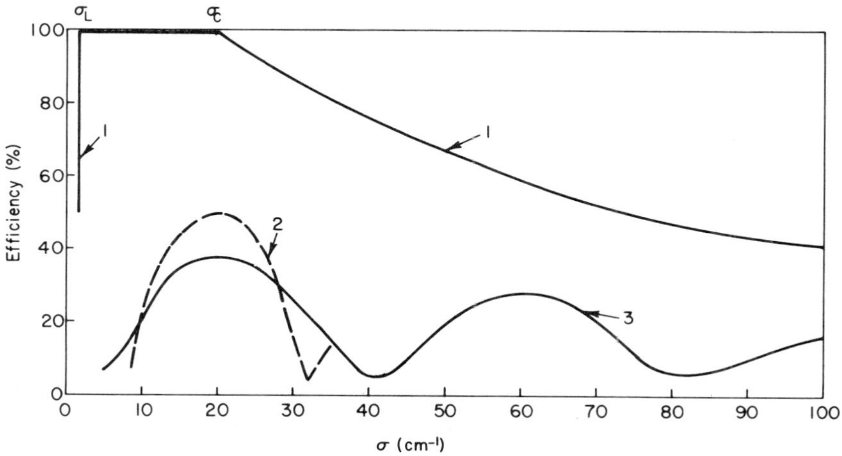


FIG. 15-4. Efficiency of introducing phase differences in (1) the lamellar grating interferometer and in Michelson interferometers using (2) a metal screen and (3) a Mylar-film beamsplitter. The efficiency of the beamsplitter in the Michelson interferometer is about one-half as large as typically reported, for the reasons cited in the text. The lower limit $\sigma_L$ is given in Eq. (15-33) and the upper limit $\sigma_c$ is given in Eq. (15-22). (After Richards [4].)

and negative arm displacements about the origin. The two-sided interferogram has several advantages, such as not having to locate the origin accurately, yielding spectra which are independent of minor linear electronic drift, etc. On the other hand, the lamellar grating interferometer is usually operated on one side of the origin. This is done to make maximum use of the available piston motion for the highest resolution possible. (see Chapter 6 on resolution). Also, the mechanical–optical difficulties increase as the piston displacement is made larger. Nevertheless, a good lamellar grating interferometer is shown in Fig. 15-5. The flux is focused at the input of a light-pipe. The beam maintains the *f*/number that it had originally, and it emerges from the exit of the light-pipe in the focal plane of the collimator mirror. The lamellar grating can be well back of the focal plane. The collimated beam returns from the lamellar grating to the large mirror and is focused in another light-pipe just above the entrance light-pipe. The use of the light-pipes into and out of the interferometer requires a minimum of space. Unless *f*/numbers above about 2 are used for the optical systems, the hole in the middle of the grating must be large, which causes large optical losses. Also, the arrangement in Fig. 15-5 keeps the beam very close to the optical axis of the collimator, reducing abberations, shadowing, and astigmatism. There are many different designs of lamellar grating interferometers, such as those in Dr. Albert Sievers's laboratory at Cornell University, Dr. Ludwig Genzel's

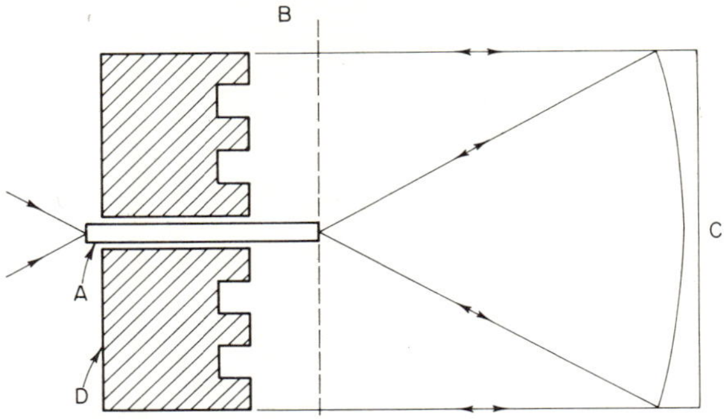


FIG. 15-5. Lamellar grating interferometer with two light-pipes (A) to bring the beams into and out of the interferometer. One light-pipe is placed over the other and their ends are in the focal plane (B) of the collimator mirror (C). The *f*/ number of the beam is maintained in the light-pipe. (D) Lamellar grating housing.

laboratory at the Max Planck Institute for Solid State Research in Stuttgart, Germany, Dr. Paul Richards's laboratory at the University of California–Berkeley, and Dr. John Strong's laboratory at the University of Massachusetts. A commercial instrument is marketed by Beckman Instruments, Inc.

## DIFFRACTION THEORY AND LAMELLAR GRATINGS

In order to explain many of the properties of the lamellar grating interferometer, the illumination of the lamellar grating must be derived. By applying scalar-Fraunhofer diffraction theory to the lamellar grating problem, we may derive the illumination or flux, which is proportional to $EE^*$, where the magnitude of the complex electric field is denoted by $E$. The diffraction theory is outlined in problems 10-3–10-5 in Strong's book [3] "Concepts of Classical Optics."† The central results are given in an article by Strong and Vanasse [2]. Chapters 7 and 9 of Stone's book [6] "Radiation and Optics" are highly recommended.

The lamellar grating interferometer can be treated as two sets of mirrors—the front facets and the back facets as shown in Fig. 15-2(b). Each set of strip mirrors can be considered as a linear array of identical long, rectangular apertures. For the regular array of apertures, the diffracted wave amplitude is given by

$$E_{\text{front}} = E_s F. \tag{15-1}$$

† In problem 10-4, $\Delta\varphi/2$ should be replaced by $\Delta\varphi$.

In Eq. (15-1), $E_s$ is the diffracted wave amplitude for a single facet and $F$ expresses the interference between the diffracted waves from the array of mirrors or facets. For a long slit of width $a/2$ with normally incident waves, the diffracted wave amplitude at an angle $\alpha$ (see Fig. 15-6) is given by

$$E_s \alpha \left\{\frac{\sin[(\pi a \sin \alpha)/2\lambda]}{(\pi a \sin \alpha)/2\lambda}\right\}. \tag{15-2}$$

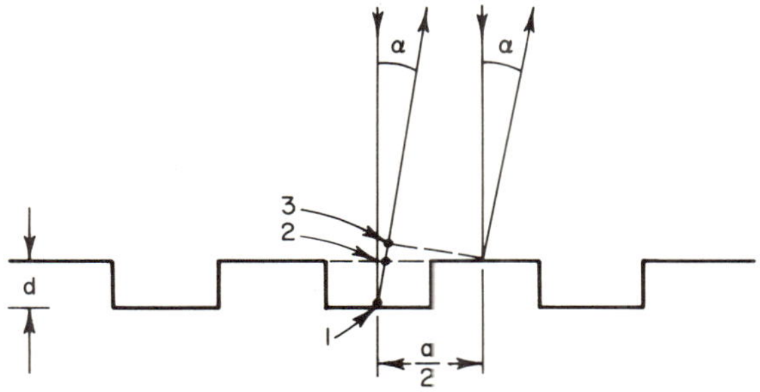


FIG. 15-6. Optical diagram of the lamellar grating with a wave at normal incidence being diffracted at an angle $\alpha$. The grating constant is $a$ and the facet displacement is $d$. The numbered points on the diffracted rays are for use in the derivation of the optical path difference.

For an array of $N$ facets at a distance $a$ aparts, one has

$$F = \frac{\sin[(N\pi a \sin \alpha)/\lambda]}{\sin[(\pi a \sin \alpha)/\lambda]} \exp -i \left[\frac{(N-1)\pi a \sin \alpha}{\lambda}\right]. \tag{15-3}$$

So far, we have only considered the front facets. In order to include the effect of the back facets, we note that the total amplitude is

$$E = E_{\text{front}} + E_{\text{back}}. \tag{15-4}$$

Since the front and back arrays are identical, the two reflected waves have only a phase difference $\varphi$ between them. Thus, Eq. (15-4) may be written as

$$E = E_s F + E_s F e^{i\varphi}. \tag{15-5}$$

In Eq. (15-5), the phase is given by $\varphi = 2\pi\delta/\lambda$. The remaining problem is to calculate $\delta$.

The total optical path difference $\delta$ is given by

$$\delta = d + l, \tag{15-6}$$

where

$$l = l_{12} + l_{23}. \tag{15-7}$$

From Fig. 15-6, one has

$$l_{12} = d/\cos\alpha \tag{15-8}$$

$$l_{23} = (\tfrac{1}{2}a - d\tan\alpha)\sin\alpha. \tag{15-9}$$

Substituting Eqs. (15-7)–(15-9) in Eq. (15-6), we have

$$\delta = d[(1 + \cos\alpha) + (a/2d)\sin\alpha]. \tag{15-10}$$

Then, the phase between the front and back facets is

$$\varphi = (2\pi/\lambda)\,\delta = (2\pi d/\lambda)[(1 + \cos\alpha) + (a/2d)\sin\alpha]. \tag{15-11}$$

Substituting Eqs. (15-2) and (15-3) in Eq. (15-5), the total amplitude of the wave diffracted from the lamellar grating is given by

$$E \propto \frac{\sin[(\pi a\sin\alpha)/2\lambda]}{(\pi a\sin\alpha)/2\lambda}\,\frac{\sin[(N\pi a\sin\alpha)/\lambda]}{\sin[(\pi a\sin\alpha)/\lambda]}\,(1 + e^{i\varphi})e^{-i\varphi_N}, \tag{15-12}$$

where

$$\varphi_N \equiv [(N-1)\pi a\sin\alpha]/\lambda. \tag{15-13}$$

So, the lamellar grating illumination is

$$\mathscr{I} \propto EE^* \propto \left\{\frac{\sin[(\pi a\sin\alpha)]/2\lambda}{(\pi a\sin\alpha)/2\lambda}\right\}^2 \left\{\frac{\sin[(N\pi a\sin\alpha)/\lambda]}{\sin[(\pi a\sin\alpha)]/\lambda}\right\}^2 \cos^2\left(\frac{\varphi}{2}\right), \tag{15-14}$$

where the definition of the cosine function and a trigonometric identity has been used. The first term is the single facet or mirror term; the second term accounts for the array of $N$ identical mirrors in either the front or back plane; and the third term results from the phase difference between the waves diffracted from the front and back mirrors.

The grating equation may be extracted from the multiple mirror term (the second term). Constructive interference between the waves from the different mirrors is found when the second term has its maxima. The largest maxima, of magnitude $N^2$, occur for the zeros in the denominator. That is, constructive interference is present for

$$(\pi a\sin\alpha)/\lambda = m\pi, \tag{15-15}$$

where $m$ is an integer (positive or negative). The grating equation immediately follows and is

$$m\lambda = a\sin\alpha, \tag{15-16}$$

where $m$ is the order of interference. For those who would prefer to derive the grating equation as is done in very elementary texts, observation of

Fig. 15-7 yields Eq. (15-16) with little effort. Using Eq. (15-16), the phase angle $\varphi$ is then given by

$$\varphi/2 = [(\pi m/2) + (\pi d/\lambda)(1 + \cos \alpha)]. \tag{15-17}$$

We would like to consider which orders of diffraction are present in the diffracted pattern and what illumination they will have. If $m \neq 0$ but is even, the first term in Eq. (15-14) is zero. Therefore, the lamellar grating has no positive or negative even orders. If one has the zeroth order, $m = 0$, the first and second terms in Eq. (15-14) are unity and $N^2$, respectively. The total illumination is then proportional to (for small angles $\alpha$)

$$\mathscr{I}(\delta) \propto N^2 \cos^2(2\pi\sigma d) \propto \cos^2(\pi\sigma\delta), \tag{15-18}$$

where $2N$ is the total number of facets. From Eq. (3-19), one has

$$B(\sigma, \delta) \propto \tfrac{1}{2}[1 + \cos(2\pi\sigma\delta)] = \cos^2(\pi\sigma\delta) \tag{15-19}$$

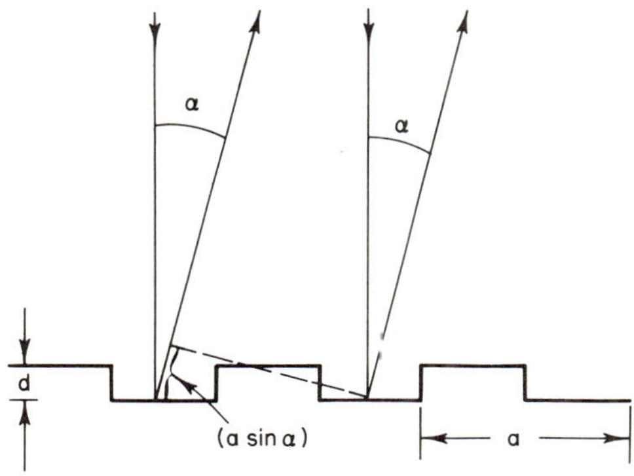


FIG. 15-7. Lamellar grating with illumination at normal incidence. The facet displacement is given by $d$, and $a$ is the grating constant. For constructive interference, one has $m\lambda = a \sin \alpha$, where $m$ is the order.

for the flux of two interfering waves after recombination. Equation (15-19) leads to Eqs. (3-25) and (3-26), which are the basic equations of Fourier transform spectroscopy. Since Eq. (15-18) from lamellar grating diffraction theory for the zeroth order and Eq. (15-19) from interferometry are the same, one uses the lamellar grating in the zeroth order for Fourier transform spectroscopy. That is, the diffraction pattern of a lamellar grating in *zeroth* order is the same as the interference pattern required in Fourier transform spectroscopy.

For $m$ equal to an odd integer, the third term in Eq. (15-14) is given by $\sin^2(\pi\sigma\delta)$. Comparing with the zeroth-order term in Eq. (15-18), one sees that the maxima of the odd orders will be displaced from the zeroth order by $\pi/2$. To summarize, one has no even orders; the $m = 0$ order is wanted for FTS; and the odd orders must be considered further.

The maximum illumination for the odd orders (positive or negative) occurs for $\pi\sigma\delta \simeq (\pi/2)$(odd integer) or $\delta \simeq$ (odd integer)$/2\sigma$. The illumination in Eq. (15-14) is then proportional to

$$\mathscr{I}_{\text{odd}} \simeq (2/\pi m)^2 N^2 \cdot 1. \tag{15-20}$$

The illumination of the zeroth order is $N^2$. Therefore, the first order is reduced to $(2/\pi)^2 \simeq 0.4$; the third order is reduced to $(2/3\pi)^2 = 0.04$; etc. Only the first orders (positive and negative) are strong. Figure 15-8 is a plot of the relative illumination of the various orders.

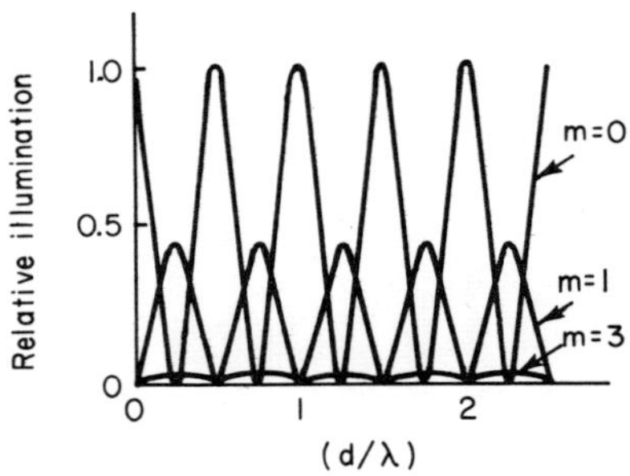


FIG. 15.8. Relative illumination in the first three grating orders for monochromatic radiation. $d$ = groove depth, $m$ = grating order number. (After Strong and Vanasse [2].)

## HIGH-ORDER DIFFRACTION PROBLEMS, $\sigma_c$, AND EFFICIENCY FOR $\sigma \geq \sigma_c$

If the first order is not diffracted at large angles, it can enter the exit slit. For a fixed wavelength, one can see from Fig. 15-8 that the first-order illumination peaks occur at the minimum of the $m = 0$ interference pattern (necessary for Fourier transform spectroscopy). If one then had the exit slit fully illuminated by the $+1$ and $-1$ orders in addition to the $m = 0$ order, the signal $I_R(\delta)$ from which the interferogram is obtained would have maxima of unity and minima of about $2(2/\pi)^2$. The interferogram signal available for Fourier analysis would be reduced because of the positive and negative first orders.

By observation of the grating equation (15-16), one sees that it is the first order ($m = \pm 1$) of the shortest wavelength in the instrument that is dispersed the least. If one has an interferometer such as shown in Fig. 15-9, he can

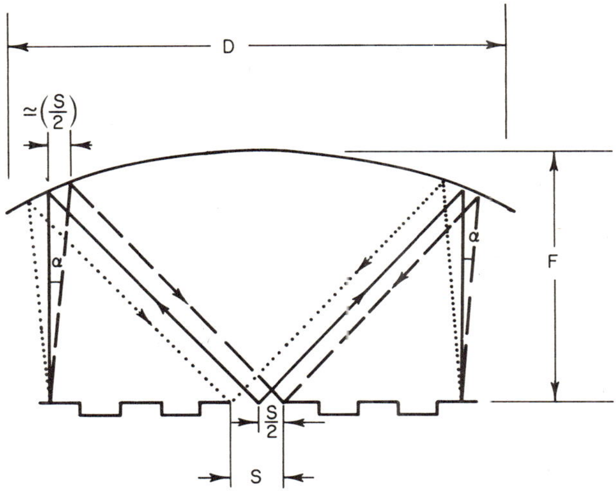


FIG. 15-9. Optical diagram of the lamellar grating interferometer showing rays diffracted at $\pm\alpha$ reaching the exit aperture. Positive $\alpha$ implies positive order. Negative $\alpha$ is equivalent to negative orders.

observe how the first-order flux may be found in the exit aperture. There is a critical wavelength for which the first order will begin entering the exit slit. By observing Fig. 15-9, one can see that a ray from the center of the entrance slit will arrive at the edge of the exit for $\alpha \simeq s/2F$. Assume the apertures are above each other. However, a ray emerging from the edge of the entrance aperture can reach the edge of the exit slit for $\alpha \simeq s/F$. For the small-angle approximation, the grating equation gives

$$m\lambda_c \simeq as/F \tag{15-21}$$

for the critical angle at which the $m$th-order flux enters the exit slit. For the important first order ($m = \pm 1$), one has

$$\sigma_c = F/as. \tag{15-22}$$

To operate the lamellar grating interferometer at large wave numbers, one sees from Eq. (15-22) that an upper limit is placed on the aperture diameter. For fixed focal length $F$, grating constant $a$, and aperture diameter $s$, the efficiency will fall for $\sigma > \sigma_c$. The efficiency falls because as the wave number is increased, the dispersion angle decreases and more high-order flux is introduced into the exit aperture. If one considers only the $m = +1$ and $m = -1$ orders, the peak illumination in the exit aperture is

$$\mathscr{I}_{|m|=1} \propto 2(2/\pi)^2 A(\sigma)/A_{\text{tot}}. \tag{15-23}$$

The area of the first-order image *overlaps* with the exit aperture by an amount $A(\sigma)$ and the total area of exit aperture is $A_{\text{tot}}$. The factor of two is to account for both the positive and negative first orders and the term $(2/\pi)^2$ is the illumination of the first order at the peak. The illumination for the zeroth order is unity. Therefore, the resultant modulation of the interferogram is

$$[I_R(\delta) - I_R(\infty)] \propto \{1 - 2(2/\pi)^2 A(\sigma)/A_{\text{tot}}\}. \tag{15-24}$$

The efficiency for the lamellar grating interferometer becomes

$$\mathscr{E} = [1 - 2(2/\pi)^2 A(\sigma)/A_{\text{tot}}] \times 100. \tag{15-25}$$

In order to calculate $A(\sigma)$ and $A_{\text{tot}}$, one must assume the shape of the apertures. If the entrance and exit apertures are circular of diameter $s$, the collimator has focal length $F$, and the grating constant is $a$, the ratio $A(\sigma)/A_{\text{tot}}$ becomes

$$\frac{A(\sigma)}{A_{\text{tot}}} \simeq \left\{1 - \frac{2(s-l)}{\pi s^2}[s^2 - (s-l)^2]^{1/2} - \frac{2}{\pi}\sin^{-1}\left(\frac{s-l}{s}\right)\right\}, \tag{15-26}$$

where

$$l = [s - (F/a\sigma)]. \tag{15-27}$$

Equation (15-25) was derived by overlapping two equal circles of diameter $s$ and noting that the overlap area is composed of the sum of two equal segments of sagitta $l/2$. Also, it was assumed that

$$\alpha \simeq (s - l)/F \tag{15-28}$$

and

$$\alpha \simeq 1/a\sigma. \tag{15-29}$$

Equations (15-25)–(15-27) enable the experimentalist to calculate the decreasing efficiency for the lamellar grating interferometer for $\sigma \geq \sigma_c = F/sa$ due to first-order overlap with the exit aperture. This cancellation in efficiency is sometimes ignored. As can be seen in Fig. 15-4, even at wave numbers such as $\sigma = 2\sigma_c$, the efficiency is still above 70%. Furthermore, the experimentalist has control over $s$, $a$, and $F$ and can determine what $\sigma_c$ is desirable. The data used to compute the curve for the lamellar grating efficiency in Fig. 15-4 were $a = 1.9$ cm, $s = 1.1$ cm, and $F = 45$ cm, which were the experimental parameters that Richards used [4].

## CAVITY EFFECT, $\sigma_L$, AND RESOLUTION

At the very lowest wave numbers, the lamellar grating interferometer cannot be used without care. The cavity effect places a long-wavelength limit on the use of a lamellar gratings. The width of the groove $a/2$ is the governing parameter. When the wave is back of the front facets, it is between what are essentially two parallel mirrors. The back facet and walls form three sides of a cavity. If one assumes that the cavity is very deep, i.e., that $d$ is large, he can make the drastic assumption that he has an open waveguide. A good reference for waveguide theory as applied to the cavity effect is Jackson [7], "Classical Electrodynamics." The assumption of a very long waveguide is too strong to be valid, but it will indicate the important parameters concerning the waveguide effect. For the component of the electric vector parallel to the walls, there are what are called transverse electric modes of wave propagation in the waveguide. The frequencies of the waves for normal modes are given by (for integers $p$ and $q$)

$$\sigma_{p,q} = (c^2\pi/2)[(p^2/b^2) + (4q^2/a^2)]^{1/2}. \qquad (15\text{-}30)$$

The width of the groove is $a/2$ and the length is $b$. The dominant mode of propagation is for $p = 0$ and $q = 1$ and this, in waveguide terminology, is called the $TE_{01}$ mode. The group velocity for the $TE_{01}$ mode is given by ($c$ is the velocity of light in a vacuum)

$$v_g = c[1 - 1/(a\sigma)^2]^{1/2}. \qquad (15\text{-}31)$$

Waves for which $\sigma$ is less than $1/a$ cannot propagate because the group velocity becomes imaginary at lower wave numbers. With the $T_{01}$ wave traveling at a velocity less than $c$, the effective optical path difference $\delta_{eff}$ increases such that

$$\delta_{eff}/\delta_{act} = c/v_g = 1/[1 - 1/(a\sigma)^2]^{1/2}, \qquad (15\text{-}32)$$

where $\delta_{act}$ is the geometric path difference. Experimentally, it has been found that $\delta_{eff}/\delta_{act} \simeq 1.05$ is consistent with good resolution as explained below. Using this 5% error in $\delta$ as the maximum allowable error, the low-wave-number limit to which a lamellar grating can be used is

$$\sigma_L \simeq 3.2/a. \qquad (15\text{-}33)$$

This lower limit for Richards's lamellar grating interferometer [4] is $\sigma_L \simeq 1.7$ cm$^{-1}$ for $a = 1.9$ cm. As one can see, widening the groove decreases the lower limit.

The ultimate spectral resolution of the lamellar grating interferometer is affected by the cavity effects. The two waves with different polarization have a relative difference in their optical paths which is about equivalent in the interferogram to having two equally bright, monochromatic lines near each other in wave number.† Then, with Eq. (15-32), one has the resolution

† The optical path difference for the polarization parallel to the grooves may be defined as $\delta_1$. For the perpendicular polarization, the path difference is $\delta_2 = 2d$. For the parallel polarization, the interferogram for a monochromatic source of frequency $\sigma_1$ would be [see Eq. (5-4)]

$$[I_R(\delta) - I_R(\infty)] = 2\cos(2\pi\delta_1\sigma_1)$$
$$= 2\cos[2\pi\delta_2(1 + \Delta\delta/\delta_2)\sigma_1],$$

where $\Delta\delta \equiv \delta_1 - \delta_2$. The interferogram would behave as though the source frequency were $\sigma_1\{1 + [(\Delta\delta)/\delta_2]\}$. For the perpendicular polarization, the interferogram would be

$$[I_R(\delta) - I_R(\infty)] = 2\cos(2\pi\delta_2\sigma_1).$$

If, on the other hand, one had two equally bright, monochromatic sources at frequencies $\sigma_1$ and $\sigma_2$, the spectrum could be expressed as

$$B(\sigma) = \delta'(\sigma - \sigma_1) + \delta'(\sigma - \sigma_2),$$

where $\delta'(\sigma - \sigma_1)$ is a Dirac delta function. The interferogram with no cavity effect, would be

$$[I_R(\delta_2) - I_R(\infty)] = 2\cos(2\pi\sigma_1\delta_2) + 2\cos(2\pi\sigma_2\delta_2).$$

By comparing the interferograms, one would have

$$\sigma_2 = \sigma_1\{1 + [(\Delta\delta)/\delta_2]\}$$

or

$$\delta\sigma/\sigma \equiv (\sigma_2 - \delta_1)/\sigma = \Delta\delta/\delta_2.$$

The difference in optical paths would make the spectrum appear as though it were produced by two close spectral lines.

Recall for that $\sigma \geq \sigma_L$ Eq. (15-32) may be expressed as

$$\delta_1/\delta_2 = \delta_{eff}/\delta_{act} \simeq 1 + \tfrac{1}{2}(1/a\sigma)^2$$
$$= 1 + [\Delta\delta/\delta_2].$$

So, the relative optical path difference is also given by

$$(\Delta\delta)/\delta_2 \simeq \tfrac{1}{2}(1/a\sigma)^2.$$

One can equate the expressions for the relative optical path differences and obtain the resolution limit

$$\delta\sigma \simeq 1/2a^2\sigma$$

for $\sigma \geq \sigma_L \simeq 3.2/a$. From Eq. (6-5), one may note that the resolution is also given by

$$\delta\sigma \simeq 1/L,$$

where $L$ is the maximum optical path difference. Therefore, there is no point in attempting to take interferograms with $L \geq 2a^2\sigma$.

limited by $\delta\sigma = 1/2a^2\sigma$ where $\sigma \geq \sigma_L$. If the wave number were at the lower limit of $\sigma_L \simeq 3.2/a$, then $\delta\sigma \simeq 0.16/a$, which for Richards's lamellar grating would yield $\delta\sigma \simeq 0.08\ \text{cm}^{-1}$. For larger wave numbers, the ultimate resolution would improve, and at $\sigma = 3$ cm, one could only have a resolution of $\delta\sigma \simeq 0.05\ \text{cm}^{-1}$ for Richards's interferometer.

It is no trouble to include the entire far-infrared region from 2 to 100 $\text{cm}^{-1}$ and stay within the limits imposed by the cavity effect and the first-order diffraction problem. However, the cavity effect does impose a resolution limitation as mentioned above.

One might suppose that the cavity effect could be overcome by using dielectric walls. In one laboratory, the strip mirrors were made of Pyrex to obtain high-quality, flat mirrors. The Pyrex furnished dielectric walls, but there is a cavity effect with dielectrics, too. In addition, the Pyrex cold-flowed in a few years and had to be replaced.

## SHADOWING

Another problem that is common to the lamellar grating interferometer is shadowing. If the waves are not perfectly collimated, there will be shadowing of the back facets by the front facets. As $d$ gets larger, more shadowing will occur. The interferogram will have dropping asymptotes for large displacements of the facets. This is not too serious a problem because the interferogram has effectively been apodized by a trapezoid function to a fair approximation. From Chapter 5, recall that one frequently introduces trapezoidal apodization anyway to correct for the effect of finite maximum optical path differences. So, additional, slight apodization only increases the width of the computed spectral lines by a small amount.

## WAVENUMBER SHIFT DUE TO OFF-AXIS OPTICAL SYSTEM

In deriving Eq. (15-18) from Eq. (15-17), it was assumed that $\alpha \simeq 0$ and $1 + \cos\alpha$ was set equal to 2. If the next-higher-order approximation of $\cos\alpha = 1 - (\alpha^2/2)$ is made, one has

$$I(\delta) \propto \cos^2\{2\pi\sigma d[1 - (\alpha^2/4)]\}. \tag{15-34}$$

If the entrance and exit slits have their centers a distance $x_0$ apart and the collimator focal length is $F$, then $\alpha^2/4 \simeq x_0{}^2/4F^2$ (see Fig. 15-9). For the computer transformation, one would use a measured displacement between the facets of $2d = \delta$. But the actual displacement is $\delta_{\text{act}} = 2d[1 - (\alpha^2/2)]$. So, one has

$$2d = \delta = \delta_{\text{act}}/(1 - \tfrac{1}{4}\alpha^2) \simeq \delta_{\text{act}}(1 + \tfrac{1}{4}\alpha^2). \tag{15-35}$$

This means that, for a monochromatic source, if one used $2d$ for the optical path difference, the spectrum from Eq. (3-25) would be given by

$$B(\sigma) = \int_0^\infty \cos^2(\pi\sigma_{\text{eff}}\delta)\cos(2\pi\sigma\delta)\,d\delta, \tag{15-36}$$

where

$$\sigma_{\text{eff}} = \sigma_0(1 - \tfrac{1}{4}\alpha^2). \tag{15-37}$$

Recalling that the Fourier transform of a Dirac delta function $\delta'(\sigma - \sigma_0)$ is $2\cos(2\pi\sigma_0\delta)$, one sees that the computed spectrum for a monochromatic source is erroneously thought to be too low by

$$\Delta\sigma = \tfrac{1}{4}\alpha^2\sigma \simeq (x_0{}^2/4F^2)\sigma. \tag{15-38}$$

So, one must increase each computed wave number by $\Delta\sigma$. For Richards's lamellar grating interferometer [4], the slit centers were 1.33 cm apart and the focal length was 45 cm and the correction was $\Delta\sigma \simeq 2.2 \times 10^{-4}\sigma$. So, all of his computed spectra taken with the lamellar grating interferometer should have each wave number increased by $2.2 \times 10^{-4}\sigma$. Only in very-high-resolution spectroscopy is this error important.

The effect of having sources of finite size on the computed wave number is discussed in Chapter 11, Field of View and Effect on Resolution. Not only is the resolution affected, but there are wavenumber corrections due to finite instrumental apertures. For Richards's lamellar grating interferometer, the aperture size correction to the wave number is negligible.† So, one has two corrections to the wave number, one due to having the entrance and exit slits a distance apart and another correction due to a finite instrumental aperture.

† The mean wavenumber shift downward for aperture size corrections is given by $\Delta\sigma \simeq h^2\sigma/16F^2$, where $h$ is the source diameter and $F$ is the focal length of the collimeter. For $h \simeq x_0$, i.e., the centers of the exit and entrance slits as close as possible, one has a shift only one-fourth as large as in Eq. (15-38). The two shifts are basically the same, except that the entended distance is twice as large for the separated apertures as for the width of the single aperture which one has for the Michelson interferometer. The width or separation distance enter in $\Delta\sigma$ as squares.

## SAMPLE SPECTRA FROM PLANE, LAMELLAR GRATING INTERFEROMETERS

Figures 15-10–15-15 show some spectra obtained with lamellar grating interferometers. Figure 15-10 demonstrates that the lamellar grating interferometer can be used to very low wave numbers with a resolving power of at least 30. Figures 15-11–15-13 are spectra obtained with the lamellar grating interferometer marketed by the Beckman Instruments, Inc. The pure rotational spectra of $^{14}N_2{}^{16}O$ and $H_2O$ are given in Figs. 15-11 and 15-12, respectively. The transmittance of hexaiodobenzene is presented in Fig. 15-13. The Beckman instrument data demonstrate that the instrument can be used from about 3.5 cm$^{-1}$ to at least 70 cm$^{-1}$ with a resolving power as high as 150. In Figs. 15-14 and 15-15, some of the capabilities of the double-beam lamellar grating interferometer made by Dr. J. Dowling are exhibited. The

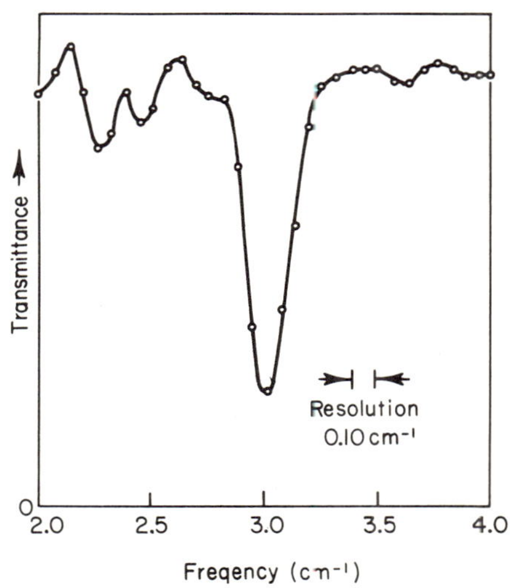


FIG. 15-10. Zero-field ferrimagnetic resonance in a 1-cm-thick powder sample of ytterbium iron garnet at 1.2°K measured using Richard's lamellar grating interferometer. This is a good example of the use of the lamellar grating interferometer at very low wave numbers with good resolution. The cavity effect limits the ultimate resolution at $\sigma = 3$ cm$^{-1}$ to about 0.05 cm$^{-1}$ (After Richards [4].)

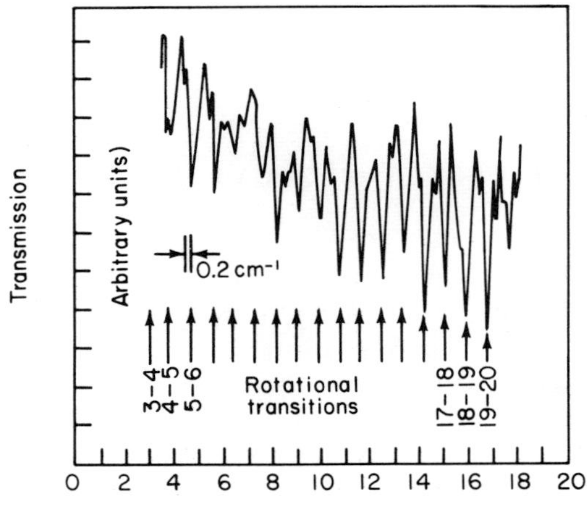


FIG. 15-11. Pure rotational spectrum of $^{14}N_2{}^{16}O$ recorded under conditions of 1 m optical path and ~10 cm Hg pressure. Recording time, 1 hr for each interferogram of ±5 cm optical path difference. Electronic time constant, 4 sec. (Beckman Instruments, Inc., Fullerton, California.)

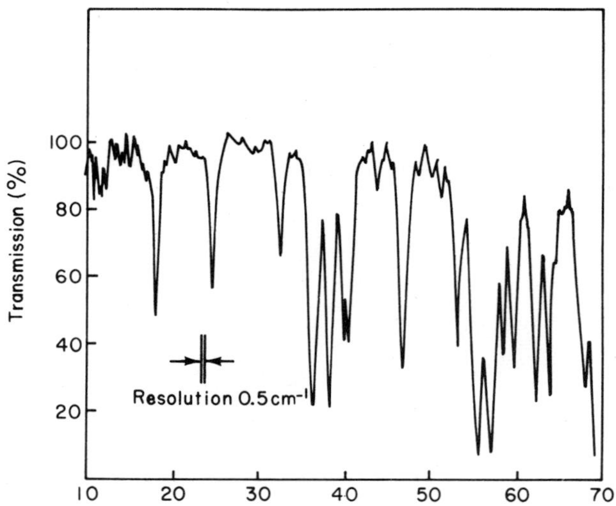


FIG. 15-12. Pure rotational spectrum of $H_2O$ vapor recorded under conditions of 1 m optical path and ~2 cm Hg pressure. Recording time, 25 min for each interferogram of ±2 cm optical path difference. Electronic time constant, 0.5 sec. (Beckman Instruments, Inc., Fullerton, California.)

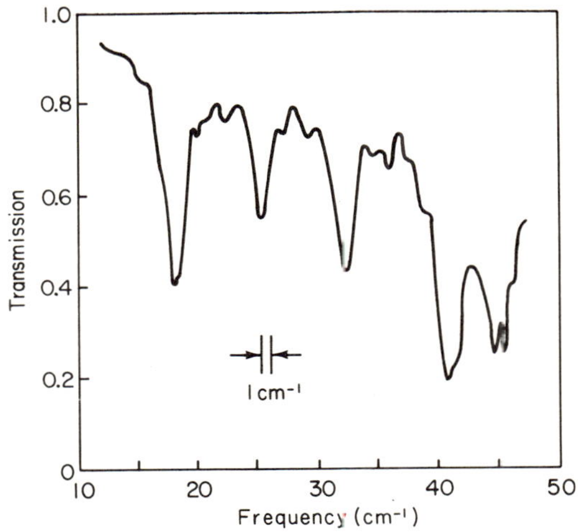


FIG. 15-13. Transmission spectrum of a 2-mm powdered disk of hexaiodobenzene. Recording time, 20 min for each interferogram of $\pm 1$ cm optical path difference. Electronic time constant, 2 sec. (Beckman Instruments, Inc., Fullerton, California.)

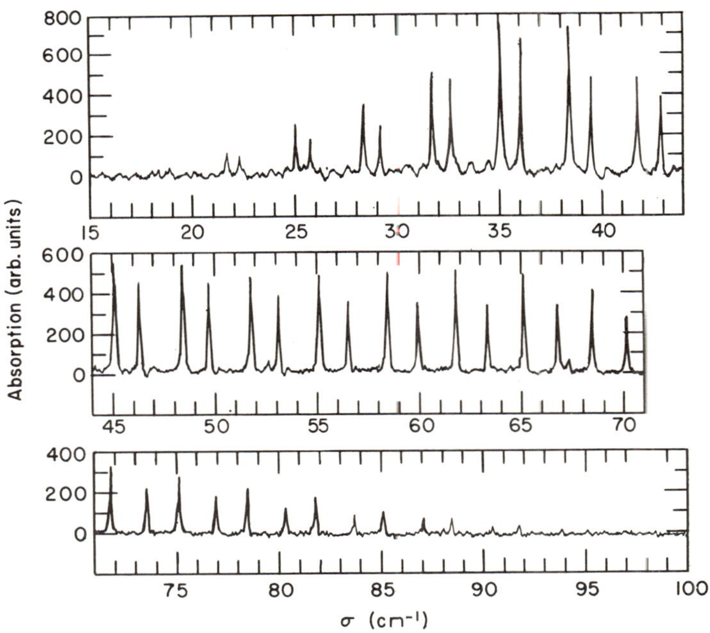


FIG. 15-14. The pure rotational spectrum of nitric oxide in the region 15–100 $cm^{-1}$ (Dowling [8]).

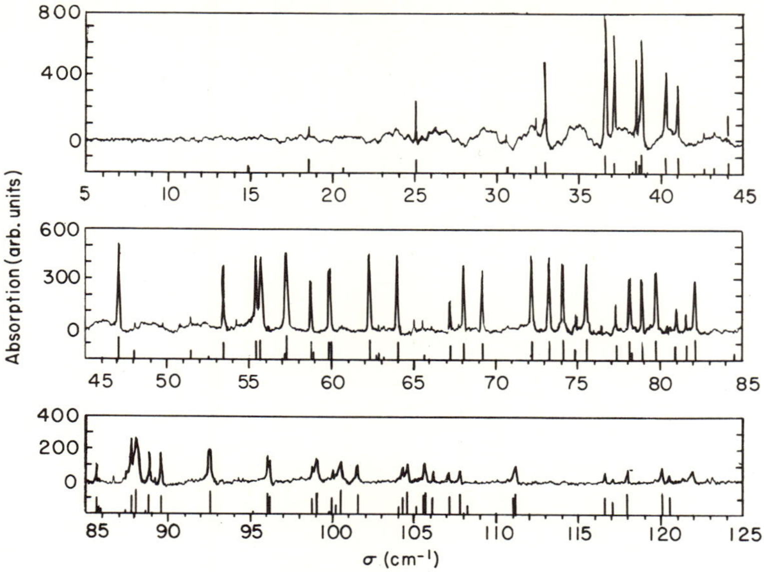


FIG. 15-15. The pure rotational spectrum of water vapor between 5 and 125 $cm^{-1}$. (After Dowling [8].)

rotational spectra of nitric oxide and water vapor are shown between 5 and 125 $cm^{-1}$ with a resolving power sometimes above 150.

Thus, it should be apparent that the plane, lamellar grating interferometer operates well at very low wave numbers. The instruments can yield reliable, high-resolution data between about 2 and 100 $cm^{-1}$. The lower limit can be easily reduced by widening the grating grooves.

## SPHERICAL, LAMELLAR GRATING INTERFEROMETERS

A recent development in lamellar grating Fourier transform spectroscopy has been the invention of the spherical, lamellar grating interferometer. The instrument is described in a paper by Hansen and Strong [9].

The basic change from the plane, lamellar grating interferometer is the use of noncollimated beams in the interferometer. The plane, lamellar grating is replaced with a spherical, lamellar grating as shown in Fig. 15-16. The entrance and exit apertures are at a distance equal to the radius of curvature from the spherical, lamellar grating. The spherical, lamellar grating is shown in Fig. 15-17. As can be seen, the facets are on a spherical surface and the

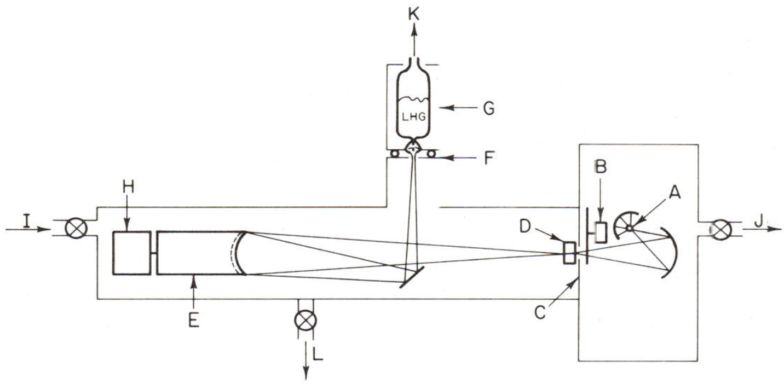


FIG. 15.16. Basic spherical lamellar grating interferometer. A: Source; B: chopper; C: entrance baffle; D: filter; E: grating; F: exit baffle; G: detector; H: drive assembly; I: gas inlet; J: gas outlet; K, L: to vacuum pump. (After Hansen and Strong [9].)

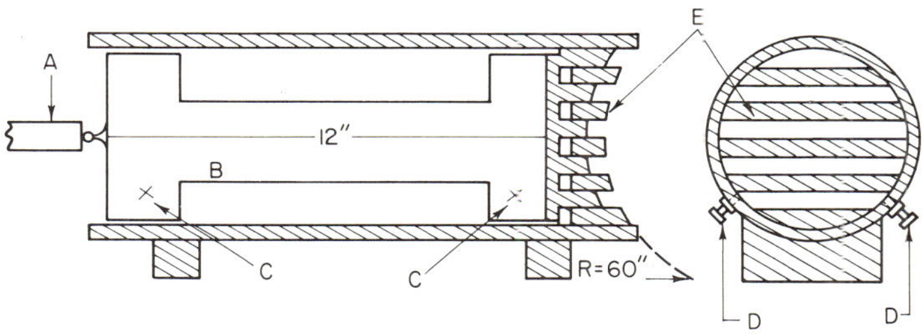


FIG. 15-17. Front and side views of assembled grating. A: Micrometer step drive; B: aluminum piston; C: highly polished surfaces ride on support screws; D: Teflon-tipped support and alignment screws; E: grating facets (28). (After Hansen and Strong [9].)

operation is very much as it is for the plane, lamellar grating interferometer except for the two spherical surfaces which are a distance $d$ apart. The images are separated a distance $2d$. The effect of this separation of the two images with noncollimated optics in interferometry will be discussed in detail.

The spherical, lamellar grating is simple to make, in that no collimation optics are needed. The spherical laminations can be ground in steel by good machine shops, and the piston drive is no more complicated than for the plane, lamellar grating. The instrument is very well suited for very-low-wave-number research and is easily mated to large-$f$/number systems. Dr. Strong's group at the University of Massachusetts has built the instrument for gas discharge studies in extremely long tubes. The instrument they built had the spherical, lamellar grating with a 152 cm radius of curvature $R$, a 9-cm grating diameter $D$, and a 0.635-cm grating constant $a$. The entrance and exit aperture diameters were $s = 2.2$ cm, and the maximum mirror displacement was $d = 1.8$ cm.

Spherical, lamellar grating interferometers may have application in millimeter-wave astronomy because no telescope is needed. The collimated incident beam would be focused a distance $R/2$, and for *f*/numbers greater than 8, the spherical mirror is a good approximation to a parabolic mirror. For very long wavelengths, the construction tolerances are eased.

The cavity effect and high-order mixing at the exit slit for the plane and spherical lamellar grating instruments are similar. The off-axis wavenumber shift will be included in the following treatment. Most importantly, one would expect the spherical, lamellar grating to produce diffraction effects similar to those for the plane, lamellar grating. Therefore, the spherical, lamellar grating also operates in the zeroth order of diffraction. One also expects that the spherical, lamellar grating interferometer would be of high efficiency, as illustrated in Fig. 15-4.

## EFFECTS OF NONCOLLIMATION ON THE COMPUTED SPECTRUM

We will now discuss the major differences between spherical and plane grating interferometers. Much of the treatment in this section is from an article by Parks and Alexander [10].

In Fig. 15-18, an optical diagram of the interferometer section of a lamellar grating interferometer is shown. It is obvious that when the two mirror surfaces are a distance $d$ apart, the images are separated a distance $2d$. One image may be focused on the detector, but the other may be focused in front of it. In order to obtain good interference or modulation in the interferogram between the two radiation beams, the *f*/number should be as large as possible so that there is maximum overlap of the images.

It is assumed that the fixed mirror focuses the source on the detector and that there is coherence over the source. One must vectorially compute the optical path length from points on the source to points on the detector for rays which have been reflected from the movable mirror. For simplicity, one assumes that the source and detector diameters are the same, that they are abutting, and that spherical aberration may be neglected. The path length may be computed for *all* source points which emit optical rays that strike the detector at a particular point. One is interested in separate rays that are reflected from the movable and the fixed mirrors which arrive at the same detector points, for these are the rays that produce the modulation in the

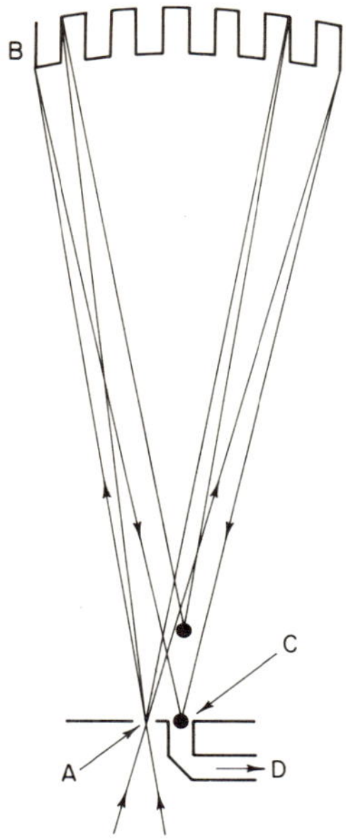


FIG. 15-18. Optical diagram of a spherical lamellar grating interferometer with separated images exhibited. A: Entrance aperture. B: Lamellar grating beamsplitter. C: Exit aperture. D: Light-pipe to detector.

interferogram. The optical path difference for these two sets of optical rays is given by

$$\delta_s = 2d(1 + \{\rho_m{}^2 - \rho_D{}^2 + 2\rho_m \rho_D[\cos(\theta - \varphi)]/2R^2\}), \qquad (15\text{-}39)$$

where the rays strike the movable mirror at $(\rho_m, \theta)$ and the detector at $(\rho_D, \varphi)$, and $R$ is the radius of curvature of the spherical mirror. These sets of rays include all rays from the source which arrive at particular mirror and detector points. For a monochromatic source, neglecting vignetting and the dependence on $d$ of the solid angle subtended by the movable mirror, the interferogram is proportional to

$$F(\delta) = (2\pi^2 r_m{}^2 r_D{}^2/R^2)(1 + \int_m d^2\rho_m \int_D d^2\rho_D\{[\cos(2\pi\sigma_0\,\delta_s)]/\pi^2 r_m{}^2 r_D{}^2\}), \qquad (15\text{-}40)$$

where $2r_m$ is the diameter of the mirror, $2r_D$ is the diameter of the source and of the detector, and $\sigma_0$ is the source wave number. The first integral is over the area of the mirror and the second integral is over the area of the detector. From the interferogram in Eq. (15-40), one may compute the spectrum (for scans of finite mirror displacements) from

$$B_{exp}(\sigma) = (L/R^2) \int_m d^2\rho_m \int_D d^2\rho_D \operatorname{sinc} \alpha, \qquad (15\text{-}41)$$

where

$$\alpha = 2\pi L\{\sigma - \sigma_0 - \sigma_0[(|\boldsymbol{\rho}_m + \boldsymbol{\rho}_D|^2 - 2\rho_D{}^2)/2R^2]\}. \tag{15-42}$$

In these equations, $\pm L$ represents the maximum optical path difference and the vectors $\boldsymbol{\rho}_m$ and $\boldsymbol{\rho}_D$ are measured radially from the optical axis to points on the mirror and detector, respectively.

Equation (15-42) can be numerically integrated using Gaussian–Legendre integration to obtain the experimental profiles. Parks and Alexander did this for the case of the spherical, lamellar grating interferometer designed by Hansen and Strong [9] ($r_m = 4.5$ cm, $r_D = 1.1$ cm, $R = 152$ cm, and $d = 1.1$ cm). The results for two values of $L\sigma_0$ are shown in Fig. 15-19. One

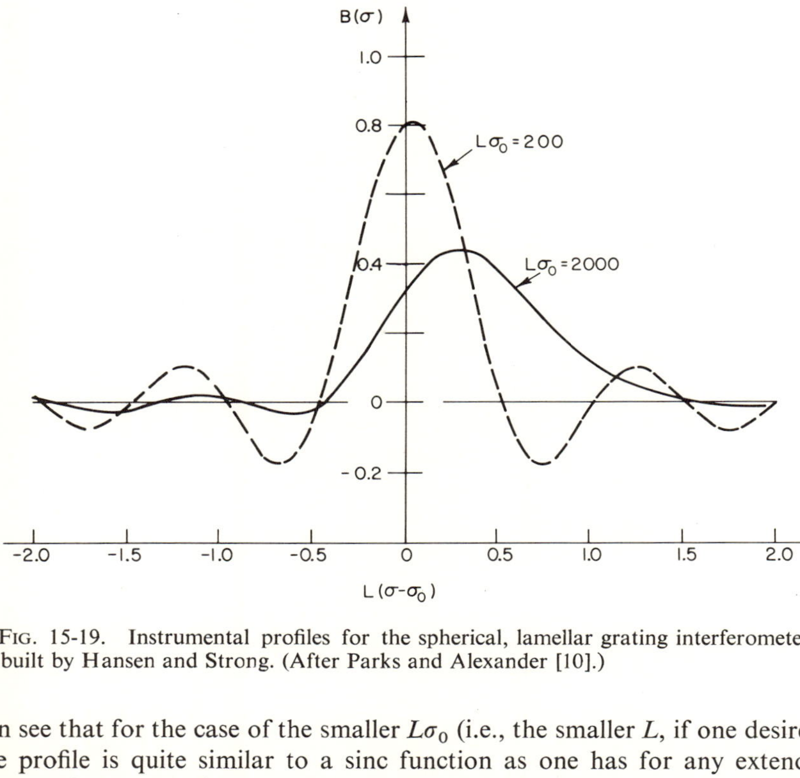


FIG. 15-19. Instrumental profiles for the spherical, lamellar grating interferometer built by Hansen and Strong. (After Parks and Alexander [10].)

can see that for the case of the smaller $L\sigma_0$ (i.e., the smaller $L$, if one desires), the profile is quite similar to a sinc function as one has for any extended source without apodization (Chapter 5). As $L$ increases, the spectrum broadens due to the increase in the dispersion of path differences. The spectral profile peaks are shifted to larger wave numbers and are not strongly skewed, in contrast to Steel's results [11]. The shift of the peak is shown in Fig. 15-20

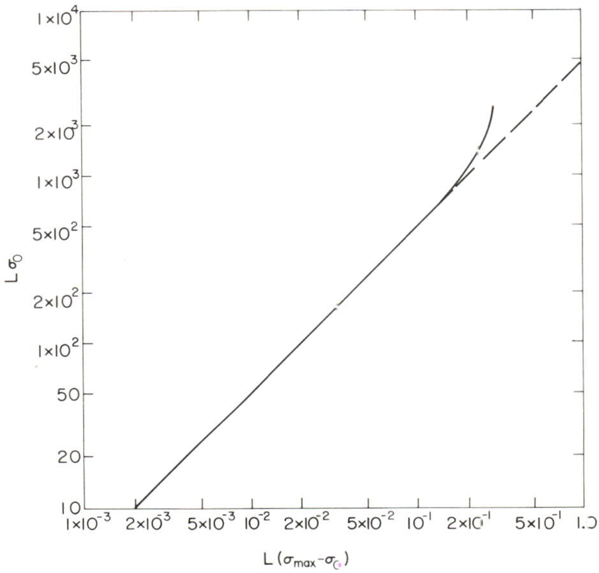


FIG. 15-20. A plot of the product of the maximum optical path difference $L$ and the source wave number $\sigma_0$ versus the product of $L$ and the wavenumber shift $\sigma_{max} - \sigma_0$ for a spherical, lamellar grating interferometer (Hansen and Strong). Note that the instrumental profile peak has been shifted up to $\sigma_{max}$ from $\sigma_0$. (After Parks and Alexander [10].)

for various values of $L\sigma_0$. In the figure, $\sigma_{max}$ is the wave number at the peak of the instrumental profile. The shift in the peak is a linear function of the product $L\sigma_0$ up to values of about $7 \times 10^2$ for the spherical, lamellar grating built by Hansen and Strong. The coefficient of the linear term for other lamellar grating instruments could be experimentally determined or calculated.

In Fig. 15-21, the full-width at one-half the peak intensity of the spectral profiles is given for various wave numbers. One can see that there is a maximum $L$ or best half-width for each wave number. The minima in the curves in Fig. 15-21 are caused by the normal sharpening of the spectra for large $L$ competing with the loss of resolution due to the dispersion in the path differences being a function of $L$, the latter finally becoming dominant.

From Fig. 15-21, one can see that, for Hansen and Strong's [9] lamellar grating interferometer, the maximum resolving power is given by

$$\sigma_0/\Gamma_{min} \leq 2.3 \times 10^3. \tag{15-43}$$

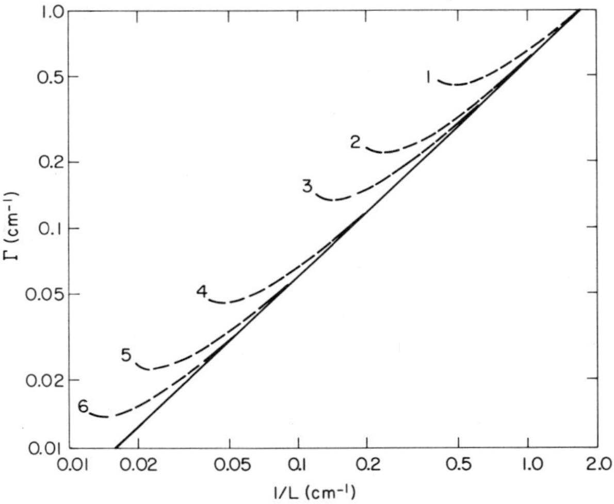


FIG. 15-21. Plots of the full-width at one-half of the peak intensity of the instrument profile, $\Gamma$, versus the reciprocal of the maximum optical path difference $1/L$ for various source wave numbers $\sigma_0$ for the spherical, lamellar grating interferometer of Hansen and Strong: (1) $\sigma_0 = 1000$, (2) 500, (3) 300, (4) 100, (5) 50, (6) 30 $cm^{-1}$. (After Parks and Alexander [10].)

For whatever $L$ and $\sigma_0$ values are chosen, there is no point in having $\sigma_0 L$ exceed $2.3 \times 10^3$. It should be emphasized that this restriction allows good resolving power for many experiments, and thus condition (15-43) is not a serious drawback.

## SAMPLE SPECTRA FROM A SPHERICAL, LAMELLAR GRATING INTERFEROMETER

Figures 15-22–15-24 show some sample interferograms and spectra taken with the spherical, lamellar grating interferometer built by Hansen and Strong. Figure 15-22 shows the spectrum of mercury arc radiation exhibiting atmospheric water vapor absorption. Channel spectra in quartz are presented in Fig. 15-23. The data in Figs. 15-22 and 15-23 show that the instrument can operate in the submillimeter wavelength range with resolving power in excess of 70. The very-low-wavenumber capabilities are illustrated in Fig. 15-24, a spectrum for which a 6.2 $cm^{-1}$ klystron source was used.

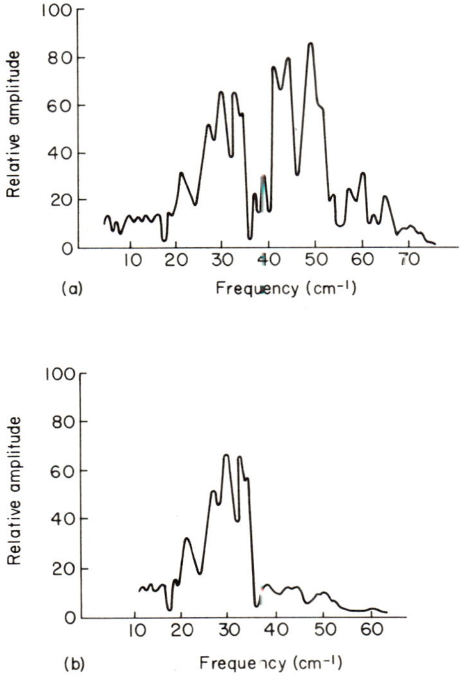


FIG. 15-22. (a) Spectrum obtained of mercury arc radiation passing through 33 m of laboratory air; (b) with a Yoshinaga filter inserted in the path of the radiation. Frequencies higher than 10 $cm^{-1}$ are attenuated. (Hansen and Strong [9].)

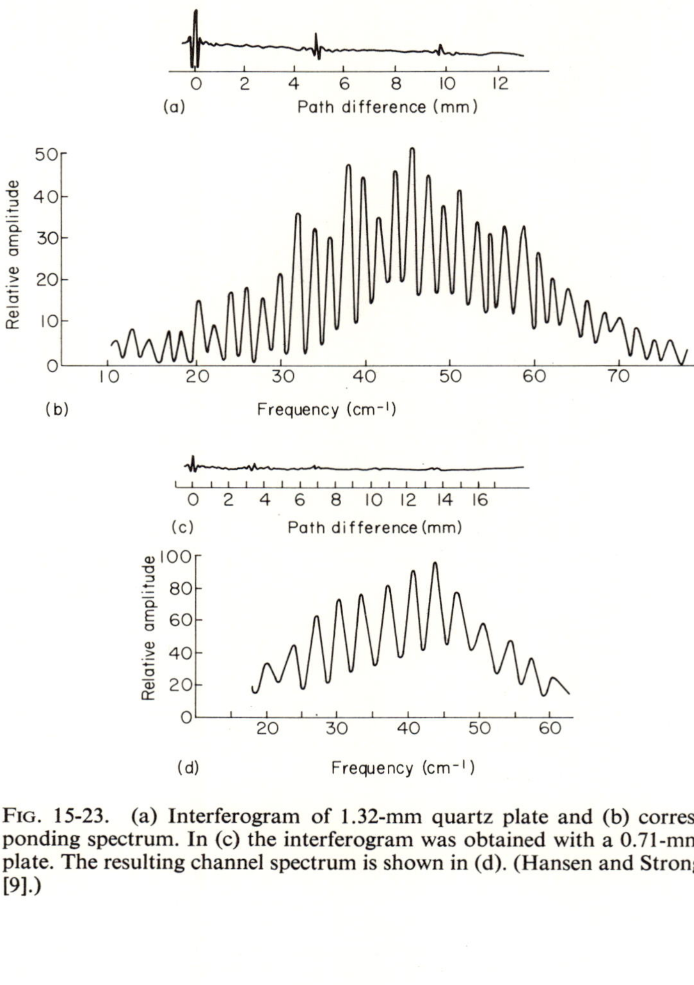


FIG. 15-23. (a) Interferogram of 1.32-mm quartz plate and (b) corresponding spectrum. In (c) the interferogram was obtained with a 0.71-mm plate. The resulting channel spectrum is shown in (d). (Hansen and Strong [9].)

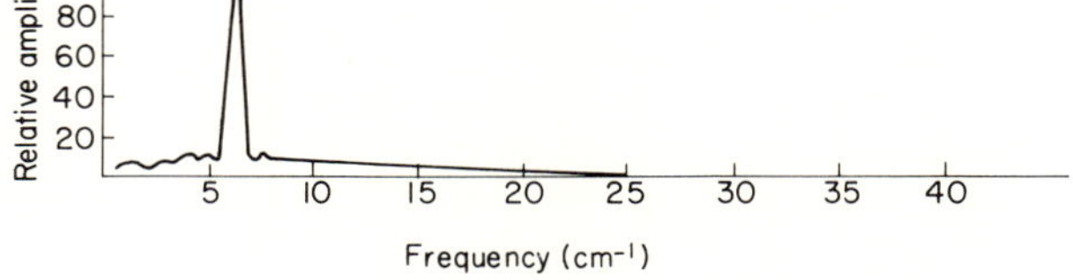


FIG. 15-24. Experimental spectrum of a monochromatic source, the klystron; peak is at 6.2 $cm^{-1}$. (Hansen and Strong [9].)

## CONCLUSIONS

In conclusion, it should be emphasized that we have shown that the lamellar grating interferometers in the zeroth order behave as Fourier transform spectrometers. Though the lamellar grating divides wavefronts and the beamsplitter in the Michelson interferometer divides amplitudes, both instruments can have the spectra computed by Eq. (3-25) or Eq. (3-26).

There are special problems associated with the lamellar grating interferometers, such as the decrease in the modulation of the interferogram due to first-order (and higher, odd-order) mixing for $\sigma \geq \sigma_c$ [Eq. (15-22)]. The decrease in efficiency is given approximately by Eq. (15-25). There is degradation of the resolution due to the cavity effect for $\sigma \geq \sigma_L$ [Eq. (15-33)]. Also, there is a frequency shift due to separated entrance and exit apertures [Eq. (15-33)]. The shifted spectra of the spherical, lamellar grating and the Michelson interferometer with noncollimated optics are illustrated in Figs. 15-19 and 15-23. Most other problems are common to the Michelson interferometer and are discussed elsewhere in this book.

In the far-infrared, the lamellar grating interferometers are more efficient than the Michelson interferometers because of the beamsplitter efficiency, as shown in Fig. 15-4 (also, there is no polarization problem at the beamsplitter). Because of this high efficiency, the lamellar grating instrument is frequently usable at low wave numbers where the source intensity is low. The new spherical, lamellar grating interferometer has a promising future because of its simplicity in construction.

### *References*


*1.* J. D. Strong and G. A. Vanasse, Modulation Interferentielle et Calculateur Analogique pour un Spectrometre Interferential, *J. Phys. Radium* **19**, 192 (1958).

*2.* J. D. Strong and G. A. Vanasse, Lamellar Grating Far-Infrared Interferometer, *J. Opt. Soc. Amer.* **50**, 113 (1960).

*3.* J. D. Strong, "Concepts of Classical Optics." Freeman, San Francisco, California, 1958. Appendix F, coauthored with G. A. Vanasse, Application of Fourier Transformations in Optics: Interferometric Spectroscopy.

*4.* P. L. Richards, High Resolution Fourier Transform Spectroscopy in the Far-Infrared, *J. Opt. Soc. Am.* **54**, 1474 (1964).

*5.* P. Connes, How Light is Analyzed, *in* "Lasers and Light," (A. Shawlow, ed.), p. 35. Freeman, San Francisco, California, 1969.

*6.* J. M. Stone, "Radiation and Optics." McGraw-Hill, New York, 1963.

*7.* J. D. Jackson, "Classical Electrodynamics," Chapter 8. Wiley, New York, 1962.

*8.* J. M. Dowling, Double Beaming in Fourier Spectroscopy, *Aspen Int. Cont. on Fourier Spectrosc., 1970*, p. 55. (G. A. Vanasse, A. T. Stair, Jr., and D. J. Baker, eds.). AFCRL-71-0019, 5 Jan. 1971, Spec. Rep. No. 114.

*9.* N. P. Hansen and J. Strong, Performance of a Simple Spherical Lamellar Grating Beamsplitter by Wavefront Division, *Aspen Int. Conf. on Fourier Spectrosc., 1970*, p. 215. (G. A. Vanasse, A. T. Stair, Jr., and D. J. Baker, eds.). AFCRL-71-0019, 5 Jan. 1971, Spec. Rep. No. 114.

*10.* W. F. Parks and Ralph W. Alexander, Interferometers with Spherical Mirrors (to be published).

*11.* W. H. Steel, Interferometers without Collimation for Fourier Spectroscopy, *J. Opt. Soc. Amer.* **54**, 151 (1964).

CHAPTER SIXTEEN

# Computation Techniques

## INTRODUCTION

The computation of the inverse Fourier transform of the interferogram to obtain the spectrum used to be one of the biggest hinderances to Fourier transform spectroscopy. With the invention and development of digital computers, Fourier spectroscopy became feasible. In this chapter, we shall discuss straightforward, conventional computations of the inverse Fourier transform. The conventional method has the advantage that one can choose the wave numbers at which the spectrum is to be computed without accepting some fixed wavenumber interval between points. If only a few wave numbers are desired in the spectrum around some special spectral feature, the conventional method is highly recommended. Mathematical filtering as described in Chapter 5 can be easily used with the straightforward computation of the transform.

First, we present two ways of computing the spectrum by conventional techniques. Then, in the next section, we count the number of computer operations involved in calculating a spectrum of $N/2$ wave numbers by the conventional techniques. This is compared with the number of operations involved in using the very fast Cooley–Tukey algorithm way of computing Fourier transforms. It will be shown that the Cooley–Tukey algorithm is approximately $0.46N/\ln_e N$ times faster than the conventional technique.

The details of the Cooley–Tukey algorithm will be left to the following chapter, real-time analysis will be considered in Chapter 18, and computer programs are given in Appendix B.

## CONVENTIONAL COMPUTATION TECHNIQUES

As is fairly obvious from the basic equation of Fourier transform spectroscopy [Eq. (3-26)],

$$B(\sigma) = \int_{-\infty}^{\infty} [I_R(\delta) - I_R(\infty)] \exp(-i2\pi\sigma\delta)\, d\delta, \tag{16-1}$$

the computational problems would be severe without a fast digital computer. In fact, most applications of Fourier transform spectroscopy were impossible until the development of modern computers.

We start with Eq. (16-1) for a double-sided interferogram sampled at equally spaced points (Chapter 7). The spectrum in a form suitable for numerical computation is

$$B(\sigma) = \sum_{j=-N/2}^{(N/2)-1} A(j\,\Delta\delta, L)[I_R(j\,\Delta\delta) - I_R(\infty)] \exp(-i2\pi\sigma j\,\Delta\delta). \tag{16-2}$$

Recall from Chapter 5 that $(\Delta\delta)N/2 = L$ is the total optical path displacement of the movable mirror from the origin and $A(j\,\Delta\delta, L)$ is the apodizing function. The first step in the calculation is to subtract $I_R(\infty)$ from each point of the measured values $I_R(j\,\Delta\delta)$ to compute the interferogram. The magnitude of $I_R(\infty)$ is usually determined in one of two ways. One is to assume that $I_R(\infty)$ is the average of the last few points of the interferogram. That is, we may compute $I_R(\infty)$ from

$$I_R(\infty) \simeq [1/(n+1)] \sum_{j=-N/2}^{-(N/2)+n} I_R(j\,\Delta\delta), \tag{16-3}$$

for some $n \ll N$.

Alternately, we may use the average value of the interferogram given by

$$I_R(\infty) \simeq (1/N) \sum_{j=-N/2}^{(N/2)-1} I(j\,\Delta\delta). \tag{16-4}$$

Equation (16-4) is a good approximation for $I_R(\infty)$ because the interferogram contains as many constructive as destructive interference points.

After we have subtracted $I_R(\infty)$ and multiplied by the apodizing function, we have

$$S(j) \equiv A(j\,\Delta\delta, L)[I_R(j\,\Delta\delta) - I_R(\infty)\ ], \tag{16-5}$$

and we wish to calculate

$$B(\sigma) = \sum_{j=-N/2}^{(N/2)-1} S(j) \exp(-i2\pi\sigma j\,\Delta\delta). \tag{16-6}$$

Clearly, $S(j)$ is real since the measured interferogram is real. Ordinarily, we have a set of wave numbers, say $\{\sigma_k\}$, for which we wish to calculate $\{B(\sigma_k)\}$ from Eq. (16-6). This can be done in two ways. We could take $S(j)$ and calculate its contribution, $S(j)\exp(-i2\pi\sigma_k j\,\Delta\delta)$ to the $B(\sigma_k)$ values for all $\sigma_k$; or we can hold $\sigma_k$ fixed and compute the sum as in Eq. (16-6) for each $\sigma_k$ one at a time. The first method is used for real-time analysis, because then we have the $S(j)$ available sequentially in time as the measurement of $I(j\,\Delta\delta)$ is made for each $j$. If the computer is directly connected to the spectrometer, it can compute the contribution of each $S(j)$ to $B(\sigma_k)$ while waiting for the measurement of the next value, i.e., $S(j+1)$, to be completed. We postpone further discussion of real-time analysis to Chapter 18.

Ordinarily, if off-line Fourier transformations are being done using the classical, conventional method, Eq. (16-6) is used for each $\sigma_k$ successively. To do this, we separate the real and imaginary parts of Eq. (16-6) as

$$B(\sigma) = \sum_{j=-N/2}^{(N/2)-1} S(j)\cos(2\pi\sigma j\,\Delta\delta) - i\sum_{j=-N/2}^{(N/2)-1} S(j)\sin(2\pi\sigma j\,\Delta\delta). \tag{16-7}$$

Now, we know that $B(\sigma)$ should be real and that if the sine transform does not vanish for all $\sigma$, it is because of instrumental errors, the co-called phase errors discussed in Chapter 12. There, it was shown that for two-sided interferograms, the linear phase errors can be eliminated by computing [Eq. (12-17)]

$$B_c(\sigma) = [B_{\exp}^*(\sigma)B_{\exp}(\sigma)]^{1/2} \tag{16-8}$$

or

$$B_c(\sigma) = \left\{\left[\sum_{j=-N/2}^{(N/2)-1} S(j)\cos(2\pi\sigma j\,\Delta\delta)\right]^2 + \left[\sum_{j=-N/2}^{(N/2)-1} S(j)\sin(2\pi\sigma j\,\Delta\delta)\right]^2\right\}^{1/2}. \tag{16-9}$$

Note that if $S(j)$ is symmetric about the origin at $j = 0$, then the sine transformation vanishes, and so $B_c(\sigma) = B(\sigma)$. For a one-sided interferogram, we assume that $S(j)$ is symmetric because we have no information about the asymmetry. Hence, we have

$$B(\sigma) = \sum_{j=0}^{(N/2)-1} S(j)\cos(2\pi\sigma j\,\Delta\delta) \tag{16-10}$$

for one-sided interferograms (with a total of $N/2$ samples in the interferogram).

Two alternatives exist for computing the required sines and cosines. One can make a table of cosines (or sines) for one period (or in just the first quadrant) and then reduce the argument $2\pi\sigma j\,\Delta\delta$ to the first period and look up the cosine (or sine) in the table. This method is usable if the $\sigma_k$ are equally spaced so $\sigma_k = k\,\Delta\sigma$.

Alternatively, one can compute the sine or cosine required for each instance. To save computation time, one uses the Chebyshev recursion formulas,

$$\sin[(n+1)\,\Delta x] = 2\sin(n\,\Delta x)\cos(\Delta x) - \sin[(n-1)\,\Delta x] \tag{16-11}$$

and

$$\cos[(n+1)\,\Delta x] = 2\cos(n\,\Delta x)\cos(\Delta x) - \cos[(n-1)\,\Delta x]. \tag{16-12}$$

Cumulative round-off errors require that one compute a new cosine (and sine) periodically using the series expansion.

## CONVENTIONAL VERSUS COOLEY–TUKEY COMPUTATIONS

Let us consider the number of computations required if we use the conventional, classical method. We shall neglect the time required to compute the sines and cosines needed. Then, for each $B(\sigma_k)$, we do $N$ multiplications for $[S(j)\cos(2\pi_k j\,\Delta\delta)]$ and $N$ multiplications for $[S(j)\sin(2\pi\sigma_k j\,\Delta\delta)]$, for a total of $2N$ multiplications. To add the values, we must do $N$ additions for the cosine transformation and the same number for the sine transformation. Hence, $4N$ operations are required for each $\sigma_k$. If we wish the entire spectrum, we must compute $B(\sigma_k)$ for at least $N/2$ wave numbers; so, $4N(N/2) = 2N^2$ operations are required for the conventional, classical method.

We should note that computing $B(\sigma_k)$ for $N/2$ equally spaced $\sigma_k$ gives us points at

$$\Delta\sigma = 1/2L. \tag{16-13}$$

Recall from Eq. (6-6) that this is the resolution. Equation (16-13) will be discussed in the text leading to Eq. (17-4). It can be shown that we can interpolate between these points if we wish to have spacings in the spectrum less than those specified by Eq. (16-13). Interpolation procedures are discussed by Connes [1]. The interpolation procedure will be faster than computing the Fourier transform for a smaller spacing.

We shall present the Cooley–Tukey algorithm in detail in the next chapter, but we now wish to present some of the reasons for comparing it with the classical calculation. Given a two-sided interferogram of $N$ points, the Cooley–Tukey algorithm will provide $B(\sigma)$ at $N/2$ points,

$$\sigma = \Delta\sigma,\quad 2\,\Delta\sigma,\quad 3\,\Delta\sigma, \ldots,\quad \sigma_{\max}, \tag{16-14}$$

where

$$\Delta\sigma = \sigma_{\max}/(N/2) \tag{16-15}$$

and

$$\sigma_{\max} = 1/(2\,\Delta\delta). \tag{16-16}$$

For the Cooley–Tukey algorithm as applied to this presentation, $N$ must be an integer power of 2, i.e., $N = 2^n$ for some integer $n$. The important feature, however, is that the number of operations required is about $2N \ln_2 N$ (as shown in Chapter 17). This should be compared to the conventional, classical method, which requires $2N^2$ operations. Thus, the computation time ratio is

$$T_{\text{classical}}/T_{\text{Cooley–Tukey}} \simeq 2N^2/3N \ln_2 N = 0.46N/\ln_e N. \tag{16-17}$$

Table 16-1 contains a comparison of this predicted ratio with that actually obtained by Forman [2]. Note that the experimental ratio is not quite as large as predicted by Eq. (16-17). In actual practice, there are other operations, such as computing $S(j) = A(j\,\Delta\delta, L)[I_R(j\,\Delta\delta) - I_R(\infty)]$, printing and/or plotting the results, ratioing sample and background spectra, etc. Such operations add terms which are proportional to $N$ to our time estimates.

It is obvious from the table that the Cooley–Tukey algorithm is orders of magnitude faster than the classical method. As the number of data points

TABLE 16-1

Comparison of Classical Procedure and Cooley–Tukey Algorithm

| $n$ | $N = 2^n$ | Theory $T_{\text{Classical}}/T_{\text{Cooley–Tukey}}$ | Experiment[a] $T_{\text{Classical}}/T_{\text{Cooley–Tukey}}$ | Time required on IBM 360/75 with 256K memory using C–T algorithm[b] (sec) |
|---|---|---|---|---|
| 7 | 128 | 12.2 | — | — |
| 8 | 256 | 21.3 | — | — |
| 9 | 512 | 38.0 | 32 | 0.22 |
| 10 | 1,024 | 68.2 | 57 | 0.44 |
| 11 | 2,048 | 124 | 103 | 0.90 |
| 12 | 4,096 | 227 | 190 | 1.85 |
| 13 | 8,192 | 420 | — | 3.82 |
| 14 | 16,384 | 780 | — | 7.86 |
| 15 | 32,768 | 1,450 | — | 16.00 |
| 16 | 65,536 | 2,520 | — | 33.44 |

[a] Forman [2].
[b] Connes [1].

to be Fourier-transformed increases, the Cooley–Tukey efficiency climbs rapidly! This table illustrates why the Cooley–Tukey algorithm means so much to Fourier transform spectroscopy.

## CONCLUSIONS

In this chapter, we have shown how the inverse Fourier transform of the interferogram is to be computed by the classical methods to obtain the spectrum. We showed how $I_R(\infty)$ is to be estimated by either Eq. (16-3) or Eq. (16-4), and then we presented Eq. (16-9) for the computation of the spectrum from a two-sided interferogram. A one-sided interferogram may be computed from Eq. (16-10). Two methods of obtaining cosine and sine values rapidly were presented.

Finally, the conventional, classical method of performing complex Fourier transforms was compared with the speed of the Cooley–Tukey algorithm. The results in Table 16-1 show that the Cooley–Tukey algorithm is orders of magnitude faster than the classical method. In addition, the Cooley–Tukey algorithm becomes more efficient with increasing numbers of data points!

### *References*

*1.* J. Connes, Computing Problems in Fourier Spectroscopy, *Aspen Int. Conf. on Fourier Spectrosc., 1970*, p. 83. (G. A. Vanasse, A. T. Stair, Jr., and D. J. Baker, eds.). AFCRL-71-0019, 5 Jan. 1971, Spec. Rep. No. 114.

*2.* M. L. Forman, Fast Fourier Transform Technique and Its Application to Fourier Spectroscopy, *J. Opt. Soc. Amer.* **56**, 978 (1966).

CHAPTER SEVENTEEN

# The Cooley–Tukey Algorithm

Ralph W. Alexander†
and
Robert J. Bell

## INTRODUCTION

The classical method of computing the discrete Fourier transform has been replaced for most purposes by the Cooley–Tukey algorithm. The reason for this, as we have seen in the last chapter, is the resultant enormous savings in computation time. This algorithm was originally discovered by Good [1]. However, its practical value was not realized until Cooley and Tukey [2] pointed out its advantages (using an arbitrary base number system). Even then, it was not until Forman [3] called attention to the tremendous savings it offered for Fourier spectroscopists that it became widely adopted. This algorithm extended the usefulness of Fourier spectroscopy into the near-infrared and visible wavenumber regions. Previously, high-resolving-power work was limited to the far-infrared partly because the computation problem was virtually impossible for higher wave numbers.

We shall begin this chapter with a short detour to review the binary number system. A brief acquaintance with the binary number system is necessary if one is to understand the details of the Cooley–Tukey algorithm. With the results of this introduction at hand, we will then turn to the Cooley–Tukey algorithm for a special case where the number of points $N$ is eight. This example is then generalized for $N = 2^n$, where $n$ is a positive integer. Because the general Cooley–Tukey algorithm yields the complex Fourier transform of a complex function, we end the chapter by considering two special cases, namely the complex Fourier transform of a real function and the real Fourier transform of a real, even function.

† Physics Department, University of Missouri-Rolla.

Another description of the Cooley–Tukey algorithm was given by Forman, as just mentioned. His version of the algorithm has the disadvantage that approximately twice as much storage space in a computer is required as for the version originally described by Cooley and Tukey. Brigham and Morrow [4] have also provided a good explanation of the algorithm, and it is somewhat similar to the method described in this chapter. The reader will find that the "tree diagrams" for Brigham's version and for the method given here are quite similar. The description given here follows one given by J. Connes [5].

In the first section, we will introduce the binary number system at a very elementary level. In the next section, we will rearrange the inverse Fourier transform of the interferogram so that it is in a form suitable for numerical computation. Then, with an example interferogram containing a total of eight data points, we demonstrate the Cooley–Tukey algorithm in the section, Cooley–Tukey Algorithm for $N = 8$. We show that the Cooley–Tukey algorithm essentially depends on repeatedly separating the interferogram sample points into two groups. The groups are identified by whether or not the index which numbers the data points is odd or even. With the separation into groups, it will be shown that many multiplications and additions are avoided in the computation of the spectrum. The material in the section is well demonstrated with tables and figures. In the next section, we will generalize from an eight-point interferogram to one with $2^n$ data points and will demonstrate the Cooley–Tukey algorithm. The storage locations of the data points will again be stressed, and it is shown that "binary bit inversion" can be used for the problem of correctly locating the entire set of interferogram points. Then, there is a section on the use of the real, inverse Fourier transform of real functions. Since the interferogram is a real function, we can use the method outlined in the section to reduce the computer time by about another factor of two.

## INTRODUCTION TO THE BINARY NUMBER SYSTEM

For the reader unfamiliar with the binary number system, this section provides a very elementary introduction. In our customary base-10 number system, the number 125.0 represents $[(1 \times 10^2) + (2 \times 10^1) + (5 \times 10^0)]$, i.e., each place in the number symbolizes a power of ten. By convention, the first place to the left of the "decimal" point has $10^0$ as a multiplier, the second place to the left has $10^1$ as a multiplier, etc. Similarly, the first place to the right of the decimal point has $10^{-1}$ for a multiplier; the second to the right has $10^{-2}$; etc. In the binary or base-2 number system, each place in the number represents a power of two. The first place to the left of the "binary" point has the multiplier $2^0$; the second place to the left has the multiplier $2^1$; etc. Again, places to the right of the binary point have negative powers; the first place to the right has the multiplier $2^{-1}$, and so forth. Thus, for example, in the binary number system, 5.75 is represented by 101.11, which is an equivalent to $[(1 \times 2^2) + (0 \times 2^1) + (1 \times 2^0) + (1 \times 2^{-1}) + (1 \times 2^{-2})]$.

In the following sections, we shall mainly be interested in integers. Let us consider an integer that is even. Since an even number is divisible by two, the $2^0 = 1$ place must contain a zero. That is, the binary representation of an even integer ends in 0. Hence, an odd integer in the binary notation ends with a 1.

## PREPARING FOR THE COOLEY–TUKEY ALGORITHM

Before considering the Cooley–Tukey algorithm, we shall further simplify the notation in Eq. (16-6). It will be seen that the Cooley–Tukey algorithm leads naturally to calculating the Fourier transform for a set of $\sigma$ that is given by $\{\sigma_{max}/(N/2), \sigma_{max}/[(N/2) - 1], \ldots, k\sigma_{max}/(N/2), \ldots, \sigma_{max}\}$. This is a set of $N/2$ values of $\sigma$. The sampling theory in Chapter 7 dictated that the maximum sampling path step that could be used is [Eq. (7-21)]

$$\Delta\delta = 1/2\sigma_{max}, \tag{17-1}$$

where $\sigma_{max}$ is the maximum wave number in the spectral range which the spectral filtering system passes. Now, Eq. (16-6) can be written (for $\sigma_k = k\,\Delta\sigma$) as

$$B(k\,\Delta\sigma) = \sum_{j=-N/2}^{(N/2)-1} S(j) \exp[-i2\pi k(\Delta\sigma) j\,\Delta\delta]. \tag{17-2}$$

Recall that $S(j) \equiv A(j\,\Delta\delta, L)[I_R(j\,\Delta\delta) - I_R(\infty)]$. With $N/2$ spectral samples, we have

$$\Delta\sigma = \sigma_{max}/(N/2). \tag{17-3}$$

[One also has $\Delta\sigma = 1/2L$ if Eq. (17-1) is substituted into Eq. (17-3), and one notes that $(N/2)\,\Delta\delta$ is equal to $L$ by the definition of $L$, the maximum optical path displacement from the origin. So, $\Delta\sigma$, by Eq. (6-6), also corresponds to the resolution.] Combining Eqs. (17-1) and (17-3), we can simplify the exponential in Eq. (17-2) with

$$(\Delta\delta)(\Delta\sigma) = 1/N. \tag{17-4}$$

Henceforth, we shall denote $B(k\,\Delta\sigma)$ by $B(k)$ and $S(j\,\Delta\delta)$ by $S(j)$, so Eq. (17-2) becomes

$$B(k) = \sum_{j=-N/2}^{(N/2)-1} S(j) \exp(-i2\pi kj/N). \tag{17-5}$$

It is convenient to eliminate negative indices, which can be done easily because of the periodicity of $\exp(-i2\pi kj/N)$. We do this by defining a set of positive $q$ given by $j + N$ for $j < 0$. Then, we have

$$\begin{aligned} B(k) = &\sum_{j=0}^{(N/2)-1} S(j) \exp(-i2\pi kj/N) \\ &+ \sum_{q=N/2}^{N-1} S(q-N) \exp[-i2\pi k(q-N)/N]. \end{aligned} \tag{17-6}$$

We can factor $\exp(-i2\pi k) = 1$ from the second sum, and on combining the two sums, we have

$$B(k) = \sum_{j=0}^{N-1} S(j) \exp(-i2\pi kj/N), \tag{17-7}$$

where we have relabeled the $S(j)$ by adding $N$ to the index of negative $j$.

Because it will be used so often, we make the following definition:

$$W_l^s = (e^{-i2\pi/l})^s. \tag{17-8}$$

## COOLEY–TUKEY ALGORITHM FOR $N=8$

We will now illustrate the Cooley–Tukey algorithm with an example before considering the general result.

Let $N = 8$, so that Eq. (17-7) becomes

$$B(k) = \sum_{j=0}^{7} S(j) \exp(-i2\pi kj/8). \tag{17-9}$$

We wish to break the set of $S(j)$ into those with even $j$ indices, which we label $Y_1(j')$, and those with odd $j$ indices, which we label $Z_1(j')$. Then, for this example, we have

$$Y_1(j') \equiv S(j) \qquad \text{for} \quad j = 2j' = 0, 2, 4, 6; \tag{17-10}$$

$$Z_1(j') \equiv S(j) \qquad \text{for} \quad j = 2j' + 1 = 1, 3, 5, 7. \tag{17-11}$$

This regrouping or relocation of the members of the set $S(j)$ is illustrated in Table 17-1.

TABLE 17-1

Division of the Set of $S(j)$ into Those $S(j)$ with Even-$j$ Indices [the $Y_1(j')$] and Those with Odd-$j$ Indices [the $Z_1(j')$]

| Location | Initial contents of location | Contents after dividing $S(j)$ into $Y_1(j')$ and $Z_1(j')$ | |
|---|---|---|---|
| $0 = 000$ | $S(000)$ | $Y_1(000) = S(000)$ | $j$ is even |
| $1 = 001$ | $S(001)$ | $Y_1(001) = S(010)$ | |
| $2 = 010$ | $S(010)$ | $Y_1(010) = S(100)$ | |
| $3 = 011$ | $S(011)$ | $Y_1(011) = S(110)$ | |
| $4 = 100$ | $S(100)$ | $Z_1(000) = S(001)$ | $j$ is odd |
| $5 = 101$ | $S(101)$ | $Z_1(001) = S(011)$ | |
| $6 = 110$ | $S(110)$ | $Z_1(010) = S(101)$ | |
| $7 = 111$ | $S(111)$ | $Z_1(011) = S(111)$ | |

We will break the set $\{S(j)\}$ into sequences which are dependent on whether $j$ is even or odd. Then, we will divide these new sets $\{Y_1(j')\}$ and $\{Z_1(j')\}$ into odd and even groups, etc. This repetitive division into odd and even groups eliminates many unnecessary additions and multiplications for the computer program. This is the essence in saving computer time with the Cooley–Tukey algorithm. With the repetitive regrouping, one will eventually find that the result is a binary bit inversion of $j$ [which will be discussed after Eq. (17-11)]. In using the Cooley–Tukey algorithm, it is *unnecessary* to go through each regrouping of the terms, as it is accomplished in its entirety by the binary bit inversion. One may relocate the $S(j)$ in the computer through a binary bit inversion of the $j$. The results are illustrated in Table 17-5 and Eq. (17-49).

We shall store the $Y_1(j')$ in locations 0–3 and the $Z_1(j')$ in locations 4–7 The $S(j)$ were initially stored in locations 0–7, and we ask how to rearrange them to have the $Y_1(j')$ in locations 0–3 and the $Z_1(j')$ in locations 4–7. If $j$ is even (say $6 = 110$ in binary), its binary representation ends in 0. But locations 0–3 begin with 0 in a three-digit binary number (three digits being required to represent the eight locations 0–7). Thus, for even $j$, we find the correct location by taking the last binary bit (a "0") and making it first and moving the remaining bits one position to the right. For $6 = 110$, we obtain 011, that is, $S(6)$ goes in location $3 = 011$.

We can now rewrite $B(k)$ in Eq. (17-9) as a sum over even $j$ terms and odd $j$ terms as

$$B(k) = \sum_{j'=0}^{3} Y_1(j') \exp[-i2\pi k(2j')/8] + \sum_{j'=0}^{3} Z_1(j') \exp[-i2\pi k(2j' + 1)/8]. \qquad (17\text{-}12)$$

Letting $G_1(k)$ be the discrete inverse Fourier transform of $Y_1(j')$ (over a set of four elements only), we may define

$$G_1(k) \equiv \sum_{j'=0}^{3} Y_1(j') \exp(-i2\pi kj'/4). \tag{17-13}$$

Similarly, we may define $H_1(k)$ as the inverse Fourier transform of $Z_1(j')$ given by

$$H_1(k) \equiv \sum_{j'=0}^{3} Z_1(j') \exp(-i2\pi kj'/4). \tag{17-14}$$

The factor of 1/4 in the exponential is present because we now have a reduced set of four data points to be transformed. By multiplying and dividing the exponential factors in Eqs. (17-13) and (17-14) by two, we may use these equations to convert Eq. (17-12) into

$$B(k) = G_1(k) + H_1(k) \exp(-i2\pi k/8). \tag{17-15}$$

Using

$$\exp[i2\pi j'(k + 4)/4] = \exp(-i2\pi j'k/4) \exp(-i2\pi j') = \exp(-i2\pi j'k/4), \tag{17-16}$$

we can show from Eq. (17-13) that

$$G_1(k + 4) = \sum_{j'=0}^{3} Y_1'(j) \exp[-i2\pi j'(k + 4)/4] = G(k). \tag{17-17}$$

Similarly, we have

$$H_1(k + 4) = H_1(k). \tag{17-18}$$

Using

$$\exp[-i2\pi(k+4)/8] = -\exp(-i2\pi k/8) \tag{17-19}$$

and Eqs. (17-15), (17-17), and (17-18), we have

$$B(k + 4) = G_1(k) - H_1(k) \exp(-i2\pi k/8). \tag{17-20}$$

Equations (17-15) and (17-20) are illustrated in Table 17-2.

Let us summarize what we have done to this point. We separated the $S(j)$ into a set with even $j$ indices [the $Y_1(j')$] and a set with odd $j$ indices [the $Z_1(j')$]. Then, we found the relation in Eq. (17-15) between $B(k)$ [the inverse Fourier transform of $S(j)$] and $G_1(k)$ [the inverse Fourier transform of $Z_1(j')$]. Also, by knowing $G_1(k)$ and $H_1(k)$, we can compute $B(k + 4)$ from Eq. (17-20). This procedure reduces the number of inverse Fourier transformations by a factor of two because we can obtain both $B(k)$ and $B(k + 4)$ from the same inverse Fourier transforms in Eqs. (17-13) and (17-14).

TABLE 17-2

The Operation of Eq. (17-15)[a]

| Location | Initial contents | Final contents |
|---|---|---|
| 000 | $G_1(000)$ | $B(000) = G_1(000) + W_8{}^0 H_1(000)$ |
| 001 | $G_1(001)$ | $B(001) = G_1(001) + W_8{}^1 H_1(001)$ |
| 010 | $G_1(010)$ | $B(010) = G_1(010) + W_8{}^2 H_1(010)$ |
| 011 | $G_1(011)$ | $B(011) = G_1(011) + W_8{}^3 H_1(011)$ |
| 100 | $H_1(000)$ | $B(100) = G_1(000) - W_8{}^0 H_1(000)$ |
| 101 | $H_1(001)$ | $B(101) = G_1(001) - W_8{}^1 H_1(001)$ |
| 110 | $H_1(010)$ | $B(110) = G_1(010) - W_8{}^2 H_1(010)$ |
| 111 | $H_1(011)$ | $B(111) = G_1(011) - W_8{}^3 H_1(011)$ |

[a] Note the relation between the first set of four $B$'s and the second set of $B$'s, as this is the relation given by Eq. (17-20). The $W_l{}^s$ are defined in Eq. (17-8). This is step 3 of Fig. 17-1.

We now perform upon $Y_1(j')$ the same process of separation into odd and even indices. The $Y_1(j')$ values with even indices become $Y_{21}(j'')$; the $Y_1(j')$ with odd indices become $Z_{21}(j'')$; we have

$$Y_{21}(j'') \equiv Y_1(j') \qquad \text{for} \quad j' = 2j'' = 0, 2, \tag{17-21}$$

$$Z_{21}(j'') \equiv Y_1(j') \qquad \text{for} \quad j' = 2j'' + 1 = 1, 3. \tag{17-22}$$

The division of $Y_1(j')$ into two sets is shown in the *upper half* of Table 17-3.

TABLE 17-3

The Division of the Set $Y_1(j')$ into Those $Y_1(j')$ with $j'$ Even [the $Y_{21}(j'')$] and Those with $j'$ Odd [the $Z_{21}(j'')$], and the Division of the Set $Z_1(j')$ into the $Y_{22}(j'')$ and $Z_{22}(j'')$

| Location | Initial contents $Y_1(j')$ | New contents after dividing $Y_1(j')$ into $Y_{21}(j'')$ and $Z_{21}(j'')$ and $Z_1(j')$ into $Y_{22}(j'')$ and $Z_{22}(j'')$ |
|---|---|---|
| 000 | $Y_1(000)$ | $Y_{21}(000) = Y_1(000)$ |
| 001 | $Y_1(001)$ | $Y_{21}(001) = Y_1(010)$ |
| 010 | $Y_1(010)$ | $Z_{21}(000) = Y_1(001)$ |
| 011 | $Y_1(011)$ | $Z_{21}(001) = Y_1(011)$ |
| 100 | $Z_1(000)$ | $Y_{22}(000) = Z_1(000)$ |
| 101 | $Z_1(001)$ | $Y_{22}(001) = Z_1(010)$ |
| 110 | $Z_1(010)$ | $Z_{22}(000) = Z_1(001)$ |
| 111 | $Z_1(011)$ | $Z_{22}(001) = Z_1(011)$ |

We have rearranged the $Y_1(j')$, in the right-hand column, only among themselves, and hence they remain in locations 0–3 (so the binary representation of the location still begins on the left with a zero). If $j'$ was even, it ended in zero and $Y_1(j')$ was relocated in location 0 or 1, in the "new contents" column, which have a zero for the middle digit. Thus, by putting the last digit on the right in $j'$ (in the "initial contents" column) in the middle and moving the middle digit to the right, we obtain the location of $Y_{21}(j'')$. This is easily verified by examining Eqs. (17-21) and (17-22) and Table 17-3. It can be seen that for $j'$ odd, we put $Y_{21}(j'')$ into the location obtained by moving the last digit of $j'$ to the middle, and the middle digit one position to the right. For example, $Y_1(3)$, which has $j' = 011$, should be moved to location $011 = 3$ to become $Z_{21}(1)$.

We now compute $G_{21}(k)$, the inverse Fourier transform of $Y_{21}(j'')$, which is defined by

$$G_{21}(k) \equiv \sum_{j''=0}^{1} Y_{21}(j'') \exp(-i2\pi kj''/2) = Y_{21}(0) + Y_{21}(1) \exp(-i2\pi k/2). \tag{17-23}$$

Likewise, the inverse Fourier transform of $Z_{21}(j'')$ is given by

$$H_{21} \equiv \sum_{j''=0}^{1} Z_{21}(j'') \exp(-i2\pi kj''/2) = Z_{21}(0) + Z_{21}(1) \exp(-i2\pi k/2). \tag{17-24}$$

A comparison of Eqs. (17-23) and (17-24) with the Eq. (17-13) for $G_1(k)$ shows that

$$G_1(k) = G_{21}(k) + H_{21}(k)e^{-i2\pi k/4}, \tag{17-25}$$

where we have also used Eqs. (17-21) and (17-22). We can easily show from the definition in Eq. (17-23) that we have the periodicity relation given by

$$G_{21}(k + 2) = G_{21}(k) \tag{17-26}$$

and from Eq. (17-24) that

$$H_{21}(k + 2) = H_{21}(k). \tag{17-27}$$

From these last three equations, we can show that

$$G_1(k + 2) = G_{21}(k) - H_{21}(k)e^{-i2\pi k/4}. \tag{17-28}$$

This last equation means only two multiplications are necessary to find $G_1(k)$ for four values of $k$ ($k = 0, 1, 2, 3$). Equations (17-23) and (17-24) also yield

$$G_{21}(k + 1) = Y_{21}(0) - Y_{21}(1)e^{-i2\pi k/2} \tag{17-29}$$

and

$$H_{21}(k + 1) = Z_{21}(0) - Z_{21}(1)e^{-i2\pi k/2}. \tag{17-30}$$

We have $G_{21}(k+1)$ available from the same set $[Y_{21}(0)$ and $Y_{21}(1)]$ that determines $G_{21}(k)$; likewise for $H_{21}(k+1)$ and $H_{21}(k)$. These last two equations again mean fewer multiplications. One multiplication only is required to obtain $G_{21}(0)$ and $G_{21}(1)$. Table 17-4 illustrates Eqs. (17-25) and (17-28).

TABLE 17-4

The Operations Required to Obtain $G_1(k)$ from $G_{21}(k)$ and $H_{21}(k)$, and $H_1(k)$ from $G_{22}(k)$ and $H_{22}(k)$[a]

| Location | Initial contents | Final contents |
|---|---|---|
| 000 | $G_{21}(000)$ | $G_1(000) = G_{21}(000) + W_4{}^0 H_{21}(000)$ |
| 001 | $G_{21}(001)$ | $G_1(001) = G_{21}(001) + W_4{}^1 H_{21}(001)$ |
| 010 | $H_{21}(000)$ | $G_1(010) = G_{21}(000) - W_4{}^0 H_{21}(000)$ |
| 011 | $H_{21}(001)$ | $G_1(011) = G_{21}(001) - W_4{}^1 H_{21}(001)$ |
| 100 | $G_{22}(000)$ | $H_1(000) = G_{22}(000) + W_4{}^0 H_{22}(000)$ |
| 101 | $G_{22}(001)$ | $H_1(001) = G_{22}(001) + W_4{}^1 H_{22}(001)$ |
| 110 | $H_{22}(000)$ | $H_1(010) = G_{22}(000) - W_4{}^0 H_{22}(000)$ |
| 111 | $H_{22}(001)$ | $H_1(011) = G_{22}(001) - W_4{}^1 H_{22}(001)$ |

[a] Note, for example, the relation between $G_1(000)$ and $G_1(010)$, as this is the relation given by Eq. (17.28) for $G_1(k)$. This table is step 2 of Fig. 17-1.

We can now devote attention to the lower half of Table 17-3. By dividing the set $Z_1(j')$ into two sets, one with even $j'$ [which we call $Y_{22}(j'')$] and one with odd $j'$ [which we call $Z_{22}(j'')$], we can do for $H_1(k)$ what we have just done for $G_1(k)$. Thus, we put

$$Y_{22}(j'') = Z_1(j') \qquad \text{for} \quad j' = 2j'' = 0, 2, \tag{17-31}$$

$$Z_{22}(j'') = Z_1(j') \qquad \text{for} \quad j' = 2j'' + 1 = 1, 3. \tag{17-32}$$

Let $G_{22}(k)$ be the Fourier transform of the $Y_{22}(j'')$. Explicitly, we can write

$$G_{22}(k) = Y_{22}(0) + Y_{22}(1)e^{-i2\pi k/2}. \tag{17-33}$$

Similarly, $H_{22}(k)$ will be the inverse Fourier transform of $Z_{22}(j')$ and

$$H_{22}(k) \equiv Z_{22}(0) + Z_{22}(1)e^{-i2\pi k/2}. \tag{17-34}$$

In the same manner as for obtaining $G_1(k)$ in Eq. (17-25), we can show from Eqs. (17-33), (17-34), and (17-14) that

$$H_1(k) = G_{22}(k) + H_{22}(k)e^{i2\pi k/4}, \tag{17-35}$$

where we have also used Eqs. (17-31) and (17-32). Again, two multiplication-saving relations exist, namely

$$G_{22}(k+1) = Y_{22}(0) - Y_{22}(1)e^{-i2\pi k/2} \tag{17-36}$$

and

$$H_{22}(k+1) = Z_{22}(0) - Z_{22}(1)e^{-i2\pi k/2}. \tag{17-37}$$

Also, from Eqs. (17-33) and (17-34), we have the periodicity relations

$$G_{22}(k+2) = G_{22}(k) \tag{17-38}$$

and

$$H_{22}(k+2) = H_{22}(k). \tag{17-39}$$

The reader is referred to Table 17-5 for an illustration of Eqs. (17-35), (17-38), and (17-39).

TABLE 17-5

The Operations Required to Obtain $G_{21}(k)$, $H_{21}(k)$, $G_{22}(k)$, and $H_{22}(k)$ from $Y_{21}(j'')$, $Z_{21}(j'')$, $Y_{22}(j'')$, and $Z_{22}(j'')$.[a]

| Location | Initial contents | Final contents |
|---|---|---|
| 000 | $Y_{21}(000) = S(000)$ | $G_{21}(0) = Y_{21}(000) + W_2{}^0 Y_{21}(001)$ |
| 001 | $Y_{21}(001) = S(100)$ | $G_{21}(1) = Y_{21}(000) - W_2{}^0 Y_{21}(001)$ |
| 010 | $Z_{21}(000) = S(010)$ | $H_{21}(0) = Z_{21}(000) + W_2{}^0 Z_{21}(001)$ |
| 011 | $Z_{21}(001) = S(110)$ | $H_{21}(1) = Z_{21}(000) - W_2{}^0 Z_{21}(001)$ |
| 100 | $Y_{22}(000) = S(001)$ | $G_{22}(0) = Y_{22}(000) + W_2{}^0 Y_{22}(001)$ |
| 101 | $Y_{22}(001) = S(101)$ | $G_{22}(1) = Y_{22}(000) - W_2{}^0 Y_{22}(001)$ |
| 110 | $Z_{22}(000) = S(011)$ | $H_{22}(0) = Z_{22}(000) + W_2{}^0 Z_{22}(001)$ |
| 111 | $Z_{22}(001) = S(111)$ | $H_{22}(1) = Z_{22}(000) - W_2{}^0 Z_{22}(001)$ |

[a] This is step 1 of Fig. 17-1.

By now, the reader may be asking why we have gone through this procedure. In answer, we note that $B(k)$ may be calculated in the following manner. We first calculate $G_{21}(k)$, $G_{22}(k)$, $H_{21}(k)$, and $H_{22}(k)$ from the $S(j)$. For example, if we wished to calculate $G_{21}(k)$, we would compute it from Eq. (17-23), which was the $Y_{21}(0)$ and $Y_{21}(1)$ defined by Eq. (17-21), which, in turn, has $Y_1(0)$ and $Y_2(2)$ defined by Eq. (17-10) in terms of $S(0)$ and $S(4)$.

That is, we can compute $G_{21}(k)$ with a simple calculation by merely locating which $S(j)$ we desire. In this example, we would have

$$G_{21}(k) = S(0) + S(4)e^{-i2\pi k/2}. \tag{17-40}$$

Then, with $G_{21}(k)$, $G_{22}(k)$, $H_{21}(k)$, and $H_{22}(k)$, we calculate $G_1(k)$ and $H_1(k)$. Finally, $B(k)$ is computed from $G_1(k)$ and $H_1(k)$.

How many operations are required? To calculate $G_{21}(0)$ from $S(0)$ and $S(4)$, we require one (complex) multiplication, i.e., $\{\exp[-i2\pi(0)/2]\}S(4)$, and one (complex) addition, i.e., $S(0) + \{\exp[-i2\pi(0)/2]\}S(4)$. We obtain $G_{21}(1)$ with only one (complex) addition, i.e., $S(0) - \{\exp[-i2\pi(0)/2]\}S(4)$. So, calculating $G_{21}(0)$ and $G_{21}(1)$ requires one multiplication and two additions. Similarly, calculating $H_{21}(k)$, $G_{22}(k)$, and $H_{22}(k)$ for $k = 0$ and $k = 1$ requires one complex multiplication and two complex additions. Hence, the first step of calculating $G_{21}(k)$, $H_{21}(k)$, $G_{22}(k)$, and $H_{22}(k)$ requires four multiplications and eight additions. This is summarized by Table 17-5 or step 1 of Fig. 17-1.

The next step in calculating $G_1(k)$ and $H_1(k)$ from $G_{21}(k)$, $H_{21}(k)$, $G_{22}(k)$, and $H_{22}(k)$ can be seen also to require four multiplications and eight additions. This step is illustrated in Table 17-4 or step 2 in Fig. 17-1. The final step in calculating $B(k)$ from $G_1(k)$ and $H_1(k)$ requires an additional four multiplications and eight additions (Table 17-2 or step 3 in Fig. 17-1). Hence, a total of 36 operations are required. If we had $N = 2^n$ (instead of $N = 8$ and $N = 3$), each of the $n$ steps would require $N$ additions and $N/2$ multiplications. Thus, $n$ steps would be required, and hence $n(N + \frac{1}{2}N) = 1.5nN$ operations would be necessary. On using $n = \log_2 N$, we have the total number of operations $M$ given by

$$M \simeq 3N \log_2 N, \tag{17-41}$$

where we have put one *complex* addition (multiplication) equal to two *real* additions (multiplications). These steps are summarized in Fig. 17-1.

Although in the above treatment we have successively divided the $S(j)$ into sets of odd or even indices, the only step in computing the Fourier transform which required the $S(j)$ was step 1 of Fig. 17-1 (or Table 17-5). For this step, the $S(j)$ were labeled as $Y_{21}(j'')$, $Z_{21}(j'')$, $Y_{22}(j'')$, and $Z_{22}(j'')$. We now summarize the process of obtaining these from the $S(j)$. In separating the set of $S(j)$ into two sets $Y_1(j')$ and $Z_1(j')$, we put $(j = j_2 j_1 j_0)$ in location $j_0 j_2 j_1$ (see Table 17-1). [Recall that $j_0 = 0$ for $Y_1(j')$ and $j_0 = 1$ for $Z_1(j')$.] When we separated the $Y_1(j')$ into two sets $Y_{21}(j'')$ and $Z_{21}(j'')$, we put $Y_1(j' = j_0 j_2 j_1)$ into location $j_0 j_1 j_2$. Thus, $S(j = j_2 j_1 j_0)$ must be initially located in $j_0 j_1 j_2$. Table 17-6 shows this in detail (see also Fig. 17-1). It can be seen that the initial location of $S(j)$ can be obtained by interchanging the binary digits of $j$ symmetrically about the middle digit. This process has been called "binary

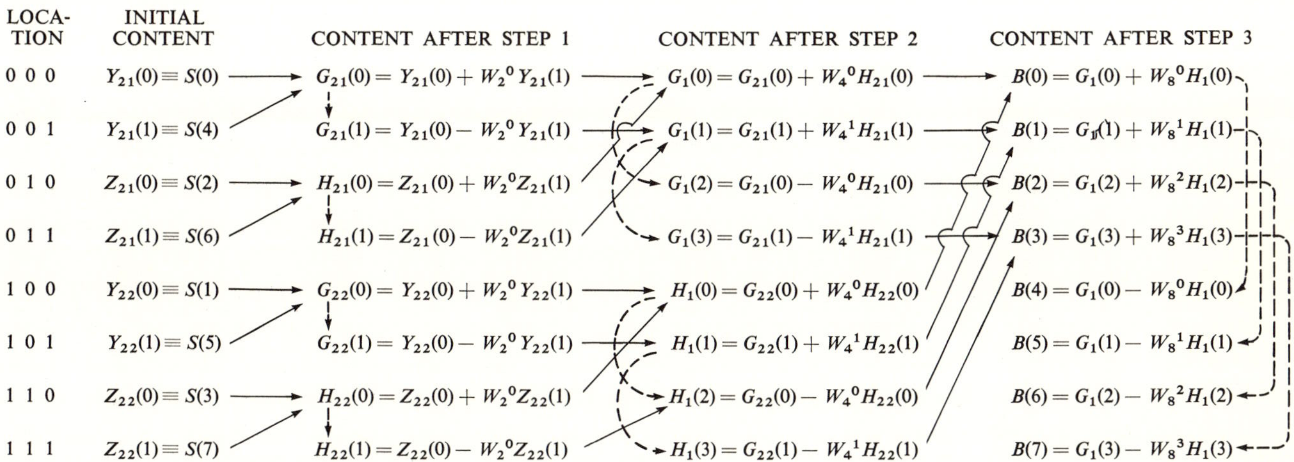


FIG. 17-1. A "tree graph" for $N = 8$. The contents of the locations 0–7 are given for each of the three steps ($N = 3$) required to compute the inverse Fourier transform using the Cooley–Tukey alogorithm. Single arrows connect locations required for each step, while dashed arrows indicate locations with a periodic relation such as $G_{21}(0) = Y_{21}(0) + W_2{}^0 Y_{21}(1)$ and $G_{21}(1) = Y_{21}(0) - W_2{}^0 Y_{21}(1)$.

TABLE 17-6

Illustration of Binary Bit Inversion

| $j$ | | Location of $S(j)$ | |
|---|---|---|---|
| Decimal | Binary | Decimal | Binary |
| 0 | 000 | 0 | 000 |
| 1 | 001 | 4 | 100 |
| 2 | 010 | 2 | 010 |
| 3 | 011 | 6 | 110 |
| 4 | 100 | 1 | 001 |
| 5 | 101 | 5 | 101 |
| 6 | 110 | 3 | 011 |
| 7 | 111 | 7 | 111 |

bit inversion" in the literature. Because we need the $S(j)$ only for step 1 (see Fig. 17-1), we may ignore the intermediate steps in the process of dividing the $S(j)$ (i.e., those in Tables 17-1 and 17-3) and initially store $S(j)$ in the location given by the binary inversion of $j$.

Notice that our method of deriving the Cooley–Tukey algorithm depended upon taking the original sum and separating it into two sums (one over even $j$, one over odd $j$). This process was repeated until the sums resulting contained only two terms [$G_{11}(k)$, $G_{22}(k)$ and $H_{11}(k)$, $H_{22}(k)$ in our $N = 8$ example]. This requires that $N = 2^n$ in our example.† This is one important restriction upon this algorithm. However, in one sense this is an unimportant restriction. If $2^{n-1} < N < 2^n$, one can put $S(j) = 0$ for $N < j \leq 2^n - 1$ and not change the resulting inverse Fourier transform. This is easily seen by considering the sum in Eq. (17-5).

It should be pointed out that this method of calculating the inverse Fourier transform is suitable for computer calculation. This is because the basic operation required is the same for all steps, as can be seen from Fig. 17-1. The figure provides a summary of what we have done above and includes Tables 17-2 and 17-4. Such a diagram as in Fig. 17-1 has been called a "tree graph" by Brigham. Note that the tree graph shows that we need $2N$ storage locations [two locations for each $S(j)$, one for the real part and one for the imaginary part]. In performing each step, we can restore the newly computed result on top of the old contents, so $4N$ additional locations are *not* needed.

† Cooley and Tukey [2] derived the algorithm for any base number system and found that the base-3 system is the most efficient. However, the base-2 system is only slightly less efficient and much more convenient on a computer.

## GENERALIZATION FOR $N=2^n$

Let us consider generalizing our example and putting $N = 2^n$ for some integer $n$. We again separate $S(j)$ into two sets, those with even $j$ indices and those with odd $j$ indices. Dropping the subscripts of 1, we have

$$Y(j') \equiv S(j) \quad \text{for} \quad j = 2j', \quad j' = 0, 1, \ldots, (N/2) - 1, \tag{17-42}$$

$$Z(j') \equiv S(j) \quad \text{for} \quad j = 2j' + 1, \quad j' = 0, 1, \ldots, (N/2) - 1. \tag{17-43}$$

As before, we let

$$G(k) \equiv \mathscr{F}^{-1}\{Y(j')\} \equiv \sum_{j'=0}^{(N/2)-1} Y(j') \exp[-i2\pi j'k/(N/2)] \tag{17-44}$$

and

$$H(k) \equiv \mathscr{F}^{-1}\{Z(j')\} \equiv \sum_{j'=0}^{(N/2)-1} Z(j') \exp[-i2\pi j'k/(N/2)]. \tag{17-45}$$

Clearly, $G(k)$ and $H(k)$ are periodic in $k$ with period $N/2$. In a manner similar to that used above for the $N = 8$ case, we find

$$B(k) = \sum_{j=0}^{N-1} S(j)e^{-i2\pi jk/N} = G(k) + W_N{}^k H(k) \tag{17-46}$$

and

$$B(k + N/2) = G(k) - W_N{}^k H(k). \tag{17-47}$$

It thus takes $N$ complex additions and $N/2$ complex multiplications to obtain $B(k)$ from $G(k)$ and $H(k)$ for a given $N$ and $k$. As before, we repeat the process on $G(k)$ and $H(k)$, again and again, until we have reduced the problem to $N/2$ pairs of equations of the form

$$\alpha(k) = \beta(0) + W_N{}^0 \beta(1) \tag{17-48}$$

and

$$\alpha(k + 1) = \beta(0) + W_N{}^0 \beta(1). \tag{17-49}$$

In our $N = 8$ example, $\alpha(k)$ corresponds to $G_{21}(k)$, $H_{21}(k)$, etc. and $\beta(j)$ to $Y_{11}(j'')$, etc. This reduction requires $n$ steps, each requiring $N$ additions and $N/2$ (complex) multiplications. Hence, as stated before, $3N \log_2 N$ operations are necessary all together.

As before, we initially break $S(j)$ into two sets with even $j$ going into locations 0 to $(N/2) - 1$, and with odd $j$ going into the locations at $N/2$ to $N - 1$.

That is, again the last binary digit is moved to the first place on the left and the rest are moved to the right one position. For example, if $j = j_{n-1}, j_{n-2}, \ldots, j_0$, after dividing $S(j)$ into odd and even $j$, $S(j)$ is in location $j' = j_0 j_{n-1} \cdots j_2 j_1$. After repeating the separation once more into odd and even $j'$, $S(j)$ is in $j'' = j_0 j_1 j_{n-1} \cdots j_2$. After repeating the process $n$ times, $S(j)$ is in location

$$j^n = j_0 j_1 \cdots j_{(n/2)-1} j_{n/2} j_{(n/2)+1} \cdots j_{n-1}, \tag{17-50}$$

*where $j^n$ means $j$ with $n$ primes*. But, this is just the result obtained by inverting the binary bits as described in the example for $N = 8$. Thus, we need not go through the process of separating $S(j)$ into two sets, etc. We need only initially relocate the $S(j)$ by putting $S(j)$ in the location given by the binary bit inversion of $j$.

Recall that $W_{2^l} = \exp(-i2\pi/2^l)$. Step 1 in Fig. 17-1 requires $W_2{}^0 = 1$. Step 2 requires $W_4{}^0$ and $W_4{}^1$. The last step (step $n$) requires $W_N{}^0$, $W_N{}^1$, $W_N{}^1, \ldots, W_N^{(N/2)-1}$[as can be seen in Eqs. (17-46) and (17-47)]. However, we need only to calculate the set $W_N{}^0$, $W_N{}^1, \ldots, W_n^{(N/2)-1}$ for the entire $n$ steps because we can generate any needed set from

$$W_{2^t}^a = [W_{2^t}^1]^a = [(W_{2^n=N}^1)^{2^{n-t}}]^a. \tag{17-51}$$

To summarize, we first reorder the $S(j)$ by putting $S(j)$ in the location obtained by inverting the binary bits of $j$. Letting $X(\gamma)$ indicate the contents of location $\gamma$, we then replace the contents (for step $t$) with

$$X(\gamma) + W_{2^n}^{\gamma \,(\mathrm{mod}\; t)} X(\gamma + t).$$

The quantity $\gamma(\mathrm{mod}\; t)$ means $\gamma - pt$, where $p$ is the integer such that

$$0 \le \gamma - pt < t,$$

with the understanding that if $\gamma(\mathrm{mod}\; t) \ge t/2$, then we use

$$X(\gamma - t) - W_{2^n}^{t-\gamma\,(\mathrm{mod}\; t)} X(\gamma).$$

In each step $t$, we start with $\gamma = 0$ and proceed until all the new $X(\gamma)$ are calculated. We return to this in more detail in Appendix B for the programming of the algorithm.

## SPECIAL CASE OF $S(j)$ REAL

So far, we have *not* required $S(j)$ to be real. However, our measured interferograms are, of course, real, and so we are wasting effort if we use the above method because we must set $\mathrm{Im}[S(j)]$ equal to zero for all $j$. This inefficiency can be avoided by the following method.

We suppose $S(j)$ is real and again separate $S(j)$ into two sets:

$$Y(j') = S(j) \qquad \text{for} \quad j = 2j', \quad j' = 0, 1, 2, \ldots, (N/2) - 1 \tag{17-52}$$

$$Z(j') = S(j) \qquad \text{for} \quad j = 2j' + 1, \quad j' = 0, 1, 2, \ldots, (N/2) - 1, \tag{17-53}$$

where now $Y(j')$ and $Z(j')$ are real. We construct a function $T(j')$ by

$$T(j') \equiv Y(j') + iZ(j') \qquad \text{for} \quad j' = 0, 2, \ldots, (N/2) - 1. \tag{17-54a}$$

Let $E(k)$ be the inverse Fourier transform of $T(j')$ which is given by

$$E(k) \equiv \sum_{j'=0}^{(N/2)-1} T(j') \exp[-i2\pi j'k/(N/2)]. \tag{17-54b}$$

We put

$$C(k) \equiv \mathscr{F}^{-1}\{Y(j')\} \equiv \sum_{j'=0}^{(N/2)-1} Y(j') \exp[-i2\pi j'k/(N/2)] \tag{17-55}$$

and

$$D(k) \equiv \mathscr{F}^{-1}\{Z(j')\}. \tag{17-56}$$

Since $Y(j')$ and $Z(j')$ are real,

$$E(k) = C(k) + iD(k), \tag{17-57}$$

$$C^*(-k) = C(k), \tag{17-58}$$

and

$$D^*(-k) = D(k). \tag{17-59}$$

Thus, we have

$$E^*(-k) = C(k) - iD(k). \tag{17-60}$$

Using Eqs. (17-54)–(17-56) to solve for $C(k)$ and $D(k)$, we have

$$C(k) = [E(k) + E^*(-k)]/2 \tag{17-61}$$

and

$$D(k) = [E(k) - E^*(-k)]/2i. \tag{17-62}$$

In a manner by now familiar to the reader, we can show that

$$B(k) = C(k) + D(k)e^{-i\pi k/N}. \tag{17-63}$$

Thus, we compute the inverse Fourier transform of $T(j') = X(j') + iY(j')$. This gives $E(k)$, and from Eqs. (17-61) and (17-62), we can find $C(k)$ and $D(k)$. Finally, Eq. (17-63) gives us $B(k)$.

The importance of this result is that for a purely real $S(j)$, we need only half the storage space and about half the computation time. This is because

$T(j')$ requires enough storage for only $N/2$ complex numbers, while the general method required storage for $N$ complex numbers. In this latter case, half the locations were initially filled with zeros [the imaginary parts of $S(j)$]. Because $E(k)$ is the inverse Fourier transform of $N/2$ complex points instead of $N$ points, we save about a factor of two in computation time.

We compute the transform of $T(j)$ and perform one more step—that given by Eq. (17-63). The use of Eq. (17-63) is formally the same as the basic step in Cooley–Tukey algorithm. Thus, the number of operations required for the inverse Fourier transform for $S(j)$ real is

$$(1.5N \log_2 N) + 1.5N,$$

where the additional factor of $1.5N$ results from the computation of Eq. (17-63). Also, only $N$ storage locations are needed instead of $2N$.

The negative indices in Eqs. (17-61) and (17-62) can be eliminated by using the fact that $E(k)$ is periodic with period $N/2$, so

$$E^*(-k) = E^*(\tfrac{1}{2}N - k). \tag{17-64}$$

We have already shown that $C(k)$ and $D(k)$ will be periodic with period $N/2$. Now, we have added the condition that $C(k)$ and $D(k)$ be real. What does this tell us about $B(k)$? Consider the real part of $B(\frac{1}{2}N + k)$. We can show that

$$\mathrm{Re}[B(\tfrac{1}{2}N + k)] = C(\tfrac{1}{2}N + k) - D(\tfrac{1}{2}N + k)\cos[2\pi(\tfrac{1}{2}N + k)/N], \tag{17-65}$$

which on using a trigonometric identity, gives

$$\mathrm{Re}[B(\tfrac{1}{2}N + k)] = C(k) + D(k)\cos(2\pi k/N) = \mathrm{Re}[B(k)]. \tag{17-66}$$

Hence, the real part of $B(k)$ is symmetric about $N/2$. Similarly, for the imaginary part of $B(k)$, we find

$$\begin{aligned}\mathrm{Im}[B(\tfrac{1}{2}N + k)] &= -D(\tfrac{1}{2}N + k)\sin[2\pi(\tfrac{1}{2}N + k)/N] \\ &= D(k)\sin(2\pi k/N),\end{aligned} \tag{17-67}$$

or

$$\mathrm{Im}[B(\tfrac{1}{2}N + k)] = -\mathrm{Im}[B(k)]. \tag{17-68}$$

So, the imaginary part of $B(k)$ is antisymmetric about $N/2$.

If we measure a double-sided interferogram, the $B(k)$ is calculated from

$$B(k) = \{[\mathrm{Re}\, B(k)]^2 + [\mathrm{Im}\, B(k)]^2\}^{1/2}. \tag{17-69}$$

Thus, from Eq. (17-66) and Eq. (17-68), we see that only $N/2$ values of $B(k)$ need be calculated for $B(k)$, as the second $N/2$ set will be identical with the first.

## SPECIAL CASE OF $S(j)$ REAL AND EVEN

A properly adjusted interferometer will produce a interferogram which is symmetric about zero displacement. That is, if a sample is taken at $\delta = 0$, then $F(-j\,\Delta\delta) = F(j\,\Delta\delta)$. Hence, if we know *a priori* that the interferogram is symmetric, then we need measure only one-half as many points, say only for $j\,\Delta\delta \geq 0$. Having done this, we could use the symmetry property to find $F(j\,\Delta\delta)$ for $j\,\Delta\delta < 0$. Recall from Chapter 12 that the phase error can be calculated from a short, two-sided interferogram, and this calculated phase error can be used to correct the spectrum computed from a long, one-sided interferogram. The method now to be described can be used to calculate the one-sided interferogram. In either case, to use the method described in the last section for a real function would require $N$ storage locations. In the following, we describe a method first published by J. Connes [5] which requires only $N/2$ storage locations. This section will contain considerably less detail than those preceding.

To keep our notation consistent with the previous sections, we shall assume that the symmetric, two-sided interferogram consists of $N$ points. After eliminating negative $j$ indices as done previously in Eq. (17-7), the symmetry property is $S(\frac{1}{2}N - j) = S(\frac{1}{2}N + j)$, where $S(j)$ has been defined by Eq. (16-5). The redundancy in the interferogram appears in the Fourier transform as a consequence of the imaginary part of the Fourier transform vanishing.

In order to save storage space and computation time, we shall define a new function $J(j)$ via an intermediate function $H(j)$. [Note that $H(j)$ differs from that used in the previous sections of this chapter.] These functions will serve to eliminate the redundant information that is contained in $S(j)$. We define

$$H(j) = S(2j) + (i/2)[S(2j+1) - S(2j-1)]. \tag{17-70}$$

$S(j)$ was a real function defined for $N$ points, while $H(j)$ is a complex function defined for $N/2$ points. In this manner, we have eliminated the zeros which made up the imaginary part of $S(j)$. $H(j)$ is a Hermitian function, i.e., $H(\frac{1}{4}N - j) = H^*(\frac{1}{4}N + j)$. This can be easily shown from the definition of $H(j)$ and the symmetry of $S(j)$. Recall that $S(j)$ is periodic in $N$, i.e., $S(N + j) = S(N - j)$ and is a real and even function, so $S(\frac{1}{2}N + j) = S(\frac{1}{2}N - j)$. $H(j)$ still contains redundant information because the real part of $H(j)$ is an even function, and the imaginary part is an odd function. Again, this redundancy is reflected in the inverse Fourier transform, which is purely real. Now, let $G(k)$ be the inverse Fourier transform of $H(j)$, so

$$G(k) = \sum_{j=0}^{(N/2)-1} H(j)e^{-i2\pi kj/(N/2)}. \tag{17-71}$$

With some algebraic manipulation, we find that

$$\begin{aligned} G(k) &= \sum_{j=0}^{(N/2)-1} \{S(2j)e^{-i2\pi kj/(N/2)} + (i/2)[S(2j+1) - S(2j-1)]e^{-i2\pi kj/(N/2)}\} \\ &= \sum_{j=0}^{(N/2)-1} \{S(2j) + (i/2)[1 - e^{-i2\pi k/(N/2)}]S(2j+1)\}e^{-i2\pi jk/(N/2)}, \end{aligned} \tag{17-72}$$

where the periodicity of $S(j)$ in $j = N$ has been used. Note that $S(2j)$ is a real, even function and its inverse Fourier transform is real; $[S(2j+1) - S(2j-1)]$ is a real, odd function from Eq. (17-70) for Hermitian functions, and thus the transform is imaginary. Thus, $G(k)$ in Eq. (17-72) is real. Because it will be useful later, we note that

$$G(\tfrac{1}{2}N - k) = \sum_{j=0}^{(N/2)-1} \{S(2j) - (i/2)[1 - e^{-i2\pi k/(N/2)}]S(2j+1)\}e^{-i2\pi jk/(N/2)} \tag{17-73}$$

Now, we will show that $B(k)$ is related to $G(k)$ and recall that

$$\begin{aligned} B(k) &\equiv \sum_{j=0}^{N-1} S(j)e^{-i2\pi jk/N} \\ &= \sum_{j=0}^{(N/2)-1} S(2j)e^{-i2\pi(2j)k/N} + \sum_{j=0}^{(N/2)-1} S(2j+1)e^{-i2\pi(2j+1)/N}. \end{aligned} \tag{17-74}$$

But also note that

$$G(\tfrac{1}{2}N - k) + G(k) = 2\sum_{j=0}^{(N/2)-1} S(2j)e^{-i2\pi k(2j)/N}. \tag{17-75}$$

and that

$$G(k) - G(\tfrac{1}{2}N - k) = -2\sin(2\pi k/N)\sum_{j=0}^{(N/2)-1} S(2j+1)e^{-i2\pi(2j+1)k/N}, \tag{17-76}$$

where we have used the relation

$$1 - e^{-i2\pi k/(N/2)} = 2ie^{-i2\pi k/N}\sin(2\pi k/N). \tag{17-77}$$

Comparing Eq. (17-74) with Eqs. (17-75) and (17-76), we find

$$B(k) = \tfrac{1}{2}\big(G(k) + G(\tfrac{1}{2}N - k) - \{[G(k) - G(\tfrac{1}{2}N - k)]/\sin(2\pi k/N)\}\big). \tag{17-78}$$

One step now remains. We do not wish to calculate directly the inverse

Fourier transform of $H(j)$, since we know $H(j)$ contains redundant information. To eliminate the redundancy, we construct a new function $J(j)$ from the relation [using $H^*(\frac{1}{4}N - j) = H(\frac{1}{4}N + j)$]

$$J(j) \equiv \tfrac{1}{2}\{H(j) + H^*(\tfrac{1}{4}N - j) + ie^{-i2\pi(2j)/N}[H(j) - H(\tfrac{1}{4}N - j)]\}. \quad (17\text{-}79)$$

$J(j)$ is a complex function defined for $N/4$ points. The inverse Fourier transform of $J(j)$ is defined as the complex function $K(k)$. With a little manipulation, it can be shown that

$$G(2k) = \text{Re } K(k) \quad (17\text{-}80)$$

and

$$G(2k + 1) = \text{Im } K(k). \quad (17\text{-}81)$$

Equation (17-78) relates $G(k)$ to $B(k)$.

Let us summarize this section. We begin with a real symmetric function $S(j)$ defined for $N$ points. From Eqs. (17-70) and (17-79), we construct a complex function $J(j)$ defined for $N/4$ points. We compute $K(k)$, the inverse Fourier transform of $J(j)$, and use Eqs. (17-78), (17-80), and (17-81) to obtain $B(k)$, the inverse Fourier transform of $S(j)$. Thus, at the price of some extra computation, we have reduced the storage space required by the computer by a factor of four for the special case of a real, symmetric function, so the method provides a savings of a factor of two for the computation of the Fourier transform of a Hermitian function.

## SPECIAL CASE OF A REAL, ODD FUNCTION

We shall only outline the method for taking the Fourier transform of a real, odd function. An interferogram which is an odd function results from a type of interferometer using internal modulation. Such an instrument produces the derivative of the interferogram obtained in the usual instrument.

The method is similar to that used for a real, even function. We first construct a Hermitian function $H(j)$ and then use the method described above for such a function. If $S(j)$ is real and odd, then $S(\frac{1}{2}N - j) = -S(\frac{1}{2}N + j)$. As customary, we assume $S(j)$ is defined for $N$ points and is related to a two-sided, odd interferogram by Eq. (16-5) and that negative-$j$ indices have been eliminated in the usual fashion. It can easily be shown that if

$$H'(j) = \tfrac{1}{2}[S(2j + 1) - S(2j - 1)] + iS(2j). \quad (17\text{-}82)$$

then $H'(j)$ is Hermitian. Hence, the method described above for a Hermitian function can be applied. Letting $G'(k)$ be the Fourier transform of $H'(j)$, we can show in a manner similar to that used for a real, even function that

$$\begin{aligned} B(k) &= \mathscr{F}^{-1}\{S(j)\} \\ &= (i/2)\big(G'(k) - G'(\tfrac{1}{2}N - k) + \{[G'(k) + G'(\tfrac{1}{2}N + k)]/\sin(2\pi k/N)\}\big) \end{aligned} \tag{17-83}$$

$B(k)$ is purely imaginary since it is the inverse Fourier transform of a real, odd function.

## CONCLUSIONS

In closing, we note that other explanations of the Cooley–Tukey algorithm have been given by other authors. Notably, there are outlines by Čooley and Tukey [2], Forman [3], Brigham and Morrow [4], and J. Connes [5]. Our explanation follows Connes's and basically is very similar to the treatment of Brigham and Morrow. It is almost obvious that instead of sorting the $S(j)$ initially, we could instead go through the algorithm and obtain the $B(k)$ in a scrambled order. Brigham and Morrow have, in fact, done this and find that $B(k)$ is in the location given by the binary digit inversion of $k$. This result perhaps is not surprising. Anyway, the reader is referred to their paper for details. Forman's version of the algorithm requires twice as many storage locations. In Appendix B, we will present a computer program that uses the method of J. Connes with a binary sort before the inverse Fourier transform.

### *References*

*1.* I. J. Good, The Interaction Algorithm and Practical Fourier Series, *J. Roy. Statist. Soc., Ser. B.* **20** 361 (1958).

*2.* J. W. Cooley and J. W. Tukey, An Algorithm for Machine Calculation and Complex Fourier Series, *Math. Comput.* **19**, 297 (1965).

*3.* M. L. Forman, Fast Fourier-Transform Technique and Its Application to Fourier Spectroscopy, *J. Opt. Soc. Amer.* **56**, 978 (1966).

*4.* E. O. Brigham and R. E. Morrow, The Fast Fourier Transform Spectrum, *IEEE Trans. Audio Electroacoust.* **AU-15(2)**, 63 (1967).

*5.* J. Connes, Computing Problems in Fourier Spectroscopy, *Aspen Int. Conf. on Fourier Spectrosc., 1970*, p. 83. (G. A. Vanasse, A. T. Stair, Jr., and D. J. Baker, eds.). AFCRL-71-0019, 5 Jan. 1971, Spec. Rep. No. 114.

CHAPTER EIGHTEEN

# *Minicomputers and Real-Time Fourier Analysis*†

Ralph W. Alexander†

## INTRODUCTION

A major drawback of Fourier transform spectroscopy can be the computation required to obtain the results of a measurement in a useful form. We have seen in the two preceding chapters that this drawback has in large measure been overcome with use of the Cooley–Tukey algorithm. However, although the time required for computation is small, a relatively long time may be spent waiting in line for a large, central batch-process computer. The development of small, relatively inexpensive digital computers allows elimination of this delay for many Fourier spectroscopists. This chapter will consider some of the possible uses of small digital computers with Fourier spectrometers.

The ideal Fourier spectrometer would have a computer associated with it which performed the Fourier analysis and displayed the spectrum continually during the measurement of the interferogram (real-time Fourier analysis). This requires either that a computer be specially designed for Fourier analysis or that a general-purpose digital computer be available during the measurement. If a general-purpose computer is to be used, the usual choice is a relatively small computer located in the same area as the spectrometer. This is because most large computers were designed for batch processing and not for real-time use. This situation may change in the future as the advantages to many users of real-time access become apparent even to computer center administrators.

A major advantage of real-time Fourier analysis, of course, is that decisions concerning the measurement can be made during the measurement.

† Physics Department, University of Missouri-Rolla.

One can see if sufficient resolution has been obtained to resolve the features of interest or if the instrument is properly adjusted. This approach can result in large savings of time.

What are the limitations of real-time analysis? First, if interferograms with large numbers of points are to be measured, the memory requirements are large. In the near-infrared and visible spectral regions, signal levels are usually relatively high, so the interferogram is sampled rapidly. This fact, often combined with the requirement for a large number of output frequencies, means an extremely fast computer is required. Extreme speed can only be obtained with special-purpose computers. For example, Connes [1] is constructing a special-purpose computer which provides the spectrum at 20,000 frequencies while accepting 50 interferogram samples per second. That is, 1 $\mu$sec is required per output point.

In the far-infrared spectral region, the rate of scanning is usually considerably slower because source intensities are so much lower. Then, a general-purpose digital computer is capable of performing the Fourier analysis in real-time. This is the case we shall be most interested in. The advantages of using a general-purpose computer are several. First, other data processing in addition to Fourier analysis can be done. For example, spectra can be ratioed or compared. Second, the computer can serve as a data processor for other types of instruments or can be used for general-purpose calculations. This can be an important advantage in terms of financing. Third, the computer can be used to control the spectrometer. In Chapter 19, a Digilab instrument is described which uses a computer to provide many control functions.

Even if the speed of a general-purpose computer is insufficient for real-time analysis, its presence may be very useful. One example would be instrument control as mentioned above; another would be providing a spectrum moments after an interferogram is complete (using the Cooley–Tukey algorithm). The Digilab FTS-14 interferometer can compute the spectra and their ratio from two 16,000-point interferograms in 2 min. This system uses external disk storage. Buijs [2] and Connes [3] have described a method for adapting the Cooley–Tukey algorithm to use with external disk or magnetic tape storage. This allows the computation of spectra from interferograms which are much larger than could be stored in core memory.

## REAL-TIME FOURIER ANALYSIS OF A ONE-SIDED INTERFEROGRAM

We first present an outline of the method and then discuss it in more detail with a sample program written in FORTRAN in Appendix B. The computer displays the spectrum during the measurement on an oscilloscope. This display consists of $M$ frequency points spaced $\Delta\sigma$ ($cm^{-1}$) apart and beginning at $\sigma_l$. During the time required to measure a point on the interferogram, the contribution of the preceding point of the interferogram to the spectrum is calculated and the updated spectrum displayed on the oscilloscope. If desired, a copy of the spectrum can be plotted on an $x$–$y$ plotter at the end of the measurement.

### INITIALIZATION

We first ask what information is required initially by the computer. To compute the $j$th value of the interferogram $S(j)$, one needs the mean value of the signal $I_R(\infty)$. The value of $I_R(\infty)$ is obtained by the operator prior to beginning a measurement. This is done by measuring the signal for a large optical path displacement from $\delta = 0$. Recall that an error in $I_R(\infty)$ means a frequency-independent error in the spectral intensity. In order to determine the frequency scale, the optical path step size $\Delta\delta$ is needed. To determine the output display, the spacing of the output points $\Delta\sigma$ ($cm^{-1}$), the number of points in the output display $M$, and finally, the wave number of the first point of the output display $\sigma_l$ are needed. Thus, our program begins by requesting the values of these parameters from the operator via the teletype.

### PARABOLIC FIT

A parabolic fit to the interferogram points near $\delta = 0$ is utilized to correct for the linear phase error (Chapter 12) that occurs if the point $\delta = 0$ is not sampled exactly. It is assumed that the operator has initially set the instrument so that the first maximum in the signal level occurs for $\delta = 0$. The

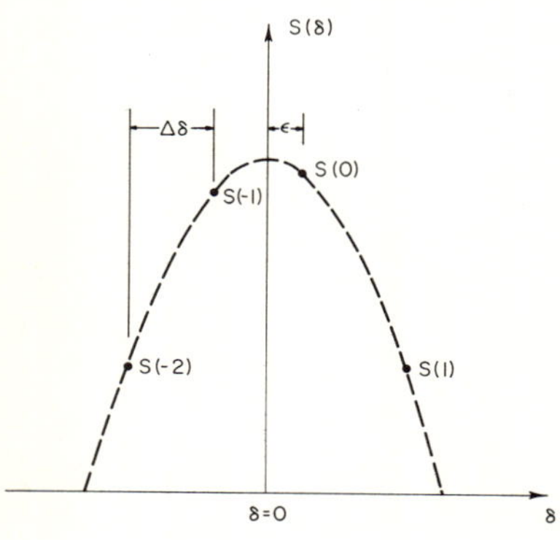


FIG. 18-1. The dots indicate points on the interferogram sampled by the computer. They are separated by an optical path difference $\Delta\delta$. The broken line indicates the parabola fitted through the maximum $S(0)$ and the first point on either side [$S(-1)$ and $S(1)$]. $\varepsilon$ is the distance between $\delta = 0$ and the value of $\delta$ for $S(0)$.

computer acquires and stores points until the maximum signal and one point beyond have been acquired. A parabola is then fitted through the maximum and one point on each side of the maximum. With $\varepsilon$ defined as in Fig. 18-1, we find using a least squares fit that

$$\varepsilon = [S(1) - S(-1)]/2[S(-1) + S(1) - 2S(0)]. \tag{18-1}$$

## COMPUTATION OF THE FOURIER TRANSFORM

Because we are calculating the Fourier transform of a one-sided interferogram (assumed to be symmetric about $\delta = 0$), we need compute only the cosine transform. Inserting the fact that the interferogram is sampled at $\delta = \varepsilon,\ \Delta\delta + \varepsilon,\ 2\,\Delta\delta + \varepsilon, \ldots$ instead of at $\delta = 0,\ \Delta\delta,\ 2\,\Delta\delta, \ldots$, the cosine Fourier transform becomes

$$B(\sigma_k) = \sum_{j=0}^{N} S(j,\Delta\delta - \varepsilon)\cos[2\pi\sigma_k(j\,\Delta\delta + \varepsilon)]. \tag{18-2}$$

Because we are not apodizing, we may compute $S(j\,\Delta\delta + \varepsilon)$ from

$$S(j) \equiv S(j\,\Delta\delta + \varepsilon) \equiv I_{\mathrm{R}}(j\,\Delta\delta + \varepsilon) - I_{\mathrm{R}}(\infty), \tag{18-3}$$

where $I_{\mathrm{R}}(\delta)$ is the measured signal at path difference $\delta$. Our procedure then will be to calculate the contribution of each $S(j)$ for a predetermined set of $\sigma_k$. This calculation will be done during the time it takes to measure the next $S(j)$. That is, if the computer acquires $I(j\,\Delta\delta + \varepsilon)$, we calculate $S(j)$ from Eq. (18-3), and then for each $\sigma_k$ we calculate

$$B_j(\sigma_k) = B_{j-1}(\sigma_k) + S(j)\cos[2\pi\sigma_k(j\,\Delta\delta + \varepsilon)], \tag{18-4}$$

where the subscript $j$ on $B_j(\sigma_k)$ denotes the maximum value of $j$ that has been used in the sum in Eq. (18-2). Of course, the first spectral value is $B_0(\sigma_k) = S(0)\cos(2\pi\sigma_k\varepsilon)$, where $S(j = 0)$ is the interferogram sample at $\delta = \varepsilon$.

## CALCULATION OF $S(j)\cos[2\pi\sigma_k(j\,\Delta\delta+\varepsilon)]$

Since we do not wish to store a table of values for $\sigma_k$, we assume $B(\sigma_k)$ is to be calculated at $M$ points spaced $\Delta\sigma$ apart, beginning with $\sigma_l$. Because considerable simplicity will result, we shall take $\sigma_l = p\,\Delta\sigma$, where $p$ is an integer. We shall let $k$ range from zero to $M - 1$, so that $\sigma_k = (p + k)\,\Delta\sigma$, and the argument of the cosine in Eq. (18-4) becomes

$$\theta(j, k) = 2\pi(p + k)(\Delta\sigma)(j\,\Delta\delta + \varepsilon) \tag{18-5}$$

We shall denote $2\pi\,\Delta\delta\,\Delta\sigma$ by $T$ and $2\pi\varepsilon\,\Delta\sigma$ by $T_0$. Both $T$ and $T_0$ are constants for a given interferogram. Hence, one has

$$\theta(j,k) = (p+k)(jT+T_0). \tag{18-6}$$

Because we shall be programming for a machine without "hardware" multiply–divide capability, we shall be interested in keeping the number of multiplications and divisions required to a minimum. For each $j$, we shall begin the calculation with $k = 0$. Hence, we find

$$\theta(j,\ k=0) = \theta(j-1, k=0) + pT. \tag{18-7}$$

These steps require only additions after the initial calculation of $pT$ and $\theta(j=0, k=0) = pT_0$. Since the calculation of the cosine requires considerable time, we prefer to store a table of cosines, or rather a table of arc cosines.

The table of arc cosines (for equally spaced values of the cosine) will be for angles in the first quadrant only. This means that each time we calculate $\theta(j,k)$, we must reduce it to the equivalent first quadrant angle if necessary. Of course, we must retain the proper sign while doing this operation. A reasonable compromise between the requirements of accuracy and storage space is to use a table of 1024 ($=2^{10}$) arc cosines for a computer with 4096 memory locations. Stored sequentially in our table are $2^{10}$ values of $\theta_i$ such that the cosine values are equally spaced with

$$\theta_i = \arccos(i2^{-10} - \tfrac{1}{2}2^{-10}). \tag{18-8}$$

We have $\theta(j, k=0)$ and require $S(j)\cos\theta(j,k=0)$. To find this product, we begin by computing $S(j)2^{-10}$. Recall that for a binary number this is a "trivial" multiplication because one must only shift the binary point in $S(j)$ 10 places to the left. Now, we search the table of arc cosines, beginning with $i = 1$. If $\theta(j,k) < \theta_1$, then we take $S(j)\cos\theta(j,k) = 0$. If not, we temporarily set $S(j)\cos\theta(j,k) = 2^{-10}S(j)$ and test to find if $\theta(j,k) < \theta_2$. If it is, we take $S(j)\cos\theta(j,k) = 2^{-10}S(j)$. If not, we temporarily set $S(j)\cos\theta(j,k) = 2^{-10}S(j) + 2^{-10}S(j) = 2\times 2^{-10}S(j)$. We continued this process until for some value of $i$, say $m$, $\theta(j,k) < \theta_m$ [and of course $\theta(j,k) \geq \theta_{m-1}$]. At this point, we have $S(j)\cos\theta(j,k) = 2^{-10}(m-1)S(j)$, the correct result to within the accuracy of our table. The important point to note is that we obtained the product very simply by successive additions because the cosines were equally spaced. It was for this reason we stored arc cosines. If cosines had been stored, then, after looking up the proper cosine, the multiplication $S(j)\cos\theta(j,k)$ would have had to be performed. Having found $S(j)\cos\theta(j,k=0)$, we add this to $B_{j-1}(\sigma_{k=0})$ to obtain $B_j(\sigma_{k=0})$, where the first value $B_0(\sigma_{k=0})$ is equal to $S(0)\cos(2\pi\sigma_{k=0}\varepsilon)$. We now set $k = 1$ and find $\theta(j, k = 1)$ from

$$\theta(j, k+1) = \theta(j, k) + (jT + T_0) \tag{18-9}$$

and proceed as for $\theta(j, k = 0)$ to find $B_j(\sigma_{k=1})$. This process is continued, incrementing $k$ by 1 each time, until $B_j(\sigma_k)$ has been found for $M$ values of $k$. Having then finished the calculation for $j$, we display the set $B(\sigma_k)$ on the oscilloscope and wait for the measurement of $I_R[(j+1)\,\Delta\delta + \varepsilon]$ to be completed. $B_{j+1}(\sigma_k)$ is then computed as above for $B_j(\sigma_k)$. This process continues until the operator indicates via the teletype that he wishes the measurement to stop.

Note that the number of points in the interferogram is not limited by the memory size, because the interferogram is not stored. Of course, round-off errors will eventually limit the number of points which can be transformed. The flow diagram in Fig. 18-2 summarizes the above program outline. A sample program written for a Hewlett-Packard computer appears in Appendix B.

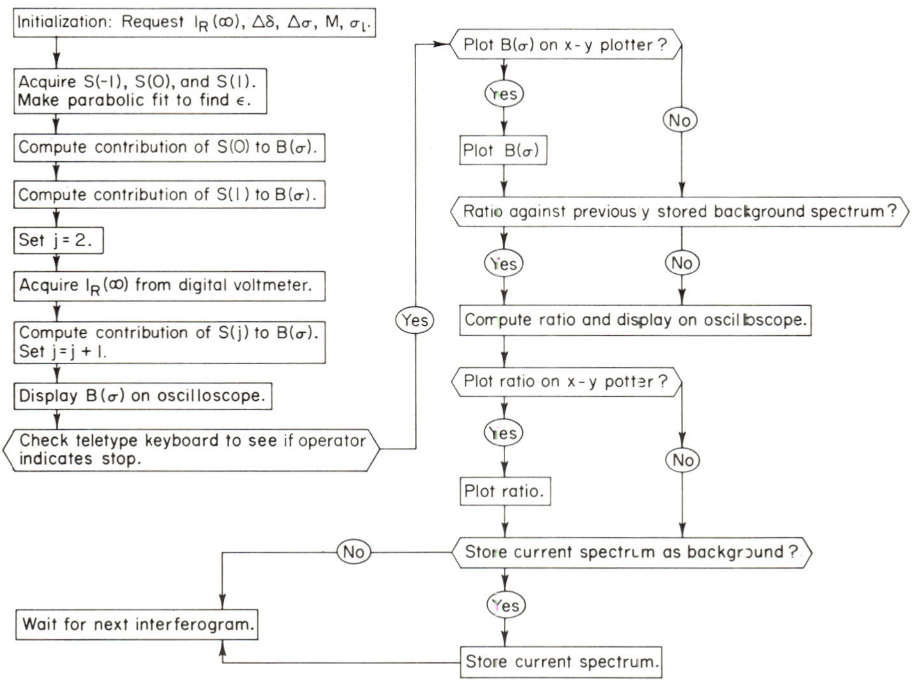


FIG. 18-2. Simplified flow chart for real-time Fourier analysis program.

## EXAMPLE: COMMERCIAL REAL-TIME SYSTEMS

To obtain an idea of what can be done with a real-time system, we consider an instrument available from Society Codery. This system, for an infrared Michelson interferometer, is described in some detail in Chapter 19 and in a paper presented at the 1970 Aspen Conference [4].

This spectrometer uses a Varian 620/i computer with a 4096-word memory of 16-bit words. It can compute the spectrum for 1500 frequency points in 1.9 sec. This is a computation rate compatible for work in the far-infrared.

The new Polytec FIR 30 is a real-time Fourier-Michelson interferometer-spectrometer which is described in Chapter 19.

## CHOICE OF A COMPUTER

A large selection of small digital computers is now available, many with the software required for Fourier analysis. For most users, a 16-bit word size will be more satisfactory than a 12-bit work size. The size of the memory and the choice of peripheral equipment depend upon the particular application. For example, the machine described above from Society Codery uses a 4096-word memory and can compute the spectrum at 1500 frequency points or, if a background spectrum is stored for ratioing, the spectrum can be computed at 750 frequency points. For many far-infrared spectrometers, this is sufficient. If the computer is required for other applications, a larger memory may be necessary. For example, the BASIC language software usually requires 6000 words of memory. BASIC is a conversational language suited for users solving problems not requiring the capacity of a large computer. For many general users, a small computer with BASIC serves as a happy compromise between a large, central computer and a small, desk-top calculator.

The peripheral equipment usually includes a teletype writer–low-speed tape punch combination. This device serves as a low-speed input–output device. An interface to a digital voltmeter is required if real-time analysis is to be performed and may be convenient for data acquisition in any case. Many users will find a high-speed tape reader convenient, particularly if off-line processing data from paper tape is going to be performed. Similarly, a high-speed punch may be necessary. Real-time analysis is really useful only if a means of immediately displaying the spectrum is available. This is usually done via a high-speed digital-to-analog converter and an oscilloscope. Some computer manufacturers offer a converter–oscilloscope package along with the appropriate software. Availability of such a unit may influence the purchaser's choice of computer. For a "hard" copy of the spectrum, the teletype may be used to produce a low-resolution plot. However, it is preferable to have a digital-to-analog converter and an $x$–$y$ recorder. Again, some manufacturers offer the converter, recorder, and software as a package.

If the computer is to be used in conjunction with an interferometer used for the near-infrared or visible regions at high resolution, external disk storage may be required. Such storage is reasonably fast and is of great convenience if much general-purpose computing is to be done.

## CONCLUSION

Small digital computers can provide real-time Fourier analysis for far-infrared interferometers and rapid Fourier analysis for other wavelengths. If extremely high-speed Fourier analysis is required, special-purpose computers may be built which operate at a speed of 1 $\mu$sec per output point. In the near future, $10^{-7}$ sec per output point may be realized. General-purpose computers, however, offer the advantage of flexibility.

## *References*

*1.* P. Connes and G. Michel, Real Time Computer for Fourier Spectroscopy, *Aspen Int. Conf. on Fourier Spectrosc., 1970*, p. 313. (G. A. Vanasse, A. T. Stair, Jr., and D. J. Baker, eds.). AFCRL-71-0019, 5 Jan. 1971, Spec. Rep. No. 114.

*2.* H. L. Buijs, Fast Fourier Transformation of Large Arrays of Data, *Appl. Opt.* **8**, 211 (1969).

*3.* J. Connes, Computing Problems in Fourier Spectroscopy, *Aspen Int. Conf. on Fourier Spectrosc., 1970*, p. 83. (G. A. Vanasse, A. T. Stair, Jr., and D. J. Baker, eds.). AFCRL-71-0019, 5 Jan. 1971, Spec. Rep. No. 114.

*4.* F. Levy, R. C. Milward, S. Bras, and R. le Toullec, "Real Time" Far Infrared Fourier Spectroscopy Using a Small Digital Computer, *Aspen Int. Conf. on Fourier Spectrosc., 1970*, p. 331. (G. A. Vanasse, A. T. Stair, Jr., and D. J. Baker, eds.). AFCRL-71-0019, 5 Jan. 1971, Spec. Rep. 114.

CHAPTER NINETEEN

# Commercial Instruments

## INTRODUCTION

In this chapter, we will discuss Fourier transform spectrometers that are made and marketed by six different companies. Each section will be headed by the name of the company and the wavenumber range which their complete line of equipment covers. Whenever available, the optical diagrams of the interferometers will be given with a brief description. Unfortunately, the author was unable to obtain detailed, exactly parallel information about all the instruments. Nonetheless, for further reliable information, the names and addresses are given in the final section for all six suppliers. At the end of each section, early-1970 prices are given for the major components. However, the reader must remember that with technical "tradeoffs against price," almost any price can be agreed upon in a competitive market.

These commercial instruments are excellent. Examples of both Michelson and lamellar grating interferometers are on the market. The asymmetric interferometers are not yet available. All of the companies have interferometers which work well up to a few hundred wave numbers.

All of the instruments are evacuable and all will operate on 60-Hz power. With the commercial choices available, it seems that the day of most "home-made" Fourier transform interferometers has past. For the well-heeled laboratories, special construction can be done to custom specifications by all of the manufacturers. In fact, it seems that Bell Telephone Laboratories, the Texas Instrument Corp., Ball Engineering Co., etc. have been involved in

Fourier transform spectrometer development. Of course NASA, Air Force Cambridge Laboratories, the University of Massachusetts, and many astronomy groups have built their own special interferometers, too.

For further reading, the reader is referred to a series of four articles by Low [1], who has done an excellent job of describing the equipment from four of the six manufacturers we consider and, in particular, has paid more attention to the computer systems than has been possible here. Ample Fourier transform computation techniques are available in Chapters 16–18 and Appendix B.

## RIIC OR BECKMAN INSTRUMENTS, INC. (3–500 $cm^{-1}$)

Beckman Instruments, Inc., sells Fourier transform spectrometers which are made in England by Research and Industrial Instruments Co. (RIIC). They have a variety of interferometers and electronic units available. They sell two Michelson interferometers (the FS-720 and FS-820) which are quite similar and a lamellar grating interferometer (LR-100).

In Fig. 19-1, an optical diagram of the FS-720 Michelson interferometer is shown. This instrument is capable of yielding spectra between 9 and 400 $cm^{-1}$ as delivered by Beckman. With very minor beamsplitter changes, the range can be extended down to 5 $cm^{-1}$.

Referring to Fig. 19-1, let us optically proceed through the FS-720. The source is a 125-W mercury lamp which is dimpled to prevent channel spectra. The lamp sits in a water-cooled housing which has three easily changed apertures of 3, 5, and 10 mm in diameter. The apertures can be used to control intensity and will affect resolution if high-resolution data are sought (Chapter 11). The chopper is a semi-disk component with a switch to furnish the reference voltage for the lock-in amplifier.

The 7.5-cm-diameter paraboloid mirror is an $f/1.5$ collimator. The entire optical system in the source housing is mounted on easily adjusted plates so that precision optical alignment may be made.

The beamsplitters are made of stretched sheets of various thicknesses of Mylar (Chapter 9). The normalized spectra from various Mylar beamsplitters are shown in Figs. 9-4–9-10. The spectral range has been extended using the Mylar sheets as specified in the captions of Figs. 9-9 and 9-10. The beamsplitter housing in the Beckman equipment is so rugged, simple, and precise that the experimentalist can change the beamsplitter in about 2 or 3 min. Spectral filters to mate with the various beamsplitters are described in Chapter 10.

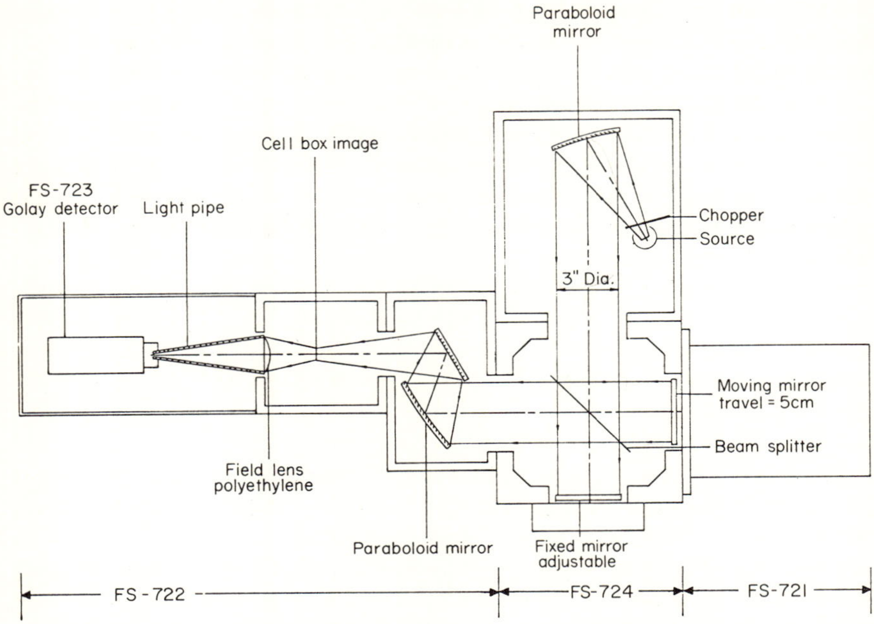


FIG. 19-1. Ray diagram for modular Fourier spectrophotometer FS-720 made by RIIC and sold by Beckman Instruments, Inc.

The fixed mirror is adjustable (when necessary) via a cantilever arrangement which is accessible under a vacuum flange which is quickly removed. The adjustment is easy to make, but one has to allow for thermal shifts for a day after the changes. Then, the instrument is stabilized and high-resolution data may be taken if there are no large-temperature ($\pm 10°$F or more) excursions in the laboratory.

The movable mirror is controlled by a micrometer screw which allows maximum optical path variations $L$ from $-10$ cm to $+10$ cm. From Chapter 6, one can see that this implies a resolution of about 0.1 $cm^{-1}$. The micrometer screw is driven by a stepping motor which is controlled by the FS-MC1 Step Control. The step control unit provides sample *path* increments $\Delta\delta$ (Chapter 7) of 5, 10, 20, 40, and 80 $\mu$m. For the high-wavenumber region at 400 $cm^{-1}$, one does not need the 5-$\mu$m step, and at the low wave numbers, a bigger step could be used. Various sample periods $\tau_s$ between 0.53 and 34 sec are available. The periods can be changed in incremental factors of two between the limits. With such a choice of sampling periods, it is not difficult to mate $\tau_s$ with the electronic time constant $\tau_e$ as described in Chapter 14. The command signal from the FS-MC1 is used to step the movable mirror and also to activate any data acquisition system for recording the interferogram.

The radiation beam is then brought to the "cell box image" position in Fig. 18-1. The radiation beam at the image has an *f*/3.4 divergence and is about the same size as the aperture in the source. The optical beam size is small enough to place wire-grid polarizers before and after samples. The polarizers may be purchased from either Beckman Instruments, Inc. or the Perkin-Elmer Corp., Norwalk, Connecticut. All sorts of liquid, gas, and solid sample cells will fit in the sample compartment. Beckman sells a fairly complete line of reflection attachments, 3–10 m gas cells, liquid cells, variable-temperature attachments, etc. One of the strong points of the FS-720 is that the sample compartment can be adapted to a very large variety of experiments. This is because the sample compartment is just a box with *O*-ring vacuum seals, and the radiation beam focus is in the center of the box. Dewars and low-temperature detectors are particularly easy to mate to the FS-720 interferometer.

Following the sample, the black polyethylene field lens helps (with the light-pipe cone) focus the radiant beam on the detector. The plane convex lens has a transmittance of 90% at 10 $cm^{-1}$ and falls monotonically such that the transmittance is 65% at 50 $cm^{-1}$, 28% at 200 $cm^{-1}$, and 15% at 500 $cm^{-1}$. Of course, it is part of the spectral filtering system. The light cone is well described in an article by Williamson [2] and is also a place to locate very small samples for transmittance studies. The copper cone merely reduces the radiation beam size so that it fits the detector.

At present, Beckman is using a Unicam Golay cell as a detector. This particular Golay cell is made in England and cannot be purchased in the United States because of patent agreements with Eppley Laboratories. However, the Unicam Golay cells are revolutionary in that they work for years without even light-bulb replacement. They are also not very microphonic. The Golay cell usually has a 1-mm-thick diamond window which is 3–5 mm in diameter. Diamond is the only material that is transparent in usable thicknesses from the vacuum-ultraviolet to microwave wave numbers. It has some minor lattice vibration absorption modes in the near-infrared wavenumber region.

The FS 720 is easy to align with a He–Ne laser as described in Appendix A. It only takes about an hour or so completely to align the instrument.

The interferometer vacuum system is good to pressures as low as $10^{-3}$ Torr. In most experiments, one may record reliable data at pressures below $5 \times 10^{-2}$ Torr and not be concerned about atmospheric water vapor absorption. The newest FS 720 instruments have a system of vacuum isolating the sample compartment. Since the mirror drive mechanism has in it something that is hygroscopic, being able to leave it under vacuum as much as possible is a distinct advantage. Some experimentalists fill the instrument with dry nitrogen before opening the interferometer to keep water vapor sorption and, later, outgassing to a minimum.

The micrometer step drive can be replaced with a continuous drive with path speeds of 5–500 m $sec^{-1}$. A moiré grating system enables one to have reference fringes for accurate mirror position determination. This system someday might be updated to rapid scanning Fourier transform spectroscopy as outlined in Appendix D.

In Fig. 19-2, an optical diagram of the FS 820 Michelson interferometer is shown. This instrument is quite similar to the FS 720. The major differences are that the drive system for the movable mirror is based on a lapped, $\frac{1}{2}$ in. diameter piston drive and that the sample compartment does not contain a well defined focal spot for samples. Usually, the instrument is sold with a crystal quartz window in the detector. This would limit the research to below 200 $cm^{-1}$. However, a diamond window is easily available at additional cost.

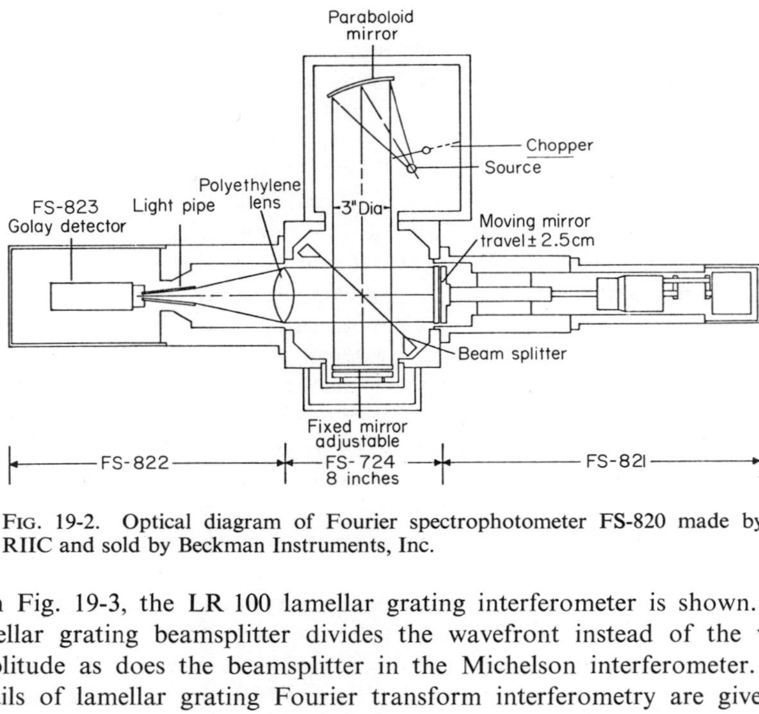


FIG. 19-2. Optical diagram of Fourier spectrophotometer FS-820 made by RIIC and sold by Beckman Instruments, Inc.

In Fig. 19-3, the LR 100 lamellar grating interferometer is shown. The lamellar grating beamsplitter divides the wavefront instead of the wave amplitude as does the beamsplitter in the Michelson interferometer. The details of lamellar grating Fourier transform interferometry are given in Chapter 15. Basically, it is a simple instrument which has about twice the radiant efficiency or throughput as a Michelson interferometer. Also, it has no instrumental polarization as introduced by the oblique incidence of the radiant beam on the beamsplitter in the Michelson interferometer. On the other hand, the lamellar grating is difficult to make with sufficient quality for research above about 100 $cm^{-1}$. (Also, there are diffraction problems at the

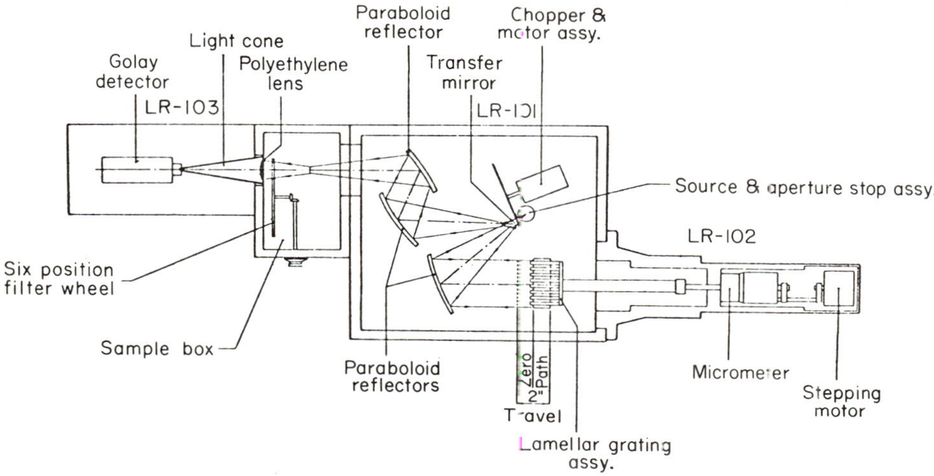


FIG. 19-3. Optical diagram of the LR-100 lamellar grating interferometer made by RIIC and sold by Beckman Instruments, Inc.

higher wave numbers, as outlined in Chapter 15.) The lamellar grating interferometer (LR 100) is thus usable between 3 and 70 $cm^{-1}$. It is best for low wavenumber research. Various grating constants from 0.64 to 1.0 cm are available, so that the highest and lowest wavenumber limits set by diffraction and cavity effects, respectively, may be controlled (see Chapter 15). The gratings are 8 cm by 8 cm and total path drives from $-5$ cm to $+5$ cm are available. The driving mechanism is a micrometer screw which is commanded by the FS MC1 Step Control. In Fig. 18-3, the "transfer mirror" is just a mirror below the source aperture to transfer the radiation beam from the lamellar grating to the sample housing. The transfer mirror is placed near the source aperture to keep the optical beam as near on axis as possible (Chapter 15).

There are various electronic systems with all of the interferometers. All of the Beckman equipment use chopped radiation beams so that ac signals may be used to discriminate against slowly drifting or constant radiant background and detector electronic noise. Then, each system is characterized by a reference voltage from the chopper to which the detector signal is "locked-in" in the phase-sensitive amplifier. For the FS-721, Beckman sells, in addition to the lock-in amplifier, a 12-bit analog-to-digital converter triggered by the Moiré grating pulses and associated electronics. For the interferometers driven by stepping motors, the analog-to-digital converter is triggered by the FS-MC1 Step Control. These systems (with the exception of the FS-MC1) can be easily replaced with lock-in amplifiers, power supplies, and data acquisition systems available from electronic supply firms such as Princeton Applied Research, the Hewlett-Packard Co., etc.

The Beckman company sells the FTC-300 Fourier transform computer. This is a hybrid digital–analog, special-purpose computer. The instrument handles up to 1024 interferogram samples in 12-bit words and gives the experimentalist his spectrum (in a plot) in a matter of minutes. The computer also will store sample-in or sample-out data and will ratio and plot the results. This is a good system, but for those readers who wish to avoid special purpose computers, Chapter 18 is called to their attention. Minicomputers which are very versatile can be purchased and used for this rapid analysis *and* other, unrelated purposes in the laboratory. The computer programs are given in Appendix B.

Some sample spectra from Beckman equipment may be found in Figs. 14-2, 14-3, 14-4. 14-7, and 14-8. These are just a few samples of the capabilities of the interferometers made by Beckman Instruments, Inc. For the prices of the major components, we give the early-1970 prices rounded off to the nearest thousand dollars:

1. FS-720 without electronics, \$8000.
2. FS-720 plus FTC-300, \$39,000.
3. FS-820 plus FTC-300 and FS-MC1, \$36,000.
4. LR-100, \$11,000.

If one purchases an interferometer, he must have high speed data handling equipment, and this usually means purchasing a data acquisition system for between \$7000 and \$20,000. If the FTC-300 system is used, one does not need the data acquisition system. Also, if direct coupling to a computer is made, the data acquisition system may be dispensed with. However, it is wise to have permanent records of the data which can be processed, and a magnetic or paper tape record are answers to the problem. Even with minicomputer in the laboratory, one may wish to have the data on record.

Using the FS-720 without electronics (\$8000), minor electronic components (\$1000), commercial lock-in amplifiers and power supplies (\$3000), and a 25 channel data acquisition system (using only one channel for the interferogram) (\$12,000), one can have (exclusive of labor costs) a complete Fourier transform spectroscopy laboratory for about \$24,000. If a minicomputer for general purposes and/or Fourier transform spectroscopy is purchased (8K memory, 16-bit words) and peripheral equipment such as $x$–$y$ plotter, high-speed tape reader, oscilloscope display, etc, are also ordered, one needs about another \$26,000. Thus, for \$50,000–\$60,000, a modern laboratory for Fourier transform spectroscopy can be established. But \$24,000 will give many researchers a satisfactory, completely functional system.

## DIGILAB, INC. (10–10,000 $cm^{-1}$)

Digilab, Inc. is a subsidiary of Block Engineering, Inc. It makes Michelson Fourier transform spectrometers which employ the latest techniques in rapid-scan spectroscopy. As outlined in Appendix D, rapid-scan Fourier transform spectroscopy is simple, but requires special electronic techniques (including data handling). Rapid-scan FTS is best performed with an on-line minicomputer—which Digilab has mated to the interferometer system. The end result of the technique using the Digilab equipment is recording interferograms in less than a second in some cases and always having them in the computer in less than 8 sec! For standard-length interferograms, the spectrum is computed and plotted in a little over 30 sec. So, from Digilab, the experimentalist has available the fastest automatic systems on the market with a mated minicomputer and plotter to display the spectra.

Digilab equipment is best referred to as complete systems. The minicomputers not only do the Fourier transforms, but are programmed to control many of the interferometer functions such as repeating scans, setting maximum optical path differences, automatically switching the radiation beam from the sample to the reference position, etc. If one purchases either the FTS-12 or FTS-14 spectrometer, he should buy the minicomputer and software programs from Digilab. Otherwise, it will take tremendous coordination with the company to mate other minicomputers with their interferometers because of the rapid-scan technique. However, Digilab is using Data General Corporation's Nova third-generation computers with a minimum of 4096 words (16 bits) and the Nova computer can be used for other laboratory purposes.

There are three basic Fourier transform spectrometers sold by Digilab:

1. The FTS-14 interferometer system is for work between 10 and 10,000 $cm^{-1}$, with iron oxide beamsplitters above 1650 $cm^{-1}$ and germanium beamsplitters below. The model 296 interferometer comprises the optical system.
2. The FTS-12 system is for research between 200 and 10,000 $cm^{-1}$. It can be used for field studies and uses the model 196 and 197 optical systems.
3. The FTS-16 interferometer is for the range between 10 and 400 cm.

The principal instrument system sold by Digilab, Inc. is the FTS-14 using the model 296 interferometer for the optical system. In Fig. 19-4, the optical diagram of the model 296 interferometer is shown. We describe the optical and computer systems separately.

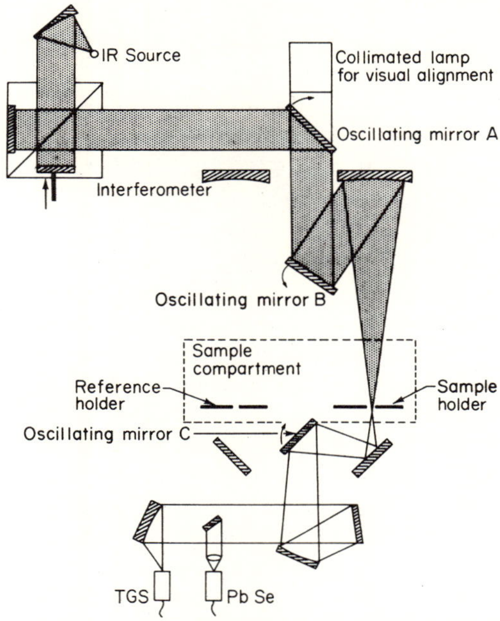


FIG. 19.4. Optical diagram of the model 296 interferometer used with the FTS-14 system made by Digilab, Inc.

Taking the optical system first and beginning with the source, one has a glower operating at 1100°C. With electronic servo-feedback systems, the source is stabilized to limit radiant fluctuations at 500 $cm^{-1}$ to less than $\pm 0.2\%$. This is necessary for repetitious, rapid-scan FTS and results in highly reproducible data. If one wishes to do research with emission spectra, the Digilab SS-IRF-250 emission kit is available. For research below 200 $cm^{-1}$ mercury arc lamps are used (Digilab has a conversion kit).

The beamsplitters are thin films deposited on supporting substrates as described in Chapter 9. The beamsplitters have relative efficiencies greater than 95% over the entire range for which they were designed. The beamsplitters and substrates are listed in Table 19-1. As described in Chapter 9, two beamsplitters of the same coating material are placed on each substrate. The center disk of the substrate has a coating on it of the proper thickness for use in the high-wavenumber range and the outer annulus coating is of the correct thickness for use as a beamsplitter at a lower, but overlapping wavenumber range. From Figure 19-4, one can see, for instance, that the center part of the radiation beam, after leaving the beamsplitter, goes to a PbSe detector and the outer portion goes to a TGS detector. The PbSe photon

TABLE 19-1

Beamsplitters for the Model 296 Interferometer

| Beamsplitter coating | Beamsplitter substrate | Wave number range (cm$^{-1}$) |
|---|---|---|
| $Fe_2O_3$ | Quartz | 3300–1,0000 |
| $Fe_2O_3$ | $CaF_2$ | 1650–6700 |
| Ge | KBr | 400–3800 |
| Ge | CsI[a] | 200–1000 |

[a] Under development.

detector is for the spectral range above 2000 cm$^{-1}$ and the new triglycine sulfate (TGS) pyroelectric bolometer is for the mid- through far-infrared. Thus, one does not have to change beamsplitters as often as he would have to with a simple beamsplitter. If one does wish to use Mylar beamsplitters, they are available in the SS-IRF-200 kit and would be used in conjunction with the TGS detector.

The movable-mirror continuous drive is at a speed of about one 1 cm sec$^{-1}$ with maximum path differences from 0.5 to 4 cm easily programmable on the computer. Under these conditions, single scans can be completed in 0.5, 1, 2, 4, or 8 sec. This also results (under standard conditions) in 2-cm$^{-1}$ resolution (Chapter 6) with apodization (Chapter 5). The instrument is controlled by the computer to make as many single scans as desired and then to average the spectra with a final spectral plot. In order to do this repetitious rapid scanning, the optical path difference must be accurately known. This is done by incorporating a He-Ne laser in the optical system and measuring optical path differences by counting laser fringes. The fringe intensity will go from a maximum to a minimum with a mirror motion of $\lambda/4$ or path difference of $\lambda/2$. With a He-Ne laser, $\lambda = 6328$ Å, so the mirror position can be easily located to within better than 0.16 $\mu$m. One must also know where the origin is located. The He-Ne laser is so near a perfect coherent, monochromatic source that it produces a near-perfect cosine interferogram. The cosine function is fine for measuring relative distances as described above, but it does not help locate the origin. The origin can be easily found by also including a simple incoherent, white-light source which will have the simple interferogram with a sharp peak at the origin and little modulation elsewhere. All of the information concerning the numbers of scans, the mirror velocity, the location of the origin, the distance from the origin, the phase-error correction (Chapter 12), and the maximum path difference (resolution, Chapter 6) can be stored, controlled, and used in the computation of the spectrum by the computer. One merely repeats the scans until he has the desired signal-to-noise ratio in the spectrum (see Fig. 14-6).

The "oscillating mirrors" in Fig 19-4 are used to switch between the sample and reference radiation beams. The switching is done after complete scans and needs to be repeated only as often as the experimentalist chooses. The operator decides when he wishes to make a reference beam scan (background) and the computer stores the complete information and employs it in ratioing (until a new background scan is wanted). In the sample compartment, there is a 5-mm image using *f*/4 optics, which is very convenient for small-sample work.

Finally, the radiation beams from the different beamsplitters are detected by the semiconductor photoconducting or pyroelectric bolometric detectors. The photoconducting detectors are amply described in the literature cited at the end of this book (the PbSe is used above 2000 $cm^{-1}$). The triglycine sulfate (TGS) detector is made by Mullard for Digilab ($850). However, these detectors may be purchased in at least $5 \times 5$ mm sizes from Barnes Engineering Co., Stamford, Connecticut for about $1300. They have a $D^*$ ("dee star" [3]) of about 4000 at 10 Hz chopping frequency and fall monotonically to 3000 at 50 Hz, 2500 at 100 Hz, 1000 at 500 Hz, and 650 at 1000 Hz. One can operate the TGS detectors as high as 2000 Hz with a $D^*$ of about 400. The detectors are available in different sizes and, with Minnesota Mining and Manufacturing Company's black velvet paint (Nextel, 101-C10 black) have a relatively flat spectral response from 10,000 $cm^{-1}$ to the far infrared.

As mentioned before, the computer is a Data General Corp. third-generation machine called the Nova. It has a minimum of 4096 words of 16 bit length and is preprogrammed to:

1. control key operational, measurement, and calibration parameters;
2. signal-average and store either single-beam or double-beam data;
3. calculate the Fourier transform, with proper apodization and phase correction in less than 20 sec either on a pair of 16,000-word interferograms to be ratioed, or on a single 32,000-word interferogram;
4. utilize any of more than a dozen available plotting routines to record single- or double-beam spectra together with a laser-derived wavenumber scale;
5. ordinate scale expansion (up to the noise limit) in per cent transmittance, absorbance, or log absorbance;
6. carry out other data manipulations, programmed by the user, such as peak selection or peak absorbance integration;
7. control other instruments or experiments around the laboratory by means of the generalized input-output.

For spectral scans between 400 and 3580 $cm^{-1}$, the Nova requires about

30 sec (2.6 $\mu$sec cycle time) to complete the calculation of the spectrum. To handle over 4096 points, Digilab also has a central processor with 16K words of memory.

The FTS-12, which is useful between 200 and 10,000 cm$^{-1}$, uses the model 196 or 197 optical system. The optical systems are so small and compact that they may be hand-held for field studies. The instrument with various detectors and resolution is partially described in Table 19-2. The instrument can make

TABLE 19-2

Detectors and Resolution for the FTS-12 System

| Wave number range (cm$^{-1}$) | Detector | Resolution (cm$^{-1}$) |
|---|---|---|
| 3,300–33,000 | Photomultiplier | 20 |
| 3,300–12,500 | PbS | 20 |
| 2,200–10,000 | PbSe | 20 |
| 625–4,000 | TGS | 4 |
| 250–2,500 | TGS | 4 |

scans in $\frac{1}{8}$, $\frac{1}{4}$, $\frac{1}{2}$, and 1 sec, which is ideal for field measurements. It has a 3 cm entrance aperture and 15 mm aperture stop. This results in a 10° full-angle field of view. The instrument has been used for emission, transmission, and reflection studies, and one can also do submicrogram solid sampling. In addition, there is an ellipsoidal mirror diffuse reflectometer available from Digilab which is useful for biological, coating, and other studies. For the FTS-12, the data handling can be managed as for the FTS-14.

In Fig. 19-5, the model FTS-16 optical system is shown. This is a Michelson Fourier transform spectrometer which uses rapid-scan techniques, and under standard conditions, is useful between 10 and 400 cm$^{-1}$. By exercising options, the experimentalist can extend the range to 5–650 cm$^{-1}$. As can be seen in Fig 19-5, the FTS-16 does not use collimated optics, but in most other respects it is an ordinary Michelson interferometer. One should note that the focii for the radiation beams from the two arms of the interferometer are at different points. However, this is a situation quite similar to that treated in Chapter 15 for the spherical, lamellar grating interferometer. For modest resolution and low wave numbers, as encountered in the far-infrared, one can afford to use noncollimated optics at virtually no penalty.

The source is a mercury arc lamp; Mylar beamsplitters are used (Chapter 9); and the stepping motor moves in 5-$\mu$m path steps (when used) with sampling times $\tau_s$ (Chapter 14) of 0.5, 1, 2, and 4 sec. Digilab can furnish either a

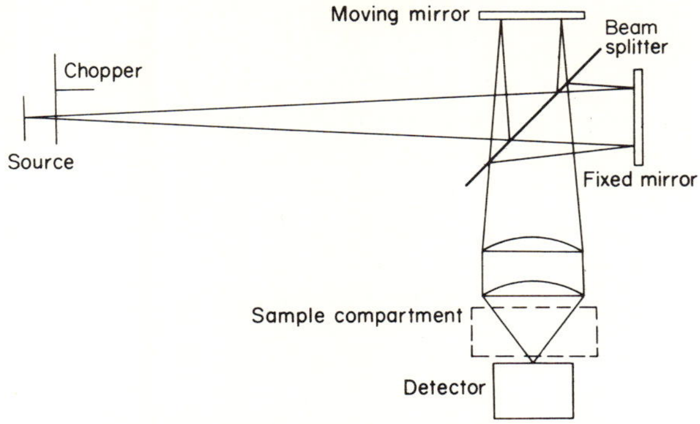


FIG. 19-5. Optical diagram of interferometer used in the FTS-16 system made by Digilab, Inc.

Golay cell or a TGS detector system with quartz or diamond windows. However, for the TGS detector, the size must be minimized if a diamond window is needed.

The prices of the Digilab equipment are as follows:

1. For the model 296, 297, or 197, the cost is $25,000.
2. For the FTS-14 including computer, the cost is $68,000.
3. For the FTS-14 far-infrared conversion kit, the cost is $2500.
4. The FTS-16 with electronics and tape readout is priced at $18,000. With the computer, the FTS-16 costs $39,000.

Sample spectra taken with Digilab equipment are shown in Figs. 14-1, 14-9, 14-10, 14-11, 14-14, 14-15, and 14-16. In addition, some rapid-scan data are presented in Fig. 14-19 for an example in the field of astronomy.

## GRUBBS-PARSONS (10–675 cm$^{-1}$)

The Grubbs-Parsons equipment is sold in the United States by Edwin Industries Corp. There are two major Michelson Fourier transform spectrometers made by Grubbs-Parsons. The model IS3 interferometer is useful in the range 10–675 cm$^{-1}$ and the Mark II is designed for the 10–200-cm$^{-1}$ range. (The latter can be modified for 10–650-cm$^{-1}$ work.) Comparatively speaking, the Grubbs-Parsons equipment is simple, small, and inexpensive.

In Fig. 19-6, an optical diagram of the IS3 spectrometer is shown on the right. The instrument uses a mercury arc source and a 20-Hz chopper. The

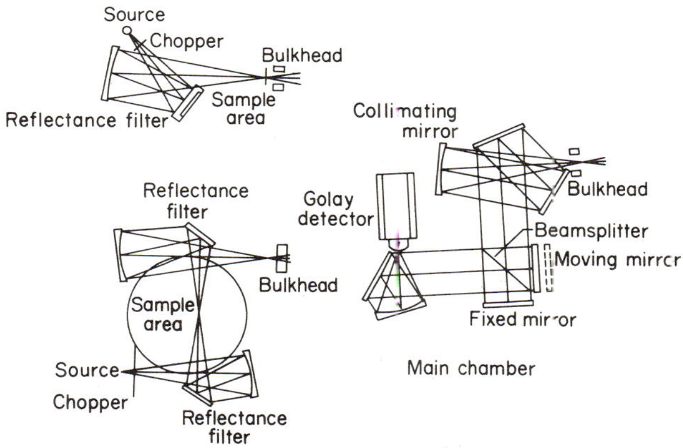


FIG. 19-6. Right: Optical diagram of the IS3 Fourier transform spectrometer made by Grubbs-Parsons. On the upper left-hand side is the SC1 sample chamber and on the lower right is the SC2 sample chamber.

beamsplitters are polyethylene terephthalate (Mylar) and are easily mounted (Chapter 9). The movable-mirror drive is in steps; starting at 5 $\mu$m, it can be increased in 5-$\mu$m increments (Chapter 7). The sampling period $\tau_s$ (Chapter 14) can be 0.5, 1, 2, or 4 sec or longer. The best resolution available is about 0.1 $cm^{-1}$.

The IS3 is a very versatile instrument, with two different sampling region kits available. The SC1 unit is shown in the upper left corner of Fig. 19-6 and the SC2 in the lower left corner. The SC1 is the smaller of the two units, but one can obtain 10-cm gas cells, six-sample solid-sample holders, liquid cells, and microsampling units to fit in it. The larger SC2 unit will hold all of the above-mentioned accessories and, in addition, accomodate larger cells and heating and cooling devices such as low-temperature Dewars.

The final optical component is the Golay cell detector. The electronic system consists of a lock-in amplifier, an analog-to-digital converter, and an Addo paper tape punch. The Fourier transforms are to be carried out on off-line computers (Chapters 16–18 and Appendix B).

The Mark II spectrometer was developed by H. A. Gebbie and J. E. Chamberlain at the National Physics Laboratory at Teddington England. It is a very inexpensive instrument for the 10–200-$cm^{-1}$ range, with an optimum resolution of 0.5 $cm^{-1}$. In Fig. 19-7, an optical diagram is shown and one can see that it has noncollimated radiation beams. It is a very simple instrument and is only 50 × 30 × 17 cm in size. The instrument functions very much as the IS3 and has accessories for refractive index measurements, Dewars, and solid, liquid, and gas samples.

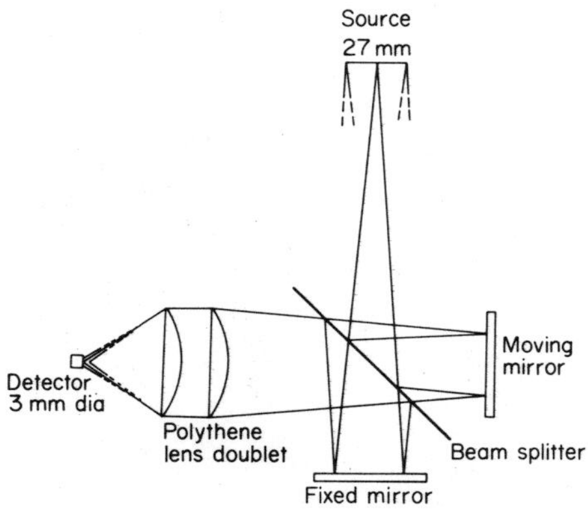


FIG. 19-7. Optical diagram of the Mark II cube interferometer made by Grubbs-Parsons.

The IS3, including electronics and tape punch, sells for \$20,000. The Mark II, with the same capabilities, retails for only \$12,000. No other manufacturer has such inexpensive equipment on the market. Therefore, one should give this equipment careful consideration.

### CODERG (10–800 $cm^{-1}$)

The CODERG (Societe de Conversions des Energies, Clichy, France) company manufacturers a single Michelson interferometer called the MIR-2. The instrument covers the range from 10 to 800 $cm^{-1}$ and can be computer controlled with real-time analysis (Chapter 18).

An optical diagram of the instrument is shown in Fig. 19-8. It uses a mercury lamp source ($f/2$ entrance aperture), a 12.5-Hz chopper, Mylar beamsplitters, and a Golay cell detector. The system is in two chambers which are vacuum coupled (capable of $10^{-6}$ Torr). The source intensity may be controlled by a variable-iris diaphragm, and the source can be easily removed to enable laser alignment of the instrument.

The instrument has three different arrangements for the sample compartment, which are shown in Fig. 19-9. In the top diagram of Fig. 19-9, the compartment is shown for obtaining reference spectra. The middle diagram illustrates the arrangement for small reflection samples located at the position *S*. In the lower diagram, a transmission sample is located at *S*. All of these arrangements are possible with the single compartment by merely moving mirrors. This is quite a convenience compared to other instruments.

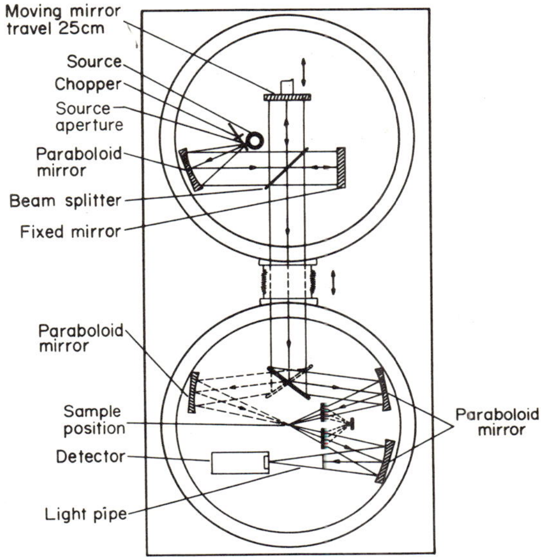


FIG. 19-8. Optical diagram of the MIR-2 Fourier transform spectrometer manufactured by CODERG.

The MIR-2 can be stepped (10-$\mu$m steps) or continuously driven, and Moiré-fringe referencing is used. One of the outstanding features of the CODERG equipment is that it, like the Digilab, Inc. spectrometers, can also be computer controlled. CODERG uses a Varian 620/i minicomputer and has the results of real-time analysis (Chapter 18) displayed on an oscilloscope. The Digilab equipment makes rapid scans and displays the computed spectrum after the scans have been repeated and averaged a prespecified number of scans. On the other hand, with one-sided interferograms and phase-error correction (Chapter 12), the CODERG system computes the entire spectrum repetitively throughout the scan. One can observe the resolution improve as the optical path difference increases and can decide visually when sufficient resolution and signal to noise have been attained.

In practice, the operator would insert the sample and, after a vacuum was established, give the computer the required data about (a) single- or double- beam operation; (b) the spectral range; (c) smallest separation between two points on the spectrum; (d) number of points (750 points maximum for double-beam, or 1500 points maximum for single-beam); (e) motor steps or motion; (f) apodizing function (a choice of ten is available) for the transform. Operation thereafter is automatic, and the instrument shuts itself off when the required resolution is obtained. Scan times for

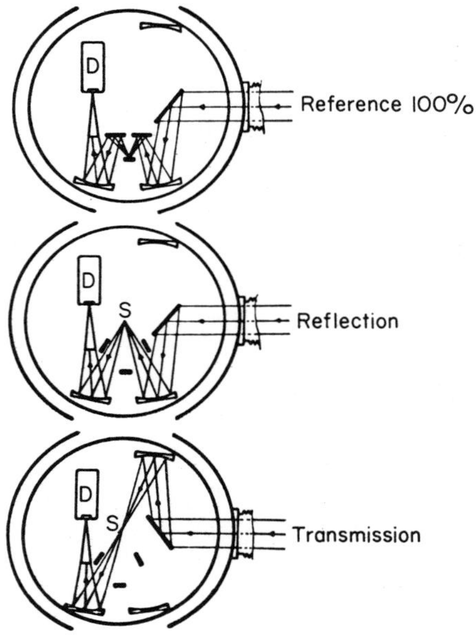


FIG. 19.9. Optical diagram of the sample compartment of the MIR-2 manufactured by CODERG. The top diagram is for reference scans; the middle diagram is for reflection samples at the position *S*; and the bottom diagram is for transmission samples at *S*. All three arrangements are possible with the one compartment by moving mirrors in and out of the radiation beam.

1500-point single-beam or 750-point double-beam spectra vary, e.g., 5 min at 1-$cm^{-1}$ resolution for a 50-$cm^{-1}$ range; 2.7 hr at 0.1-$cm^{-1}$ resolution for a 250-$cm^{-1}$ range.

If one does not wish to purchase the minicomputer, CODERG can furnish the MIR-2 with just a punched-paper-tape output.

With the electronics system including the tape readout the MIR-2 costs \$20,000. The data system incorporating the computer is \$26,000, so that the computerized system costs \$46,000.

### IDEALAB (10–10,000 $cm^{-1}$)

The Idealab Company sells two Michelson Fourier transform spectrometers, the IF-3 and the IF-6. The essential difference between the two is that the IF-3 affords a best resolution of 0.5 $cm^{-1}$, whereas the IF-6 can provide a best resolution of 0.1 $cm^{-1}$. The instrument is a continuous-drive interferometer, but the manufacturer does not sell a minicomputer for on-line analysis.

The IF-3 has a 2-cm maximum optical path difference and the IF-6 has a 10-cm maximum path difference. Otherwise, they are the same, so the following description applies to both instruments.

The IF-3 has various sources for the different spectral ranges: a glower for the near-infrared and a mercury arc for the far-infrared. Since it is a continuous-drive instrument, one has possible rapid-scan capability (Appendix D) with mirror velocities between 0.0167 and 1 cm $sec^{-1}$. The interferometer mirror movement is kept linear by use of servocontrol to provide insensitivity to vibration. The total scan intervals can be as short as 1 sec or as long as 10 min (standard). At the slower scan speeds, spectral filtering is not accomplished by electronic means as described in Appendix D. Also, if one wishes, he can purchase from Idealab a step control unit for mirror movement.

The beamsplitters for greater than 250 $cm^{-1}$ are dielectric films on transparent substrates (Chapter 9). For less than 250 $cm^{-1}$, Mylar beamsplitters are used (Chapter 9).

In addition to providing interferogram data, the IF-3 system has a white-light fiducial mark to locate the origin and a He–Ne laser for measurement of optical path differences. All three channels of information are available at the output.

The IF-3 is well suited for military and space science investigations because it can function in ambient vibration accelerations as great as 0.2 g.

The IF-3 drive unit cost \$18,000; the IF-6 drive unit sells for \$32,000; and the cube holding the beamsplitters and mirror retails at \$2000. The reference channels (complete with lasers and white-light sources) are \$1500.

## POLYTEC GmbH (10–1000 $cm^{-1}$)

This new German firm has just marketed a Fourier–Michelson interferometer–spectrometer with real-time analysis. The instrument is called the FIR-30 and is a fully integrated optical–electrical–mechanical–computer system.

At the sample, the system is *f*/3, and without braking vacuum ($5 \times 10^{-2}$ Torr), independent sample and reference scans can be made. Reflection or transmission studies are possible and by changing one mirror, the sample chamber can be used as a gas cell with optical paths up to 5 meters long.

In Fig. 19-10, the optical diagram is shown. The source is a high pressure

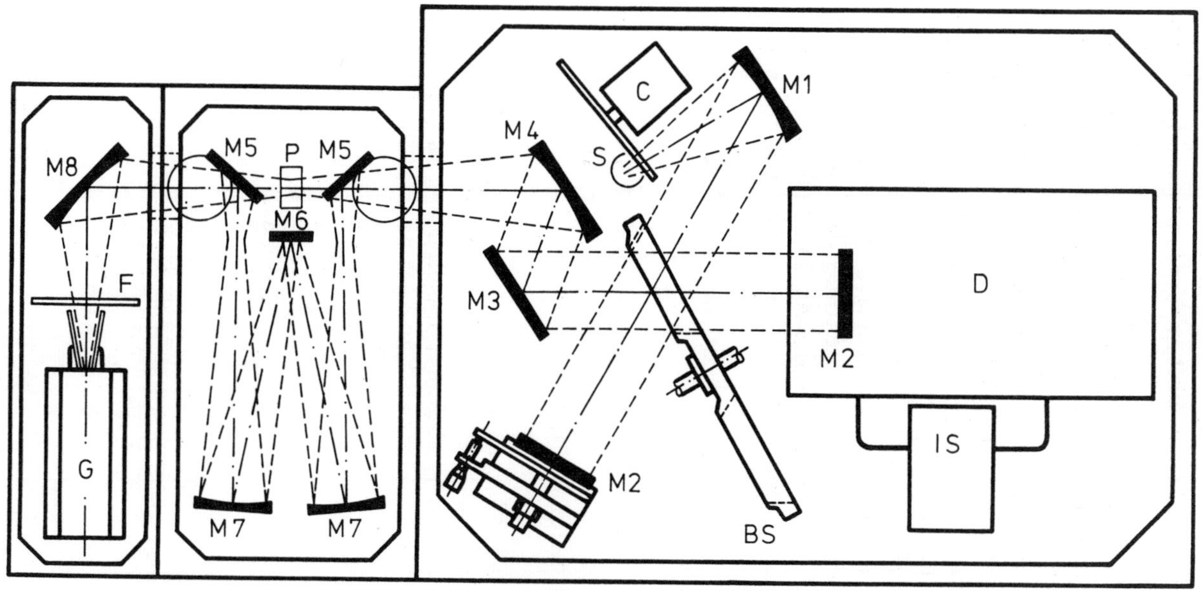


FIG. 19-10. Optical diagram of the Polytec FIR-30, M1 parabolic collimator, M2 movable Michelson mirror, M3 plane mirror, M4 parabolic condensor, M5 plane mirror (rotatable out of beam), M6 plane mirror, M7 spherical mirror, M8 elliptical mirror, C modulator, S Hg high pressure lamp, BS beamsplitter changer, D mirror drive, IS optical path incrementor control (Moiré control), F filter changer, G Golay detector, and P sample holder or sample cell.

dc mercury arc (closed-cycle cooling system) followed by a 12.5-Hz chopper. The beamsplitters (the radiation beam is incident at 30°) are mounted on a ring so that they can be changed automatically without disturbing the vacuum. There are four mylar and polypropylene beamsplitters available at 2.5-, 6-, 15-, and 50-$\mu$m thicknesses. The continuous motion can be monitored by a Moiré system. It is easy externally to change mirrors in the sample region by remote switches, and samples between 2- and 15-mm diameter are easily inserted. The instrument comes with a Golay cell detector which has an elliptical focusing mirror and a filter wheel just before it.

The computer is a Varian 620/i and soon will be a Nova 1200. All Fourier transforms are furnished, and the experimentalist can store other spectra and extract differences, ratios, etc. The output after each step is displayed on an accompanying oscilloscope (Tektronix 601) and so the instrument is truly a real-time interferometer–spectrometer. After the entire spectrum has been obtained, the spectra or ratio are traced on a strip-plotter having a wavenumber marker. All of the equipment can be operated, by the buyer's choice, on 250 V, 3.5 A, 50 Hz or on 115 V, 6 A, 60 Hz power.

The wavenumber range is about 10 to 1000 cm$^{-1}$ but can be extended to lower wave numbers with different filters and beamsplitters.

The entire instrument sells for about $57,000 (182,000 DM) in Germany. The first instrument was installed at L. Genzel's laboratory at the Max Planck Institüt für Festkörperforschung, Stuttgart, West Germany.

## CONCLUSIONS

We have discussed Fourier transform spectrometers made by:

1. Research and Industrial Instruments Company (RIIC), a division of Beckman–RIIC Ltd., Worsley Bridge Road, London SE26, England. U.S. distributor: Beckman Instruments, Inc., 2500 Harbor Boulevard, Fullerton, California 92634.
2. Digilab, Inc., a subsidiary of Block Engineering, Inc., 19 Blackstone Street, Cambridge, Massachusetts 02139.
3. Sir Howard Grubbs-Parsons and Co., Ltd., Walkergate, Newcastle-upon-Tyne 6, England NE62YB. U.S. distributor: Edwin Industries Corp., 11933 Tech Road, Silver Spring, Maryland 20904.
4. Societé de Conversion des Énergies (CODERG), 15 Impasse Barbier 92 Clichy, France. U.S. distributor: Scientific Instrumentation, Inc., P.O. Box 1345, Palo Alto, California 94302.
5. Idealab, Inc., Franklin, Massachusetts 02038.
6. Polytec GmbH, 7501 Grunwetter–Karlsruhe, W. Germany. U.S. distributor: Molectron Corp., 930 Thompson Place, Sunnyvale, Calif. 94088.

Only the Beckman or RIIC equipment include a lamellar grating interferometer. All of the companies sell Michelson interferometers. All of the interferometers are either step-driven or continuous-drive instruments with conventional analysis except for some of the Digilab equipment. Digilab does sell a true rapid-scan interferometer, but the capability is latent in the Idealab instrument. The CODERG and Digilab instruments have been mated with minicomputers for real-time or rapid spectral analysis, respectively.

All of the instruments are vacuum equipment, and all manufacturers have instruments with the capability of going as low as 10 $cm^{-1}$. The Digilab and Idealab equipment have the highest wavenumber capability, 10,000 $cm^{-1}$.

It is treacherous to comment on commercial products; so, the author suggests that the reader *contact* the manufacturers for the latest prices, innovations, and capabilities.

## *References*

*1.* M. F. D. Low, Fourier Transform Spectroscopy, *J. Chem. Ed.* **47**, A163, A255, A349, A415 (1970).

*2.* D. E. Williamson, Cone Channel Condenser Optics, *J. Opt. Soc. Amer.* **42**, 712 (1952).

*3.* R. D. Hudson, "Infrared System Engineering," p. 270. Wiley, New York, 1969.

APPENDIX A

# Optical Alignment of A Michelson Interferometer

## INTRODUCTION

It is very critical to have the movable and fixed mirrors each perpendicular to the incident radiation beams so that the radiation beams follow the same path back to the beamsplitter. After leaving the plane mirrors, the optical beams must be brought to a single focus if maximum modulation in the interferogram and minimum wavenumber spread in the computed spectrum are to be achieved. (The spherical mirror or spherical, lamellar grating interferometers are exceptions as described in Chapter 15.) So, mirror tilt and lack of alignment of one mirror relative to the other can seriously affect the efficiency of the interferometer.

The material in this appendix is given as an illustration of how one might align a Michelson interferometer using a laser. The laser offers convenience and speed in adjusting the position of the mirrors. It is an easier and quicker tool for adequate positioning of the mirrors than the old method of trying to locate fringes without any aids. This particular example applies to a Michelson interferometer (Beckman FS-720 interferometer) which used Mylar beamsplitters and a mercury arc source in the far-infrared. We are also dealing with an interferometer which has the movable and fixed mirrors positioned at right angles to each other.

The criterion for a perfectly aligned interferometer is that the radiant signal at the origin of the interferogram $I_R(0)$ be twice the radiant signal at large path differences $I_R(\infty)$. One can use this test as a means of deciding when an interferometer is in need of readjustment.

Before beginning the discussion of the various stages of alignment (coarse, intermediate, and fine), the cat's-eye retroreflector should be mentioned. In Appendix C, two types of retroreflector are described. The cat's-eye retroreflector components replace the movable and fixed mirrors and make optical alignment trivial even for asymmetric interferometers described in Chapter 8. The initial, alignment of the two mirrors in the cat's eye is critical, but the placement of the retroreflectors in the interferometer does not have stringent conditions.

## COARSE ADJUSTMENT

If the optical system of a Michelson interferometer is badly out of alignment such as for an initial step-up, the following procedure may be used. If only minor changes are needed, the next section should be read.

For the particular interferometer in mind, one has readily available adjustments on the fixed mirror and the beamsplitter and not on the movable mirror. The general procedure will be to locate the optical axis, then to adjust the beamsplitter relative to the movable mirror, to position the fixed mirror relative to the beamsplitter and movable mirror, and, finally, to focus the radiation beams on the center of the sample holder.

With an inexpensive He–Ne laser, one projects a visible beam down the axis of the interferometer. The radiation beam from the laser can be introduced into the interferometer at a point *after* the collimator and *before* the beamsplitter with glass prisms mounted in a suitable holder. The holder may take a few hours to make; but the time is well spent. One should remove the beamsplitter. Fine wire or string cross-hairs are strung before *and* after the beamsplitter position to locate approximately the central axis. (In the Beckman FS-720, the fixed mirror is entirely removed and one set of cross-hairs may be strung on bolts in its stead.) A card suitably placed in the path of the laser beam behind the beamsplitter module will show the shadow of *both* pairs of cross-hairs. The prisms are then carefully moved until the *shadows* of the two sets of cross-hairs coincide. The laser beam does not have to be right on the cross-hairs, just parallel to the line between the cross-hairs. After some practice, this alignment only takes a few minutes. Now the laser beam is near and parallel to the optical axis.

Now, put the beamsplitter in its holder. Adjust the beamsplitter so that the reflected radiation beam going to and the radiation beam reflected from the movable mirror coincide on the beamsplitter. There are two checks of coincidence. A little atmospheric water vapor chilled by the blow-off from a small liquid nitrogen flask makes the laser beam very visible (smoke will also do, but has cancerous drawbacks). Also, the laser beam, on returning to the beamsplitter, produces a slightly diffused spot of light on the beamsplitter. The beamsplitter can be adjusted until the primary and returning laser beam spots coincide. (Now, the beamsplitter is never to be moved again, as this is its final adjustment.)

Remove the cross-hairs at the fixed mirror and return the fixed mirror to its holder, firmly setting the assembly. Block the optical beam to the movable mirror and align the fixed mirror. As was done for the movable mirror, one now adjusts the fixed mirror until the incident and reflected optical beams for the fixed mirror coincide on the beamsplitter. (The coarse, intermediate, and fine adjustment screws on the back of the fixed mirror are utilized in the FS-720.)

Unblock the laser beam and observe the optical beams from the mirrors in the arms. With only the intermediate and fine adjustment controls of the fixed mirror, make the recombined optical beams from the two arms coincide just after the beamsplitter. Observe the optical beam in the sample chamber and move the fixed mirror until the brightest spots completely merge.

The beamsplitter and fixed and movable mirrors are now coarsely in position. The sample-chamber focusing optical components should next be adjusted to move the single laser beam spot to the approximate middle of the sample holder. Just moving the last mirror before the sample holder will usually do this. Having completed these phases of the adjustment, the prisms and laser beams should be removed.

Turn on the mercury lamp source (and be very cautious of cornea damage from the ultraviolet radiation). Put wire cross-hairs over the center of the source. Move both the source and the parabolic mirror used for collimation so that the image of the source is back on the source and is approximately *in focus*. The radiation beam goes from the source, is collimated by the parabolic mirror, is reflected back from the fixed mirror, and is focused on the source. This assures that the beam from the source and collimator is (1) collimated and (2) collimated parallel to the axis established by the laser beam. Therefore, the source–collimator and beamsplitter–mirror assemblies are *both* aligned.

Since the source may now be in such a position that the image is not at the center of the sample chamber, the last plane mirror must be adjusted again. Bringing the image to the center completes the coarse adjustment of the entire interferometer. Do not be concerned if the image at the sample holder is slightly blurred.

## INTERMEDIATE AND FINE ADJUSTMENT

Insert a stop with about a 1-in.-diameter hole in the center of the collimated optical beam *before* the beamsplitter and turn off the chopper if it was running. At the sample chamber, two separate images of the source wire cross-hairs will be seen. (If the instrument is very nearly aligned at this point, the two images will be very close together and may appear as a single, blurred image.) One image comes from each of the two arms of the interferometer. By using the *fine* adjustments on the fixed mirror, bring these two images into coincidence. When this is done, there will be a single, sharp image of the cross-hairs at the sampling point.

Now, if one places an index card in the collimated radiation beam *after* the beamsplitter, interference fringes are visible in a darkened room. The contrast of these fringes can be improved with the fine adjustments on the fixed mirror. If oil films form on the mirrors and beamsplitters due to the vacuum pumps, the fringes will be grossly distorted.

Lock all set-screws and nuts, being careful to hold adjustment, seal the interferometer, and begin operation. If the adjustments are good, then the ratio

$$[I_R(0) - I_R(\infty)]/I_R(\infty)$$

should be greater than 0.6. After careful adjustment, the ratio will be over 0.9 even for large-wavenumber radiation beams.

APPENDIX B

# *Computer Programs*

Ralph W. Alexander†
and
Harold V. Romero‡

This appendix contains two programs, both written for the HP 2114B computer in FORTRAN. Only a few modifications, such as changes to input–output statements, should be necessary for use with other machines.

The first program utilizes the Cooley–Tukey algorithm to transform a double-sided interferogram of up to 512 points. It is designed to read tapes from a data acquisition system using IBM paper tape code. It would be simpler if ASCII code had been used. Most small-computer manufacturers offer fast Fourier transform subroutines. If available, such a subroutine could replace the fast Fourier transform section of this program. The program plots the transformed interferogram on an HP 2700A plotter. This section would of course need modification for other plotters.

If the ratio of two spectra is desired, this can be done with a minor modification, but then only 256 points can be transformed.

## PROGRAM USING COOLEY–TUKEY ALGORITHM

```
FTN, B
C   COOLEY-TUKEY ALGORITHM FOR HP 2114B
C   THIS PROGRAM CAN DO UP TO 512 OF
C   A DOUBLE-SIDED INTERFEROGRAM. UP TO 530 POINTS
```

† Physics Department, University Missouri-Rolla.
‡ Los Alamos Scientific Laboratory, Los Alamos, New Mexico.

```
C   OF DATA CAN BE READ. PROGRAM USES ONLY
C   N POINTS ON EITHER SIDE OF THE MAXIMUM.
        DIMENSIONS S(530), LL(9)
        EQUIVALENCE (L13, LL(1)), (L12, LL(2)), (L11, LL(3)),
        1 (L10, LL(4)), (L9, LL(5)), (L8, LL(6)), (L7, LL(7)),
        2 (L6, LL(8)), (L5, LL(9))
        EQUIVALENCE (NP, N), (DENOM, AVE), (R, REAL), (FI, FIMAG),
        1 (SUBW, DENOM), (ILOC, N1), (LEND, N3), (JEND, N4), (BR DR),
        2 (BI, DI), (REAL, CR), (FIMAG, CI), (JMAX, JPP),
        3 (M2, N3), (M11, N4), (M21, JMAX), (FK, FL), (KN, N3), (KI, N4),
        4 (HOLD1, WREAL), (HOLD2, WIMAG), (RES, FK), (SIGI, WREAL),
        5 (SIG2, WIMAG), (K2, L8), (K1, N1), (K3, JMAX), (MASK, N1),
        6 (FMULT, FK), (SUM, BR), (DATA, BI), (SMAX, BR), (L8, M),
        7 (SMIN, REAL), (DNU, FIMAG), (FK, FNU), (FNUMX, BI),
        8 (N1, P), (N3, NU)
        PI=3.14159
C
C   REQUEST OPTICAL PATH DIFFERENCE – STEP
C
    1000 WRITE (2, 3)
       3 FORMAT ("STEP")
         READ (1, 4) STEP
       4 FORMAT (F15.6)
C
C   READ INTERFEROGRAM AFTER ASKING FOR NUMBER OF POINTS IN
C   INTERFEROGRAM
C
         WRITE (2, 5)
       5 FORMAT ("NP")
       6 FORMAT (14)
C   THIS SECTION READS THE IBM-CODE PAPER
C   TAPE FROM DATA ACQUISITION SYSTEM AND
C   CONVERTS DATA TO FORTRAN FORMAT. SUB-
C   ROUTINE READ READS ONE LINE FROM TAPE
C   (IT IS WRITTEN IN THE H-P ASSEMBLY
C   LANGUAGE).
         MASK=17B
         DO 7 K=1, NP
         FMULT=100
         SUM=0
         DO 500 J-1, 11
         CALL READ(IDATA)
         IF(J-3)500, 500, 501
     501 IF(J-10)503,500, 500
     503 IDATA=IAND(IDATA, MASK)
         DATA=FLOAT(IDATA)*FMULT
         SUM=SUM+DATA
         FMULT=FMULT*.1
     500 CONTINUE
       7 S(K)=SUM
```

```
C
C   FIND THE MAXIMUM VALUE OF S (SMAX) AND ITS SUBSCRIPT
C   (JMAX).
C
          SMAX=0
          DO 9 J=1, NP
          IF(S(J)-SMAX) 9, 9, 8
        8 SMAX=S(J)
          JMAX=J
        9 CONTINUE
C
C   DETERMINE N2POW SUCH THAT NP/2<2**N2POW<NP. THAT IS,
C   THE INTERFEROGRAM IS TO BE TRUNCATED TO N=2**N2POW
C   POINTS ON EITHER SIDE OF △=0.
C
          M=NP
          N2POW=-1
       10 M=M/2
          IF(M)12, 12, 11
       11 N2POW=N2POW+1
          GO TO 10
       12 CONTINUE
C
C   TRUNCATE INTERFEROGRAM. S(N1) IS THE FIRST POINT OF THE
C   TRUNCATED INTERFEROGRAM AND S(N4) THE LAST. N3 AND N2
C   GIVE THE LIMIT OF THE FLAT TOP OF THE TRAPEZOID USED FOR
C   APODIZATION.
C
          N=2**N2POW
          N1=JMAX-N
          N2=JMAX-3
          N3=JMAX+3
          N4=JMAX+N-1
          FN=FLOAT(N)
C
C   COMPUTE AVERAGE VALUE OF INTERFEROGRAM, I(∞), HERE
C   CALLED AVE.
C
       30 DO 22 J=N1, N4
       22 AVE=AVE+S(J)
          AVE=AVE/FN
C
C   SUBTRACT I(∞)=AVE FROM ALL POINTS OF INTERFEROGRAM.
C
          DO 23 J=N1, N4
       23 S(J)=S(J)-AVE
C
C   NOW PERFORM TRAPEZOIDAL APODIZATION.
C
          DENOM=FLOAT(N2-N1)
```

```
      DENOM=1./DENOM
      DO 40 J=N1, N2
   40 S(J)=S(J)*DENOM*FLOAT(J-N1)
      DENOM=FLOAT(N4-N3)
      DENOM=1./DENOM
      DO 41 J=N3, N4
   41 S(J)=S(J)*DENOM*FLOAT(N4-J)
C  RECALL THAT TO COMPUTE THE TRANSFORM OF A REAL FUNCTION,
C  WE CONSTRUCTED A NEW FUNCTION
C
C          T(J′) = Y(J′) + i*Z(J′)
C  WHERE
C          Y(J′) + S(J)   FOR   J=2J′
C  AND
           Z(J′) = S(J)   FOR   J=2J′+1
C
C  SO WE NOW SET UP THE REAL AND IMAGINARY PARTS OF T. WE SHALL
C  CONSIDER THE REAL PART OF T(J′) TO BE IN THE LOCATION INDEXED
C  BY 2J-1 AND THE IMAGINARY PART TO BE IN THE LOCATION INDEXED
C  BY 2J. THAT IS, THE IMAGINARY PART OF T(J′) IS STORED IN THE
C  LOCATION IMMEDIATELY AFTER THE LOCATION OF THE REAL PART.
C  THIS REQUIRES ONLY A SHIFTING OF THE S ARRAY TO A NEW ORIGIN.
C
      DO 51 J=1, N
      M1=2J+N1-2
      S(2*J-1)=S(M1)
      M1=M1+1
   51 S(2*J)=S(M1)
C
C  NOW PERFORM 'BINARY BIT INVERSION' SORT. FIRST SET UP THE
C  REQUIRED PARAMETERS.
C
      DO 68 J=1, 9
      LL(J)=1
      IF(J-N2POW)69, 69, 68
   69 LL(J)=FLOAT(2**(N2POW+1-J))
   68 CONTINUE
C
C  NOTE THAT LI=LL(14-I) VIA THE EQUIVALENCE STATEMENT. DO
C  'BINARY BIT INVERSION' SORT.
C
      J5=1
      IJ=1
      DO 100 J6=J5, L6, L5
      DO 100 J7=J6, L7, L6
      DO 100 J8=J7, L8, L7
      DO 100 J9=J8, L9, L8
      DO 100 J10=J9, L10, L9
      DO 100 J11=J10, L11, L10
      DO 100 J12=J11, L12, L11
```

```
      DO 100 J13=J12, L13, L12
      IF(IJ-JI)102, 101, 101
  102 R=S(2*IJ-1)
      FI=S(2*IJ)
      S(2*IJ-1)=S(2*JI-1)
      S(2*IJ)=S(2*JI)
      S(2*JI-1)=R
      S(2*JI)=FI
  101 IJ=IJ+1
  100 CONTINUE
C
C   NOW READY TO USE THE COMPLEX COOLEY-TUKEY ALGORITHM FOR
C   THE N COMPLEX VALUES OF T(J').
      SUBW=2.
      ILOC=2
      LEND=1
      JEND-N/2
      DO 220 K=1, N2POW
      ARG=2.*PI/SUBW
      DO 210 J=1, JEND
      DO 200 L=1, LEND
      FL=FLOAT(L-1)
      JP=2*ILOC*(J-1)+2*(L-1)+1
      JPP=JP+ILOC
      WREAL=COS(FL*ARG)
      WIMAG=-SIN(FL*ARG)
      REAL=S(JPP)*WREAL-S(JPP+1)*WIMAG
      FIMAG=S(JPP)*WIMAG+S(JPP+1)*WREAL
      S(JPP)+S(JP)-REAL
      S(JPP+1)=S(JP+1)-FIMAG
      S(JP)=S(JP)+REAL
  200 S(JP+1)=S(JP+1)+FIMAG
C
  210 CONTINUE
      LEND=LEND*2
      JEND=JEND/2
      ILOC=ILOC*2
      SUBW=SUBW*2
C
  220 CONTINUE
C
C   END OF COOLEY-TUKEY ALGORITHM FOR T(J'). NOW MUST FIND
C   C(K) AND D(K) FROM EQ. 17-61 AND 17-62.
C
      N2=N/2
      DO 250 M=2, N2
      M1=2*N+3-2*M
      M2=2*M-1
      M11=M1+1
      M21=M2+1
```

```
      DR=(S(M21)+S(M11))*.5
      DI=(S(M1)-S(M2))*.5
      CR=(S(M2)+S(M1))*.5
      CI=(S(M21)-S(M11))*.5
      S(M1)=DR
      S(M11)=DI
      S(M2)=CR
  250 S(M21)=CI
C
C  NOTE THAT THE C(K) ARE STORED IN THE FIRST N LOCATIONS AND
C  THE D(K) IN THE SECOND N LOCATIONS, IN REVERSE ORDER. NOW
C  USE EQ. 17-63 TO COMPUTE B(K) AND EQ. 17-69 TO GET CORRECTED
C  B(K).
C
      ARG=PI/FN
      DO 250 K=2, N2
      FK=FLOAT(K-1)
      WREAL=COS(FK*ARG)
      WIMAG=-SIN(FK*ARG)
      KN=2*N+3-2*K
      KI=2*K-1
      REAL=S(KN)*WREAL-S(KN+1)*WIMAG
      FIMAG=S(KN)*WIMAG+S(KN+1)*WREAL
      BR=S(KI)+REAL
      BI=S(KI+1)+FIMAG
      HOLD1=SQRT(BR*BR+BI*BI)
      BR=S(KI)-REAL
      BI=S(KI+1)-FIMAG
      S(KI)=HOLD1
  350 S(KN)=SQRT(BR*BR+BI*BI)
C
C  IF N IS ODD, THEN ONE POINT WAS MISSED. DO THIS NOW. S(O) ALSO
C  HAS BEEN SKIPPED.
C
      IF((N/2+1)-N/2)300, 300, 260
  260 S(N+1)=SQRT(S(N+1)*S(N+1)+S(N+2)*S(N+2))
C
C  NOW WE HAVE B(K) (EXCEPT FOR K=O) STORED AS S(2) THROUGH
C  S(N).
C
C  NOW PRINT OUT SPECTRUM. RES IS SPACING IN CM-1 OF OUTPUT
C  POINTS.
C
      RES=1./(2.*FN*STEP*1.E-4)
  420 FORMAT3(F9.3, E12.4)
      SIG1=0
      SIG2=FLOAT(N2)*RES
      DO 410 K=1, N3
      K1=2*K-1
      K2=K1+N2
```

```
      WRITE(2, 420)SIG1, S(K1), SIG2, S(K2)
      SIG1=SIG1+RES
  410 SIG2=SIG2+RES
C
C  NOW PLOT SPECTRUM USING HP 2700 PLOTTER. THIS PLOTTER IS IN
C  PARALLEL WITH TELETYPE. FIRST FIND MIN AND MAX OF TRANS-
C  FORMED INTERFEROGRAM (SMIN AND SMAX), EXCLUDING FIRST
C  TEN POINTS. FOR PLOTTING, SPECTRUM IS NORMALIZED TO RUN
C  BETWEEN 0 AND 9999.
C
      SMAX=0.
      SMIN=100000.
      FNU=8.*DNU
      FNUMX=DNU*FLOAT(N)
      DO 550 K=10, N
      J=2*K-1
      IF(S(J)-SMAX)530, 530, 540
  530 IF(S(J)-SMIN)535, 535, 550
  535 SMIN=S(J)
      GO TO 550
  540 SMAX=S(J)
  550 CONTINUE
      WRITE(2, 600)
  600 FORMAT("PLTL")
      DENOM=SMAX-SMIN
      DO 650 K=10, N
      FNU=FNU+DNU
      J=2*K-1
      HOLD1=(S(J)-SMIN)/DENOM
      HOLD1=HOLD1*9999.
      P=IFIX(HOLD1)
      NU=IFIX((FNU/FNUMX)*9999.)
  650 WRITE(2, 700)NU, P
  700 FORMAT(I4,1X, I4)
      END
      ENDS
```

SUBROUTINE READ is written in the Hewlett–Packard assembly language. It reads one tape frame.

```
ASMB,   B, L, R
        NAM READ
        EXT .ENTR
        ENT READ
IDATA   NOP
HOLD    NOP
```

```
READ    NOP
        JSB .ENTR         .ENTR is a library routine providing address of para-
        DEF IDATA         meter IDATA in call.
        CLA               Tape reader is in IO slot 10 octal.
        LDA 10B
        STA HOLD          Disable interrupt.
        CLA
        STA 10B
        STC 10B, C
        SFS 10B
        JMP *-1
        LIA 10B
        STA IDATA, I
        LDA HOLD
        STA 10B
        CLC 10B
        JMP READ, I
        END
```

TABLE OF SYMBOLS

| | |
|---|---|
| PI | 3.14159. |
| STEP | Optical path difference between points in interferogram. |
| NP | Number of points in interferogram (before truncation). This is the number of data points to be read by program. |
| MASK | 000017 binary used in conversion from IBM code to internal FORTRAN format. |
| FMULT | Decimal place value used in IBM to internal FORTRAN format. |
| SUM | Temporary storage. |
| S(K) | Storage for interferogram and its Fourier transform. |
| SMAX | Maximum value of interferogram. |
| JMAX | Subscript of maximum value of interferogram. |
| M | Temporary storage. |
| N2POW | Satisfies relation: NP/2 < 2**N2POW ≤ NP, i.e., after truncation, interferogram 8 + AS N = 2**N2POW points on either side of maximum. |
| N | 2**N2POW. |
| N2 | Lower limit of flat top of trapezoidal apodizing function. |
| N3 | Upper limit of flat top of trapezoidal apodizing function. |
| N1 | Index of first point of truncated interferogram. |
| N4 | Index of last point of truncated interferogram. |
| FN | FLOAT(N). |
| AVE | Temporary storage. |
| DENOM | Temporary storage for denominator of apodizing function. |
| M1 | Temporary storage for indices. |
| J1 | Temporary storage for indices. |
| LL(J) | Indices used in "binary sort." Related to LI by LI = LL(14-I). |
| L1 | Index used in binary sort. |
| R | Temporary storage (for real part of a complex number). |

| | |
|---|---|
| FI | Temporary storage (for imaginary part of complex number). |
| JI | Index used in binary sort. |
| IJ | Index used in binary sort. |
| SUBW | Parameter used Cooley–Tukey algorithm. In the quantity $W_l^J$, SUBW corresponds to $l$. |
| ILOC | Index used in Cooley–Tukey algorithm. |
| WREAL | Real part of $W_K^J$. |
| WIMAG | Imaginary part of $W_K^J$. |
| REAL | Temporary storage (for real part of complex number). |
| FIMAG | Temporary storage (for imaginary part of complex number). |
| N2 | $N/2$. |
| M2 | Temporary storage for index. |
| M11, M21 | Temporary storage for index. |
| BR, CR, DR | Temporary storage (for real part of complex number). |
| BI, CI, DI | Temporary storage (for imaginary part of complex number). |
| RES | Spacing in wave numbers between points of the transformed interferogram. |
| S1G1, S1G2 | Output wave number in form column table. |
| SMAX, SMIN | Maximum and minimum, respectively, of transformed interferogram. |
| FNU | Output wave number for plot (X-AXIS). |
| FNUMX | Maximum wave number in output. |

## PROGRAM FOR REAL-TIME ANALYSIS

This is a simple HP–FORTRAN program to do Fourier analysis of a one-sided interferogram while the interferogram is being measured. A parabolic fit to the three points nearest $\delta = 0$ is used to provide an approximate phase-error correction as described in Chapter 18. Computation for 128 output points requires about 4 sec. This is considerably slower than the single precision assembly-language program suggested in Chapter 18.

```
FTN, B
      PROGRAM RLTM
      DIMENSION X(3), B(128), IBUFF(128)
      COMMON B
      BMAX=.00001
      TPI=6.28318
C  B(K) WILL BE THE COMPUTED SPECTRAL POINTS AND BMAX THE
C  MAXIMUM VALUE OF THIS SET. IBUFF WILL BE THE B(K) NOR-
C  MALIZED TO A MAXIMUM VALUE OF 250 FOR THE PURPOSE OUTPUT
C  TO THE OSCILLOSCOPE DISPLAY
      WRITE(2, 1)
    1 FORMAT("IAVE=")
C  IAVE IS VALUE OF INTERFEROGRAM FOR LARGE δ (HERE CALL FO).
```

```
    2 READ(1, 6)FO
      WRITE(2, 2)
    2 FORMAT("DDEL=")
      READ(1, 6)DDEL
C   DDEL=△δ
      WRITE(2, 3)
    3 FORMAT("DSIG=")
      READ(1, 6)DSIG
C   DSIG IS THE WAVENUMBER INCREMENT DESIRED FOR OUTPUT
    POINTS.
      WRITE(2, 4)
    4 FORMAT("SIGL=")
      READ(1, 6)SIGL
C   SIGL IS WAVENUMBER FOR FIRST POINT IN OUTPUT SPECTRUM.
      WRITE(1, 5)
    5 FORMAT("NPT=")
      READ(1, 7)NPT
      FNPT=FLOAT(NPT)
    6 FORMAT(E15.9)
    7 FORMAT(I3)
C   NPT IS NUMBER OF POINTS IN INTERFEROGRAM. IT IS USED FOR
C   THE TRIANGULAR APODIZATION FUNCTION.
      DO 15 K=1, 128
   15 B(K)=0.
      CALL DVM(DATA)
C   DVM IS AN ASSEMBLY LANGUAGE SUBROUTINE TO READ A HEWLETT-
C   PACKARD 2401C DVM AND CONVERT READING TO A FLOATING
C   POINT NUMBER.
      X(1)=DATA-FO
      CALL DVM(DATA)
      X(2)=DATA-FO
  100 CALL DVM(DATA)
      X(3)=DATA-FO
      IF(X(2)-X(3))110, 111, 111
C   IF SECOND POINT WAS NOT THE MAXIMUM OF INTERFEROGRAM,
C   ACQUIRE ANOTHER POINT.
  110 X(1)=X(2)
      X(2)=X(3)
      GO TO 100
  111 EPSLN=3.*(X(1)+X(3))
      EPSLN=EPSLN-2.*(X(1)+X(2)+X(3))
      EPSLN=(X(3)-X(1))/EPSLN
      EPSLN=EPSLN*DDEL
C   CALCULATE ε USING PARABOLIC FIT.
      T=TPI*DSIG
      TO=T*EPSLN
   25 IF(TO-TPI)26, 27, 27
   27 TO=TO-TPI
      GO TO 25
   26 T=T*DDEL
```

```
   40 IF(T-TPI)30, 30, 35
   35 T=T-TPI
      GO TO 40
   30 N=IFIX(SIGL/DSIG)
      P=FLOAT(N)
      THETA=P*TO
      PT=P*T
      S=X(2)
      FJT=TO
C  HAVE NOW CALCULATED THE PARAMETERS To, Tp, THETA=
C  θ(j, K=0) AND FJT=TO+jpT AS DEFINED IN CHAPTER 18.
      CALL TRNSF(S, THETA, FJT, IBUFF, BMAX)
      WRITE(2, 4)THETA
      FJT=FJT+T
      THETA=THETA+PT
      S=X(3)
      CALL TRNSF(S, THETA, FJT, IBUFF, BMAX)
C  USING SUBROUTINE TRNSF, CALCULATE THE B(K) FOR j=0 AND
C  j=1. NOW REPEAT FOR NPT POINTS. NOTE TRIANGULAR APODIZA-
C  TION IS USED. CALL TO SUBROUTINE SPOT DISPLAYS B(K) ON
C  OSCILLOSCOPE.
      DO 20 K=1, NIN
      CALL DVM(DATA)
      S=(DATA-FO)*(FNPT-FLOAT(K))/FNPT
      FJT=FJT+T
      THETA=THETA+PT
      CALL TRNSF(S, THETA, FJT, IBUFF, BMAX)
      DO 45 J=1, 200
   45 CALL SPOT(IBUFF)
   20 CONTINUE
```

At this point, hard copy plot and printout similar to that of proceding program is made.

**SUBROUTINE DVM.** This routine is written in the H–P assembly language and acquires reading from a 2401C DVM.

```
ASMB, B, R
        NAM DVM
        ENT DVM
        EXT .ENTR, FLOAT      .ENTR and FLOAT are HP library routines.
.1110   EQU 13B
DATA    BSS 1
DVM     NOP
        JSB .ENTR             .ENTR is used to transfer parameter address.
        DEF DATA
        CLA
```

```
        CLB
        STA .1110           Disable interrupt.
        STC .1110, C        Wait for completion of reading.
        SFS .1110
        JMP *-1
INPUT   CLC . 1110
        LIA .1110           Read DVM output.
        LIB . 1110
        ALF, ALF            Convert to floating point.
        AND MSXK37
        STA IDATA
        BLF, BLF
        LDA 1
        AND MSK17
        ADA IDATA
        CLB
        JSB BCDTB
        JSB FLOAT
        DST DATA, I
        JMP DVM, I
BCDTB   NOP
        LDB DM16
        STB CTR
        LDB DTABL
        STB PTR
        CLB
LOOP    SSA
        ADB PTR, I
        RAL
        ISZ PTR
        ISZ CTR
        JMP LOOP
        LDA 1
        JMP BCDTB, I
TABLE   DEC 4000, 2000, 2000, 1000, 400, 200, 200, 100
        DEC 40, 20, 20, 10, 4, 2, 2, 1
        ORB
DTABL   DEF TABLE
PTR     BSS 1
MSK37   OCT 377
MSK17   OCT 177400
CTR     BSS 1
```

SUBROUTINE SPOT is written in the H–P assembly language and outputs 128 values of $B(K)$ to oscilloscope display. Output of form (X, IBUFF), where X is $x$ coordinate (eight bits) and IBUFF is $y$ coordinate (eight bits). Both axes run from 0 to 256.

```
ASMB, R, B
        NAM SPOT
        ENT SPOT
        EXT .ENTR
OSC     EQU 15B
ORD     BSS 1
SPOT    NOP
        JSB .ENTR
        DEF ORD
        CLF OSC
        LDA LIM
        STA CTR             Set loop counter
        CLB
RPT     LDA ORD, I          Bring IBUFF(K)
        ADA 1B              Add X coordinate
        OTA OSC             Display
        ADB INX             Increment X cocrdinate
        ISZ ORD             Increment K
        ISZ CTR             Finished?
        JMP RPT
        JMP SPOT, I
LIM     DEC -128
INX     OCT 1000
CTR     BSS 1
        END
```

SUBROUTINE TRNSF computes B(K) + S*COS$\theta$(j, k) for current $j$ and $k = 1$ to 128. IBUFF are the $B(K)$ normalized to a maximum value of 256.

```
          FTN, B
          SUBROUTINE TRANSF (S, THETA, FJT, IBUFF, BMAX)
          DIMENSION B (128), IBUFF(128)
          COMMON B
          P12=6.28318
      200 IF (THETA-P12) 210, 210, 205
      205 THETA=THETA-P12
          GO TO 200
      210 IF(FJT-P12)230, 230, 220
      220 FJT=FJT-P12
          GO TO 210
      230 TJK=THETA—FJT
C     TJK IS θ(j, k)
          DO 102 K=1, 128
          TJK=TJK+FJT
          B(K)=B(K)+S*COS(TJK)
          WRITE(2, 25)B(K)
       25 FORMAT(E15.9)
```

```
      IF(B(K))105, 105, 110
  105 IBUFF(K)=0
      GO TO 102
  110 IF(B(K)-BMAX))120, 120, 115
  115 BMAX=B(K)
  120 IBUFF(K)=IFIX(250.*B(K))/BMAX)
  102 CONTINUE
      RETURN
      END
      ENDS
```

APPENDIX C

# *Mirror Tilt, The Cat's-Eye Retroreflector*

## INTRODUCTION

If one or both of the fixed and movable mirrors in an interferometer are not perfectly aligned, the wavefronts emerging from the beamsplitter will not be parallel. This causes a loss in modulation in the interferogram, introduces mean displacements in the optical path difference, and results in a loss in resolution in the computed spectrum.

The problems can be overcome by replacing the plane mirrors with retroreflectors. There are many types of retroreflector, such as the corner cubes placed on the moon by astronauts Armstrong and Aldrin in 1969, and several are described by Steel [1]. The corner cube reflector, with three reflecting surfaces, has six axes of rotation to align, which are very difficult and expensive to position. The cat's-eye retroreflector has only two mirrors and, as will be shown, only one very critical dimension—the distance between the two mirrors in the retroreflector.

For the above reasons, the cat's eye is a superior retroreflector and can be made with either a small, convex or concave mirror used in conjunction with a large, concave (parabolic) mirror. It turns out experimentally that the cat's-eye retroreflector using the convex secondary mirror produces the largest throughput when extended sources are used as in Fourier transform spectroscopy [2]. Systems with convex secondary mirrors are described in a theoretical article by Cuisenier and Pinard [3]. An excellent example of the use of the cat's-eye retroreflector is in the asymmetric interferometer described by Sanderson and Scott [2].

For cat's-eye retroreflectors with concave secondary mirrors, the reader is referred to an article by Beer and Marjaniemi [4]. It is claimed by Beer and Marjaniemi that a concave secondary produces less distortion of the wavefronts than does a convex secondary. However, this statement is true only if *plane* wavefronts are considered to be incident upon the retroreflector. As mentioned above, Sanderson found experimentally that if there are skew rays in the incident beam (as is the case for an extended source), the use of a plane or concave secondary considerably reduces the throughput [2]. Evidently, some skew rays have been left out of the computer ray-tracing treatment of the cat's eye by Beer and Marjaniemi. Nonetheless, the cat's eye with restricted fields of view yields excellent results for the 2000–8000-$cm^{-1}$ region, as shown by Schindler [5].

Before proceeding with the discussion, it should be stated explicitly that the entire retroreflector is moved in order to make changes in the optical path. Also, it should be mentioned that the radiation beam entering on the bottom of a retroreflector leaves at the top of the system. That is, the radiation beam is laterally displaced, but the radiation beam can enter both sides of the retroreflector simultaneously so that nearly all the *total* area is fully utilized.

The retroreflector is excellent for changing the direction of propagation of a radiation beam. It can change the direction by 180° with only a few seconds of arc error. However, one must closely examine whether the emerging wavefront is distorted. Much of the following discussion is concerned with the flatness of the reflected wavefront.

## CAT'S-EYE RETROREFLECTOR (CONVEX SECONDARY)

The cat's eye retroreflector with a convex secondary mirror is shown in Fig. C-1. The primary mirror is usually a parabolic surface for best results, as shown by Cuisenier and Pinard (the discussion of the spherical primary mirror may be found in their article [3]). The primary and secondary mirrors are paraxial with about the same radii of curvature. The focal point $F_1$ of the parabolic mirror is at the vertex of the convex secondary mirror.

As long as the optical rays are incident at no more than 1° off the optical axis formed by the line between $C_1$ and $C_2$, the cat's-eye retroreflector will turn the radiation beams by exactly 180°. The deviations of the returned wavefront will vary from the plane wave by an amount $z$ given by

$$z = [(2y^6/R_1{}^6) - (D_1{}^2 y^4/3R_1{}^6)]\delta, \qquad \text{(C-1)}$$

where $y$ is the distance from the optical axis to any point in the aperture

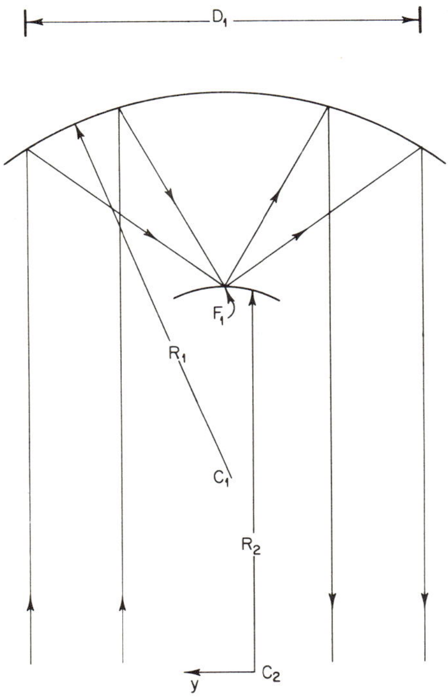


FIG. C-1. Optical diagram of a cat's-eye retroreflector with a convex secondary mirror.

and $\delta$ is the optical *path* difference of the radiation beams in the two arms of the interferometer. The worst deviation is for $y = D_1/2$, and is given by

$$z_{\max} = (\delta/96)(D_1/R_1)^6 = \delta/[6144(f/\text{No.})^6], \tag{C-2}$$

where

$$(f/\text{No.}) \equiv F_1/D_1 \simeq R_1/2D_1. \tag{C-3}$$

Equation (C-2) indicates that the deviations can be minimized with large $f$/numbers. Since the resolution $\delta\sigma$ and maximum path difference $L$ of an interferometer are related by $\delta\sigma = 1/L$ [Eq. (6-5)], we have the maximum $z$ given by

$$z_{\max} = R/[6144(f/\text{No.})^6], \tag{C-4}$$

where $R$ is the resolving power of the interferometer scan and $\lambda = 1/\sigma$ is the wavelength. For an $f/4$ primary mirror in the cat's eye, Eq. (C-4) indicates that the deviations will be of the order of $\lambda/25$ for resolving powers even as high as $10^6$. However, this assumes that the two retroreflectors are identical.

If the distance $e$ between the primary and secondary mirrors is not the same for the two retroreflectors, serious wavefront deviations appear. The variation in the wavefront due to a lack of perfect mating of the distances $e$ is by far the most important error in the retroreflector systems. If the two distances differ by a distance $de$, the maximum deviation in the wavefront becomes

$$z_{\max} = de/[4(f/\text{No.})^2] \tag{C-5}$$

For an $f/4$ primary mirror, $z_{\max}$ is $de/64$; so, for visible light, $de$ should be reduced to less than 6$\mu$m if the wavefront is to deviate by less than $\lambda/5$. This is a realizable, but stringent condition.

So the important consideration is to mate the two retroreflectors very carefully. Considering the two retroreflectors in the Michelson interferometer, the maximum optical path difference used does affect the wavefront deviations, but not nearly as much as the spacing between the primary and secondary mirrors of *both* of the retroreflectors.

In Fig. C-2, an optical diagram of an asymmetric Michelson interferometer built by Sanderson and Scott [4] is shown. The collimating mirror, telescope mirror, and cat's eye primaries are all identical paraboloids. They

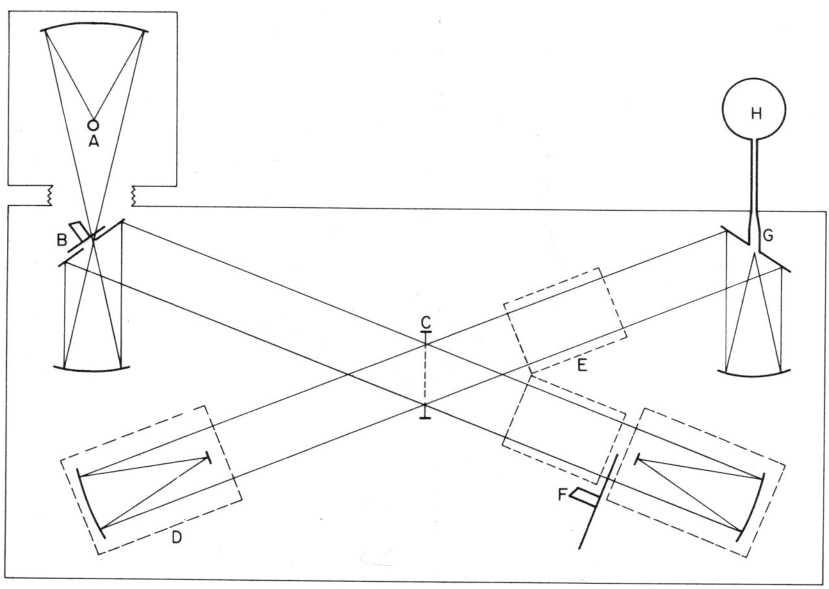


FIG. C-2. Optical diagram of asymmetric interferometer using cat's-eye retroreflectors. A: Source; B: chopper; C: beamsplitter; D: movable mirror (20 cm); E: cell; F: chopper; G: cone; H: detector.

have a diameter of 15 cm and a focal length of 25 cm. The primary of the cat's eye forms an image of the source on the small, convex secondary mirror. The curvature of the secondary mirror ($R_2 \simeq 2F_1$) is chosen so that the beamsplitter is imaged back upon itself by the cat's-eye assembly. This arrangement optimizes the throughput without causing appreciable loss of resolution. Use of the convex secondary improves the signal by a factor of two over that obtained using a plane secondary [2].

The use of the cat's eye greatly facilitates alignment. For the instrument built by Sanderson and Scott, the interferometer arms are about 70 cm long. With plane end mirrors, the adjustment of these mirrors and of the beamsplitter for asymmetric interferometry would be very critical. With the cat's eye, alignment is done by the following simple procedure. Cross-hairs are placed in the hole in the entrance flat. When the cross-hairs are simultaneously imaged on both secondaries and at the hole in the exit flat, it is possible to observe interference fringes. Final alignment is done with the secondary mirrors of the cat's-eye retroreflectors. An important advantage of this arrangement is the insensitivity of the alignment to the orientation of the beamsplitter. This makes it possible to change beamsplitters without re-aligning optics.

It should be pointed out that for the interferometer built by Sanderson and Scott, the fractional correction to the observed wave numbers due to the extended source was about $-4 \times 10^{-4}$ if the input solid angle were uniformly filled. Thus, from Eq. (11-25), the solid angle is about $16\pi \times 10^{-4}$ ster. From Eq. (11-5), the maximum angle of divergence $\theta$ is about 2.3° and so the field of view is about 4.6° for Sanderson and Scott's asymmetric interferometer.

In making the mirrors for use in the near infrared, the primary mirrors should be figured to 1/10 of a visible wavelength and have nearly equal radii. However, the secondary mirrors can be made at lower optician shop tolerances.

## CAT'S-EYE RETROREFLECTOR (CONCAVE SECONDARY)

The cat's-eye retroreflector with a concave mirror for the secondary reflector was described in an article by Beer and Marjaniemi [4]. In Fig. C-3, an optical diagram of a retroreflector is shown with the pertinent parameters defined. In the article by Beer and Marjaniemi, two types of cat's-eye retroreflector were considered—one with a parabolic mirror for the primary reflector and one with a spherical mirror for the primary reflector. They

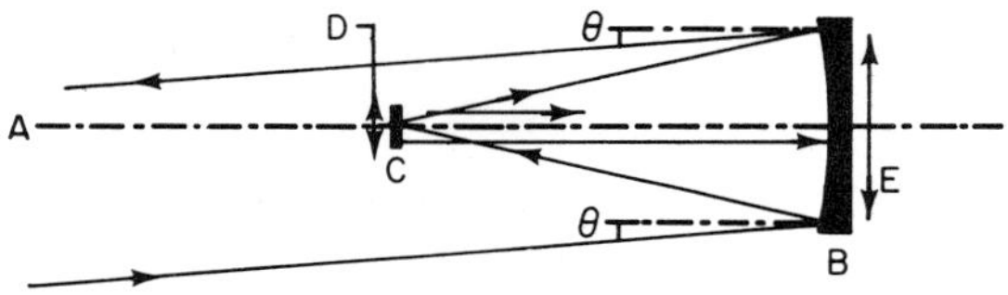


FIG. C-3. The cat's eye retroreflector. A: Optical axis; B: primary reflector; C: secondary reflector; D: aperture $2A_2$; E: aperture $2A_1$. Within the limitations of Gaussian optics, emergent rays are parallel to those incident. The secondary lies in the focal plane of the primary (nominally).

concluded, using computer ray-tracing techniques, that the parabolic primary mirror makes the best retroreflector. Also, using the planeness of the reflected wavefront as a criterion, they prove that a concave secondary mirror is superior to a convex or plane secondary mirror. With only plane waves incident and using concave secondary mirrors in the retroreflectors, one gets circular fringes in an interferometer, whereas the convex secondary mirrors yield hyperbolic fringe patterns. Also, with the concave secondary mirror and parabolic primary reflector, they found that the reflected ray deviates from the perfect back-reflection direction (perfect retroreflector) by less than $6''$ of arc for a $10^\circ$ field ($\theta = 5^\circ$). This greatly reduces alignment problems. Nonetheless, the reader must remember that Beer and Marjaniemi evidentally did not include all possible skew rays. Thus, we rely on the experiments of Sanderson and Scott [2] to decide the merits of the different cat's eyes with extended sources.

Considering plane wavefronts incident on the retroreflector, there are two criteria which can be used to determine if a retroreflector is optimized with respect to its design parameters. One criterion is that the wavefront, after being reflected by the retroreflector, have minimum deviations from the parallel, plane wavefronts. The other criterion is somewhat more involved. It is that the *derivative* of the function describing the emergent wavefronts from the retroreflector should be a minimum. The retroreflector cannot be considered by itself. With a thick beamsplitter, there is a lateral shear (or displacement as shown in Fig. C-4) of the radiation beams. If the wavefronts from the retroreflectors incident on the beamsplitter are plane, parallel, and normal to the optical axis, there is no effect on the optical path difference. However if the wavefronts are not plane, there will be a variation in the optical path difference across the aperture and a change in the mean path difference. If one has a poor retroreflector, this variation can occur, and one will have loss of resolution and shifts in the computed spectrum. Also, there can be considerable loss in the modulation in the interferogram. The magnitudes of the departures of the wavefronts from planeness are a function of

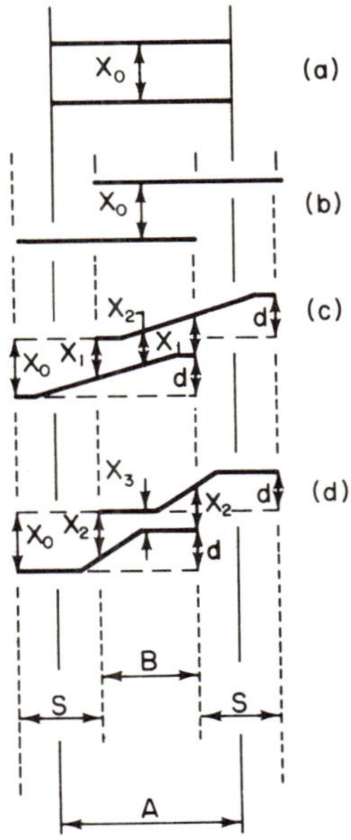


FIG. C-4. Illustration of the interference between two identical and parallel, but sheared and distorted, wavefronts. A: Overall aperture; B: interfering region; S: shear $S$. It is clear that in order to keep the variation in path difference across the interfering region minimal, the derivative of the wavefront function, as well as its absolute value, should be a minimum. (a) Wavefronts plane parallel, no shear. Path difference $x_0$ across entire aperture. (b) Wavefronts plane parallel, shear $S$ introduced. Path difference $x_0$ across interfering region. (c) Wavefronts parallel, distorted by maximum value $d$, shear $S$ introduced. Path difference variable across interfering region and mean path difference $\neq x_0$. (d) Wavefronts parallel, distorted by maximum value $d$ with steeper slope than in (c), shear $S$ introduced. Path difference more strongly variable across interfering region than in (c), but mean path difference almost unchanged. (After Beer and Marjaniemi, [4].)

shear. While the change in mean path difference is mainly a function of the *deviation* of the wavefront from an ideal plane wave, the variation across the aperture is a function of the *derivative*. These variations are illustrated in Fig. C-4 and again explained in the figure caption.

Beer and Marjaniemi studied the cat's-eye retroreflector by computer ray tracing and presented their results in graphic form. Their results are valid for any wave number by the choice of normalized parameters. In this appendix, only one case will be illustrated: parabolic primary mirror and 1° field of view. Other examples are in their article. In order to understand their curves, we present the relevant definitions and equations.

Beer and Marjaniemi defined:

$Y/A_1 \equiv$ relative semiaperture $\equiv$ radial distance to a point in the aperture from the optical axis normalized in units of the radius of the primary mirror, $A_1 \equiv D_1/2$, unitless. (C-6)

focal ratio $\equiv$ focal length of the primary mirror divided by the radial distance to a point in the aperture from the optical axis in units of $D_1$, unitless. Also called the $f$/No. by Beer and Marjaniemi. (C-7) normalized phase error ≡ departure distance, in microns, of a wavefront from a plane divided by the product of the relative semiaperture and $A_1$, in centimeters. (C-8)

ray deviation from the direction of the reflected beam of a perfect retroreflector in microradians ≡ slope of the normalized phase error versus relative semi-aperture curve $\times 10^2$. (C-9)

$\delta x \equiv$ the axial displacement of the secondary mirror from the paraxial focal point of the primary mirror. (C-10)

The factor of $10^2$ in Eq. (C-9) was not explained by Beer and Marjaniemi.

For a 1° field of view ($\theta = 0.5°$) and a parabolic primary mirror, Beer and Marjaniemi computed the effect on wavefronts of changing the curvature of the secondary mirror. In Fig. C-5, the results are presented for various radii

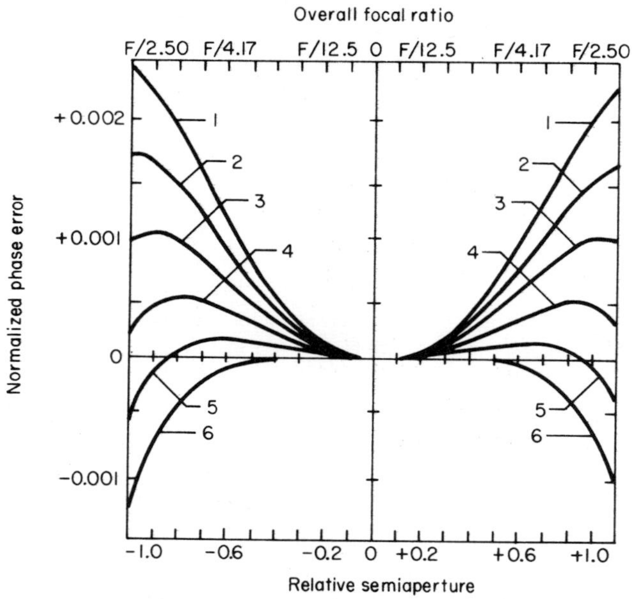


FIG. C-5. Parabolic primary, spherical secondary. Effect on wavefronts of changing the curvature of the secondary, narrow-angle (1° field) case. Parameter $\delta x = 0$. $R_2$ value: (1) $-2.105F_1$, (2) $-2.083F_1$, (3) $-2.062F_1$, (4) $-2.041F_1$, (5) $-2.020F_1$, (6) $-2.000F_1$. (After Beer and Marjaniemi [4].)

of curvature $R_2$ of the secondary mirror in terms of the focal length of the primary mirror $F_1$. The minimum derivative condition indicates that $R_2$ should be $-2.062F_2$. (The minus sign indicates that the secondary mirror should be concave.)

In Fig. C-6, the effect on wavefronts of displacing the secondary mirror from the focal point of the primary mirror for various $\delta x$ as defined in Eq. (C-10) is presented. The optimum radius of the secondary mirror of $-2.072F_1$ was chosen for the curves in Fig. C-6. It should be noted that the normalized phase errors for the values of $x$ used are two orders of magnitude larger than

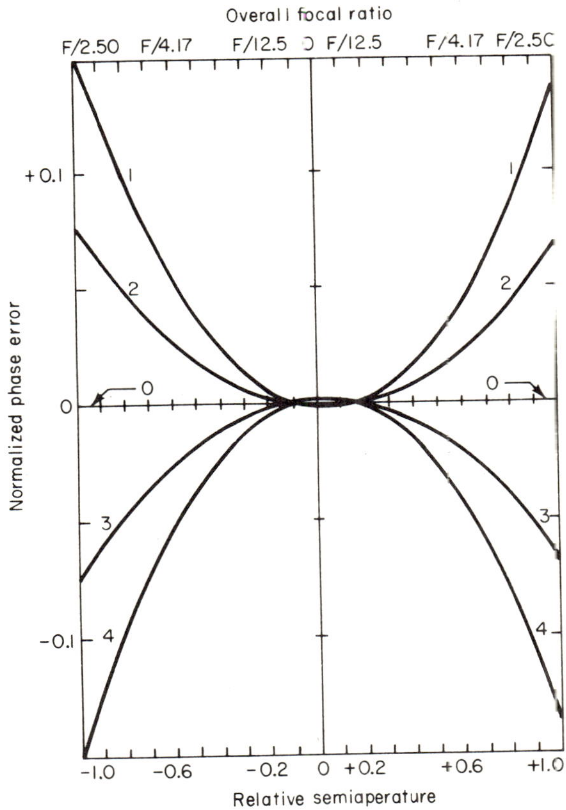


FIG. C-6. Parabolic primary, spherical secondary. Effect on wavefronts of displacing the secondary mirror from the primary mirror focal point, narrow angle (1° field) case. Parameter $R_2 = -2.062F_1$. $\delta x$ values: (1) $-0.0004A_1$, (2) $-0.0002\ A_1$, (3) $+0.0002A_1$, (4) $+0.0004A_1$. (After Beer and Marjaniemi [4].)

chosen in selecting the optimum $R_2$ in Fig. C-5. We shall return to this point later

In Fig. C-7, the optimum radius $R_2$ and zero axial displacement of the secondary mirror have been assumed in order to study the effect of the variation of the field of view on the normalized phase error. The angles listed on the curves are for various $\theta$, with

$$\text{field of view} = 2\theta. \tag{C-11}$$

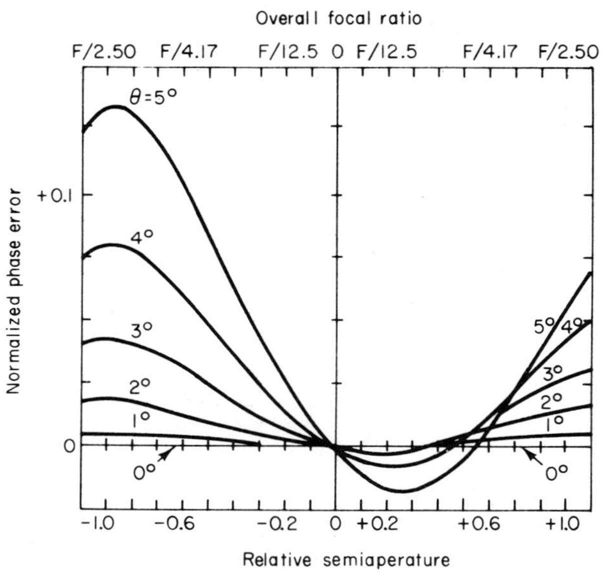


FIG. C-7. Parabolic primary, spherical secondary. Wavefronts for angles of incidence 0–5° and parameters $R_2 = -2.062F_1$ and $\delta x = 0$. (After Beer and Marjaniemi [4].)

With Figs. C-5–C-7, one can analyze a cat's-eye retroreflector with a parabolic primary mirror, a concave secondary mirror, and a 1° field of view. We have indicated that the derivative of the normalized phase error versus relative semiaperture curve will be minimized to obtain the optimum design parameters. However, we also need a limit on allowable errors. The allowable errors are based on the departure of the wavefront from the best *possible* wavefront (not always a plane wavefront) by amounts *equal* to the deviations of the best possible wavefront from a plane wave. (The best possible wavefront is the one determined by requiring that the derivative be a minimum.) Beer and Marjaniemi called these allowable errors the "100% errors in the departure of the wavefront from a plane wavefront."

In Fig. C-5, one has the smallest positive and negative slopes or derivatives for $R_2 = -2.062F_1$. The slopes can be about doubled in going to the $R_2 = -2.083F_1$ curve, so we state that the optimum value of the radius of curvature of the secondary mirror is $R_2 = -(2.06 \pm 0.02)F_1$. That is, the secondary mirror should have a focal length nearly, but not quite, equal to the focal length of the primary reflector.

It is claimed by Beer and Marjaniemi that the optimum value of the displacement of the secondary mirror from the focal point of the primary mirror is $\delta x = (0 \pm 2 \times 10^{-6})A_1$ (see Fig. C-6). For a 10-cm-diameter mirror, this means that the tolerances on the location of the secondary mirror $\delta x$ would be $\pm 10^{-5}$ cm or $\pm 1000$ Å. This is too stringent to be practical, Therefore. instead of asking what is the optimum, we shall ask what $\delta x$ can be tolerated and what the resulting departures from plane waves are. Suppose the experimentalist can align the secondary mirror such that $\delta x = (0 \pm 5)$ $\mu$m. If the primary mirror diameter $D_1$ is 10 cm, then $\delta x = (0 \pm 1 \times 10^{-4})A_1$ cm. From Fig. C-6, one might suppose for a relative semiaperture of 1.0 (implying $f/2$) that the normalized phase error is thus about $\pm 2 \times 10^{-2}$. Then, using the definition in Eq. (C-8), the distance of departure from a plane wave is about $\pm 2 \times 10^{-2} A_1$ $\mu$m, and for the 10-cm diameter mirror, is $\pm 0.1$ $\mu$m $= \pm 1000$ Å or $\pm 1/5$ of a visible wavelength. It is usually very satisfactory to have an optical component of such high quality. Normalized phase error tolerances of $\pm 2 \times 10^{-2}$, introduced by the tolerances on the location of the secondary mirror, are much larger than the limits of $R_2$ as shown in Fig. C-5. Therefore, we may regress and claim that within safe limits $R_2 = -(2.06 \pm 0.1)F_1$.

We might inquire about the ray deviations from perfect back reflection from the retroreflector. The restrictions of a 10° field of view ($\theta = 5°$) and an $f/5$ focal ratio, from Fig. C-7 and using Eq. (C-9), yield a ray deviation of about $200 \times 10^{-6}$ rad or 5″. That is, even if the field of view is 10 times wider than anticipated (1°), the experimentalist has a retroreflector that only deviates from perfection by 5″ of arc!

Thus, the design parameters for a retroreflector which is good enough for visible wavelength Fourier transform spectroscopy have been specified by Beer and Marjaniemi. Since the experimentalist in the infrared region only needs a wavefront plane within about 1/10 of the longer, infrared wavelengths, he may ease the tolerances on $\delta x$ even more than described above. For instance, at 5 $\mu$m wavelength or 2000 cm$^{-1}$, it would only be necessary for 1/10 wavelength tolerances to locate the secondary mirror at $\delta x = 0 \pm 25$ $\mu$m.

As claimed by Beer and Marjaniemi, although without proof, the lateral positioning of the mirror is not at all critical. So, with the cat's-eye retroreflector, the alignment for retroreflection with 1° fields of view is trivial. One has back reflection for even a 10° field of view deviating by about only 5″ of arc. If $\delta x$ is accurate to 5 $\mu$m, then the wavefront is flat and departs from a plane wave by less than 1/5 of a visible wavelength.

The practical system shown in Fig. C-8 was described by Schindler [5]. The key feature of the interferometer is the holding of the secondary mirror relative to the primary mirror with a fused quartz tube so that thermal

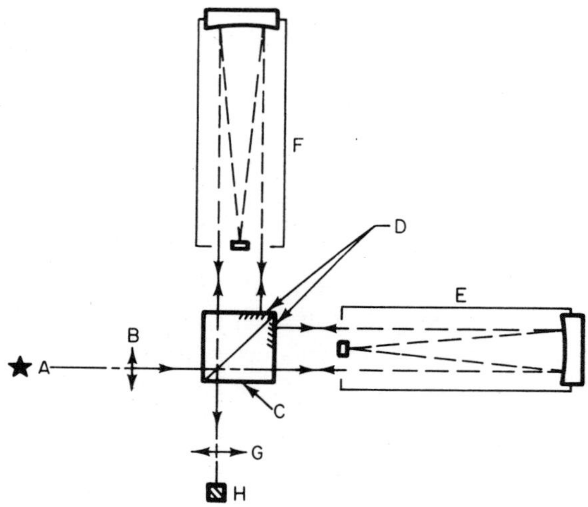


FIG. C-8. Optical diagram of interferometer using cat's-eye retroreflectors. A: Source; B: collimator; C: beamsplitter cube; D: fully reflecting faces; E, F: cats'-eye retroreflectors; G: condenser; H: detector. (After Schindler [5].)

expansion (of $\delta x$) could be limited. (A cer-vit glass-ceramic tube made by Owens-Illinois of Toledo, Ohio has an expansion coefficient seven times smaller.) The quartz tube is large and is kept out of the optical beams. Also, Schindler make a double pass through the cat's-eye retroreflector to make the instrument insensitive to the lateral displacement of the moving retroreflector optical axis from that of the fixed one. This also made the change in optical path difference four times the distance of motion of the retroreflector system. In essence, Schindler used only one-half of each retroreflector to accept radiant flux from the beamsplitter as shown in Fig. C-8.

## *References*


*1.* W. H. Steel, Interferometers for Fourier Spectroscopy, *Aspen Int. Conf. on Fourier Spectrosc., 1970*, p. 43. (G. A. Vanasse, A. T. Stair, Jr., and D. J. Baker, eds.). AFCRL-71-0019, 5 Jan. 1971, Spec. Rep. No. 114.

*2.* R. B. Sanderson and H. E. Scott, A High Resolution Far Infrared Interferometer, *Appl. Opt.* **10**, 1097 (1971).

*3.* M. Cuisenier and J. Pinard, Apectrometre de Fourier a ' Oeils de Chatè ' et a Balayage Rapide, *J. Phys. Coll. C-2* **28**, C-297 (1967).

*4.* R. Beer and D. Marjaniemi, Wavefronts and Construction Tolerances for a Cat's-Eye Retroreflector, *Appl. Opt.* **7**, 1191 (1966).

*5.* R. A. Schindler, A Small, High Speed Interferometer for Aircraft, Balloon, and Spacecraft Applications, *Appl. Opt.* **9**, 301 (1970).

APPENDIX D

# *Rapid-Scan Fourier Transform Spectroscopy*

In usual practice, the interferogram is recorded by moving the mirror in steps through the path distance $L$. However, one may move the mirror very rapidly at constant velocity and record the interferogram on magnetic rape. Rapid scanning also means moving the mirror at a rate such that the interferometric fringes act as their own chopper. The modulation in the interferogram furnishes an ac signal which can be processed for spectral filtering and electronic noise reduction before the spectrum is computed. The techniques of rapid-scan Fourier spectroscopy make it possible to record single interferograms in under 1 sec! Repetition is used to improve the signal-to-noise ratio whenever necessary (see Fig. 14-1).

Rapid-scan Fourier transform spectroscopy was developed by Mertz [1] and described in an article "Rapid Scanning Fourier Transform Spectrometry" and his book [2] "Transformations in Optics." Early interferometric modulation references may be found in articles by Madden and Strong [3]; Genzel and Weber [4]; Genzel and Weber [5]; and Genzel [6].

If one has a spectral signal $B(\sigma, \delta)$ for a given wave number $\sigma$ and optical path $\delta$, it may be expressed as [Eq. (3-19)]

$$B(\sigma, \delta) = \mathscr{E}^2(\sigma)[1 + \cos(2\pi\sigma\delta)]. \tag{D-1}$$

If the mirror is moved at a constant velocity $v$, then the optical path difference as a function of time $t$ is given by

$$\delta = 2vt. \tag{D-2}$$

The maximum path displacement $L$ from the origin determines the total time for a single interferogram scan, with

$$-\tfrac{1}{2}L/v \leq t \leq \tfrac{1}{2}L/v. \tag{D-3}$$

With the optical path given by Eq. (D-2), the recorded signal versus time [by Eq. (3-26)] becomes

$$I_{\mathrm{R}}(t) = \int_0^\infty \mathscr{E}^2(\sigma)\, d\sigma + \int_0^\infty \mathscr{E}^2(\sigma) \cos(4\pi\sigma v t)\, d\sigma, \tag{D-4}$$

or the interferogram as a function of time is

$$F(t) \equiv I_{\mathrm{R}}(t) - I_{\mathrm{R}}(\infty) = \int_0^\infty \mathscr{E}^2(\sigma) \cos(\omega t)\, d\sigma \tag{D-5}$$

where the angular frequency at any $\sigma$ is

$$\omega = 2\pi\nu = 4\pi\sigma v. \tag{D-6}$$

From Eqs. (D-5) and (D-6), one has the interferogram composed of the sum of many spectral components, $\mathscr{E}^2(\sigma)$, whose ac signal frequency is determined by the wave number. The spectral content at each wave number $\sigma$ will appear at an electronic frequency $\nu = 2\sigma v$. Continuing the derivation that lead to Eq. (3-26), one can compute the spectrum from

$$B(\sigma) = \int_{-L/2v}^{L/2v} (\text{ac signal}) \exp(-i\omega t)\, dt. \tag{D-7}$$

This simple idea of recording an interferogram rapidly and as an ac signal has many consequences. The results have been capitalized on in astronomy and in the Digilab Fourier transform spectrometers.

If it is desired to compute a spectrum between wave numbers $\sigma_1$ and $\sigma_2$, electronic band-pass filters can be used to isolate the spectral range in the interferogram. The desired range of frequencies in the interferogram is determined by

$$2v\sigma_1 \leq \nu \leq 2v\sigma_2. \tag{D-8}$$

One can reduce or eliminate electronic noise by accepting only ac signal voltages above a certain level. The voltage level is set low enough to pass spectral information but too high for noise. Since the system requires no chopper, one has the entire signal on the detector all the time. Therefore, it is twice as efficient as spectrometers with choppers.

With mirror velocities of the order of $10^{-2}$ cm sec$^{-1}$ at visible range wave numbers of $2 \times 10^4$ cm$^{-1}$, one would expect fringe frequencies in the interferogram of the order of $2 \times 10^2$ Hz. That is, one can record the interferogram at frequencies which are easy to handle electronically. But, best of all, the experimentalist can choose the exact frequencies he wishes by choosing $v$ for a given spectral range. In astronomy, the fringe frequency is set above scintillation frequencies.

Weak spectral signals in rapid-scan Fourier transform spectroscopy require that the interferogram be recorded repetitiously. One can average the interferograms *or* the computed spectra to get the required signal-to-noise ratio in the final averaged spectrum. The Cooley–Tukey algorithm is so fast that most experimentalists prefer to average the spectra.

The observational procedure may be to record the repetitive signals on one channel of an ordinary tape recorder. Simultaneous timing information is recorded on the second channel, thereby avoiding any problems of tape speed variations. The timing information consists of a bipolar pulse (simple pulses do not record well) at the beginning of each scan and a low-level sinusoidal signal of the desired sampling frequency (Chapter 7). The timing channel may be derived directly from the interferometer by monochromatic reference fringes or by Moiré fringes, thereby avoiding the need for accurate constant velocity in the mirror drive.

Source signals may be drifting in time, but not in spectral content. The repetitive scans enable the experimentalist to get accurate spectra in less than 1 sec. In ordinary Fourier transform spectroscopy, the slowly recorded interferogram would be badly distorted. By the repetitive scan method, one can also identify and discard bad spectra which grossly deviate from the mean.

The Digilab Fourier transform spectrometers described in Chapter 19 use the principle of rapid scanning and in Fig. 14-1, one has an example of the reproducibility and signal-to-noise ratios to be had with their equipment. In Fig. 14-19, some astronomical data taken by rapid-scan techniques are shown.

## *References*

*1.* L. Mertz, Rapid Scanning Fourier Transform Spectroscopy, *J. Phys. Coll. C2, Suppl. 3–4* **28**, 88 (1967).

*2.* L. Mertz, "Transformations in Optics," p. 44. Wiley, New York, 1965.

*3.* R. Madden and J. Strong, Cylindrical Interference Filter of Variable Thickness, *J. Opt. Soc. Amer.* **44**, 352 (1954).

*4.* L. Genzel and R. Weber, On the Theory of Interference Modulation in Two-Beam Interferometry, *Z. Angew. Phys.* **10**, 127 (1958).

*5.* L. Genzel and R. Weber, Spectroscopy in the Far Infrared by Interference Modulation, *Z. Angew. Phys.* **10**, 195 (1958).

*6.* L. Genzel, Aperiodic and Periodic Interference Modulation for Spectrographic Purposes, *J. Mol. Spectrosc.* **4**, 241 (1960).

# *Author Bibliography*


E. Archbold and H. A. Gebbie, The Use of a Michelson Interferometer to Determine Luminescence Spectra of Optical Maser Materials, *Proc. Phys. Soc.* **80**, 793 (1962).

R. Bachman, H. D. Kirsch, and T. H. Geballe, Low Temperature Silicon Thermometer and Bolometer, *Rev. Sci. Instrum.* **41**, 547 (1970).

J. M. Ballontyne, Double Beam Operation of a Fourier Spectrometer, *Appl. Opt.* **6**, 587 (1967).

Barnes Engineering Company, The Pyroelectric Detector, Bulletin 2-220A, 30 Commerce Road, Stamford, Connecticut (1969).

R. Beer and D. Marjaniemi, Wavefronts and Construction Tolerances for a Cat's-Eye Retroreflector, *Appl. Opt.* **7**, 1191 (1966).

H. A. Beggie and N. W. B. Stone, A Michelson Interferometer for Far Infrared Spectroscopy of Gases, *Infrared Phys.* **4**, 85 (1964).

E. E. Bell, Measurement of Spectral Transmittance and Reflectance with a Far Infrared Michelson Interferometer, *Jap. J. Appl. Phys., Suppl. 1* **4**, 412 (1965).

E. E. Bell, Measurement of the Far Infrared Optical Properties of Solids with a Michelson Interferometer Used in the Asymmetric Mode: Part I, Mathematical Formulation, *Infrared Phys.* **6**, 57 (1966).

E. E. Bell, "Optical Constants and Their Measurement" (L. Genzel, ed.) Bd XXV/2a. Springer-Verlag, Heidelberg, 1967.

E. E. Bell, The Use of Asymmetric Interferograms in Transmittance Measurements, *J. Phys., Paris, Colloq. C-2* **28**, C2-18 (1967).

E. E. Bell, Amplitude Spectroscopy, *Aspen Int. Conf. on Fourier Spectrosc., 1970*, p. 71. (G. A. Vanasse, A. T. Stair, Jr., and D. J. Baker, eds.). AFCRL-71-0019, 5 Jan. 1971, Spec. Rep. No. 114.

R. J. Bell and T. E. Gilmer, Jr., A New Radiation Chopper Principle, *Appl. Opt.* **4**, 45 (1965).

R. J. Bell and G. M. Goldman, Optical Constants of Black Polyethylene, *J. Opt. Soc. Amer.* **57**, 1552 (1967).

R. J. Bell and H. V. Romero, A Study of an Array of Square Openings, *Appl. Opt.* **9**, 2341 (1970).

R. J. Bell, H. V. Romero, and J. M. Blea, Theory and Experiments for Multi-element Grid Filters in a Dielectric, *Appl. Opt.* **9**, 2350 (1970).

G. D. Bergland, The Fast Fourier Transform Recursive Equations for Arbitrary Length Records, *Math. Comput.* **21**, 236 (1967).

G. D. Bergland, A Fast Fourier Transform Algorithm Using Base 8 Iterations, *Math. Comput.* **22**, 275 (1968).

D. W. Berreman, Low Pass Filters for Far Infrared Spectroscopy, *Rev. Sci. Instrum.* **37**, 513 (1966).

J. M. Blea, W. F. Parks, P. A. R. Ade, and R. J. Bell, Optical Properties of Black Polyethylene from 3 to 4000 $cm^{-1}$, *J. Opt. Soc. Amer.* **60**, 603 (1970).

D. Bloor, Bibliography of Far Infrared Spectroscopy, *Infrared Phys.* **10**, 1 (1970).

M. Born and E. Wolf, "Principles of Optics," 2nd ed., p. 61. Pergamon Press, Macmillan, New York, 1964.

D. R. Bosomworth and H. D. Gush, Far Infrared Spectroscopy of Compressed Gases Using a Michelson Interferometer, *Jap. J. Appl. Phys., Suppl. 1* **4**, 588 (1965).

M. Bottema and H. J. Bolle, An Interferometer with Spherical Mirrors for Fourier Spectroscopy, *Aspen Int. Conf. on Fourier Spectrosc., 1970*, p. 210. (G. A. Vanasse, A. T. Stair, Jr., and D. J. Baker, eds.). AFCRL-71-0019, 5 Jan. 1971, Spec. Rep. No. 114.

H. Bowers, Atmospheric Turbulence and Interferometric Spectroscopy, *Appl. Opt.* **3**, 627 (1964).

W. Boyle and K. Rodgers, Performance Characteristics of New Low-Temperature Bolometer, *J. Opt. Soc. Amer.* **52**, 859 (1962).

R. Bracewell, "The Fourier Transform and Its Applications." McGraw-Hill, New York, 1965.

E. O. Brigham and R. E. Morrow, The Fast Fourier Spectrum, *IEEE Trans. Audio Electroacoust.* **AU-15(2)**, 63 (1967).

H. L. Buijs, Fast Fourier Transformation of Large Arrays of Data, *Appl. Opt.* **8**, 211 (1969).

R. Cano and M. Mattioli, Far Infrared Spectroscopy with a Michelson Interferometer and an InSb Detector, *Infrared Phys.* **7**, 25 (1967).

J. E. Chamberlain, On a Relation Between Absorption Strength and Refractive Index, *Infrared Phys.* **5**, 175 (1965).

J. E. Chamberlain, On a Convergence Correction for Dispersive Fourier Spectrometry, *Appl. Opt.* **6**, 980 (1967).

J. E. Chamberlain, On the Calculation of Integrated Absorption Strengths From Refraction Data, *J. Quant. Spectrosc. Radiat. Transfer* **7**, 151 (1967).

J. E. Chamberlain and H. A. Gebbie, Sub-Millimetre Dispersion and Rotational Line Strengths of the Hydrogen Halides, *Nature* **208**, 480 (1965).

J. E. Chamberlain and H. A. Gebbie, Dispersion Measurements on Polytetrafluorethylene in The Far Infrared, *Appl. Opt.* **5**, 393 (1966).

J. E. Chamberlain, J. E. Gibbs, and H. A. Gebbie, Refractometry in the Far Infrared Using a Two Beam Interferometer, *Nature* **198**, 874 (1963).

J. E. Chamberlain, G. W. Chantry, F. D. Findlay, H. A. Gebbie, J. E. Gibbs, N. W. B. Stone, and A. J. Wright, The Spectra Transmission at Infra-Red Wavelengths of Michelson Interferometers with Dielectric Film Beam Splitters, *Infrared Phys.* **6**, 195 (1966).

J. E. Chamberlain, A. E. Costley, and H. A. Gebbie, Submillimeter Dispersion of Liquid Tetrabromoethane, *Spectrochim. Acta* **23A**, 2255 (1967).

J. E. Chamberlain, H. A. Gebbie, G. W. F. Pardoc, and M. Davies, Millimetre-Wave Fourier Transform Spectroscopy for Studies of the Liquid State, *Chem. Phys. Lett.* **1**, 523 (1968).

J. E. Chamberlain, J. E. Gibbs, and H. A. Gebbie, The Determination of Refractive Index Spectra by Fourier Spectroscopy, *Infrared Phys.* **9**, 185 (1969).

L. W. Chaney, S. R. Drayson, and C. Young, Fourier Transform Spectrometer—Radiative Measurements and Temperature Inversion, *Appl. Opt.* **6**, 347 (1967).

G. W. Chantry, H. M. Evans, J. Chamberlain, and H. A. Gebbie, Absorption and Dispersive Studies in the Range 10–10,000 $\mu$m Using a Modular Michelson Interferometer, *Infrared Phys.* **9**, 85 (1969).

W. T. Cochran, James W. Cooley, David L. Favin, Howard D. Helms, R. A. Kaenel, W. W. Lang, G. C. Maling, Jr., D. E. Nelson, C. M. Rader, and P. D. Welch, What is the Fast Fourier Transform, *Proc. IEEE* **55**, 1664 (1967).

J. Connes, Domaine d'Utilisation de la Methode par Transformee Fourier, *J. Phys. Radium* **19**, 197 (1958).

J. Connes, Recherches sur la Spectroscopie par Transformation de Fourier, *Rev. Opt.* **40**, 45–78, 116–140, 171–190, 231–265 (1961).

J. Connes, Computing Problems in Fourier Spectroscopy, *Aspen Int. Conf. on Fourier Spectrosc., 1970*, p. 83. (G. A. Vanasse, A. T. Stair, Jr., and D. J. Baker, eds.). AFCRL-71-0019, 5 Jan. 1971, Spec. Rep. No. 114.

J. Connes and P. Connes, Near-Infrared Planetary Spectra by Fourier Spectroscopy. I. Instruments and Results, *J. Opt. Soc. Amer.* **56**, 896 (1966).

J. Connes and V. Nozal, Mathematical Filtering in Spectroscopy by the Fourier Transformation. *Phys. Radium* **22**, 359 (1961).

J. Connes, P. Connes, and J. P. Maillard, "Atlas des Spectres dans le Proche Infrarouge de Venus, Mars, Jupiter, et Saturne" Centre National de la Recherche Scientique, Paris, 1969.

P. Connes, Increase of the Product of Luminosity × Resolution with Interferometers by Employing Path Differences Independent of the Incidence, *Rev. Opt.*, **35**, 37 (1956).

P. Connes, Spectrometre Interferentiel a Selection par 1; Amplitude de Modulation, *J. Phys. Radium* **19**, 215 (1958).

P. Connes, How Light is Analyzed, *in* "Lasers and Light" (A. Shawlow, ed.), p. 35, Freeman, San Francisco, California, 1969.

P. Connes, High Resolution and High Information Fourier Spectroscopy, *Aspen Int. Conf. on Fourier Spectrosc., 1970*, p. 121. (G. A. Vanasse, A. T. Stair, Jr., and D. J. Baker, eds.). AFCRL-71-0019, 5 Jan. 1971, Spec. Rep. No. 114.

P. Connes, ICO Conference, "Optical Instruments and Techniques." Oriel Press, Newcastle-upon-Tyne (to be published).

P. Connes and G. Michel, Real Time Computer for Fourier Spectroscopy, *Aspen Int. Conf. on Fourier Spectrosc., 1970*, p. 313. (G. A. Vanasse, A. T. Stair, Jr., and D. J. Baker, eds.). AFCRL-71-0019, 5 Jan. 1971, Spec. Rep. No. 114.

J. W. Cooley and J. W. Tukey, An Algorithm for the Machine Calculation of Complex Fourier Series, *Math. Computat.* **19**, 297 (1965).

M. Cuisenier, Etude et realisation d'un interferometre de Michelson permetland l'utilisation de la spectroscopie par transformation de Fourier a haute resolution sur un intervalle spectral etendu, These de 3[e] cycle, Paris.

M. Cuisenier and J. Pinard, Apectrometre de Fourier a 'Oeils de Chat' et a Balayage Rapide, *J. Phys. Colloq. C-2* **28**, C-297 (1967).

R. Curbelo and C. Foskett, An Automated Interferometer Spectrometer: Real Time Data Acquisition and On-Line Computation to 0.5 cm, *Aspen Int. Conf. on Fourier Spectrosc., 1970*, p. 221. (G. A. Vanasse, A. T. Stair, Jr., and P. J. Baker, eds.). AFCRL-71-0019, 5 Jan. 1971, Spec. Rep. No. 114.

J. Dowling, Direct Interpretation of Far-Infrared Interferograms with Application to Diatomic and Linear Molecules, *J. Opt. Soc. Amer.*, **54**, 663 (1964).

J. M. Dowling and R. T. Hall, Far Infrared Interferometry: Upper Limit for Line Widths for the Pure Rotational Band of Carbon Monoxide, *J. Mol. Spectrosc.* **19**, 108 (1966).

J. M. Dowling, Signal and Noise in Two-Beam Interferometry, *Appl. Opt.*, **6**, 1580 (1967).

J. M. Dowling, Double Beaming in Fourier Spectroscopy, *Aspen Int. Conf. on Fourier Spectrosc., 1970*, p. 55. (G. A. Vanasse, A. T. Stair, Jr., and D. J. Baker, eds.). AFCRL-71-0019, 5 Jan. 1971, Spec. Rep. No. 114.

J. M. Dowling and R. T. Hall, The Techniques of High Resolution, Double-Beam, Two-Beam Interferometry with a Far Infrared Lamellar Grating Interferometer, *J. Phys. Colloq. C2 Suppl. are M° 3–4* **28**, C2-156 (1967).

J. M. Dowling and R. T. Hall, Comments on 'High Resolution Grating Spectrometer for the Far Infrared,' *J. Opt. Soc. Amer.* **57**, 269 (1967).

J. M. Dowling, Investigations in the Far Infrared with a Lamellar Grating Interferometer, Air Force Rep. No. SSD-TR-67-30; Aerospace Rep. No. TR-1001 (9260-01)7, March, 1967); Space Phys. Lab., Laboratories Div., Lab. Operations, Aerospace Corp., Los Angeles, California.

S. I. Drasky and R. J. Bell, Radiation Chopper for the Submillimeter Wavelength Region, *Infrared Phys.* **5**, 137 (1965).

H. Y. Fan, M. L. Shepherd, and W. Spitzer, "Photoconductivity Conference." Wiley, New York, 1956.

D. W. Faries, K. A. Gehring, P. L. Richards, and Y. R. Shen, Tunable Far-Infrared Radiation Generated from the Difference Frequency Between Two Ruby Lasers, *Phys. Rev.* **180**, 363 (1969).

Fast Computational Techniques Issue of *IEEE Trans. Audio Electroacoust.* **AU-15(2)** (1967).

P. Fellgett, thesis, Cambridge Univ. (1951).

P. Fellgett, A Propos de la Theorie du Spectrometre Interferentiel Multiplex, *J. Phys. Radium* **19**, 187 (1958).

P. Fellgett, Spectrometry Interferentiel Multiplex Pour Measures Infrared Sur Les Etoiles, *J. Phys. Radium* **19**, 237 (1958).

A. Filler, Apodization and Interpolation in Fourier Transform Spectroscopy, *J. Opt. Soc. Amer.* **54**, 762 (1964).

M. L. Forman, Fast Fourier-Transform Technique and Its Application to Fourier Spectroscopy, *J. Opt. Soc. Amer.* **56**, 978 (1966).

M. L. Forman *et al.*, Nonlinear Phase Correction of Interferograms Obtained in Fourier Spectroscopy, Utah State Univ., Logan, Utah (1 July 1965); Defense Documentation Center, Defense Supply Agency, AD 624 086.

M. L. Forman, W. H. Steel, and G. A. Vanasse, Correction of Asymmetric Interferograms Obtained in Fourier Spectroscopy, *J. Opt. Soc. Amer.* **56**, 59 (1966).

M. Francon, "Optical Interferometry," (translated by I. Wilmanns). Academic Press, New York, 1966.

H. Gebbie and G. Vanasse, Interferometric Spectroscopy in the Far-Infrared, *Nature* **178**, 432 (1956).

H. Gebbie, Spectres D'Absorption Atmospherique dans l'Infrared Lointain par Interferometrie a Deux Ondes, *J. Phys. Radium* **19**, 197 (1958).

H. A. Gebbie, Submillimeter Wave Spectroscopy Using a Michelson Interferometer, *in* "Advances in Quantum Electronics." Columbia Univ. Press, New York, 1961.

H. A. Gebbie, Fourier Transform versus Grating Spectroscopy, *Appl. Opt.* **8**, 501 (1969).

H. A. Gebbie and N. W. B. Stone, A Michelson Interferometer for Far-Infrared Spectroscopy of Gases, *Infrared Phys.* **4**, 85 (1964).

H. A. Gebbie and R. Q. Twiss, Two-Beam Interferometric Spectroscopy, *Rep. Progr. Phys.* **29**, 729 (1966).

H. Gebbie, N. Stone, and D. Williams, A Interferometric Study of the Far-Infrared Spectrum of Compressed Nitrogen, *Mol. Phys.* **6**, 215 (1963).

R. Geick, Der Brechungsindex Von Kristallinem Und Geschmolzenem Quarz Im Spektralbereich Um 100 $\mu$m, *Z. Phys.* **161**, 116 (1961).

L. Genzel, Aperiodic and Periodic Interference Modulation for Spectrographic Purposes, *J. Mol. Spectrosc.* **4**, 241 (1960).

L. Genzel, Application of Frustrated Total Reflection as a Filter in the Far Infrared, *Infrared Phys.* **4**, 253 (1964).

L. Genzel, Fourier-Spectroskopie, Plenarvortrage, p. 128, 33 Physiker-Tagung 1968, Karlsruhe, W. Germany, B. G. Teubner, Stuttgart, 1969.

L. Genzel, Far-Infrared Sources, "Far Infrared Properties of Solids," p. 51 (S. S. Mitra and S. Nudelman, eds.). Plenum, New York 1970.

L. Genzel and R. Weber, Zur Theorie de Interference-Modulation fur Zweistrahl-Interferenzen, *Z. angew. Phys.* **10**, 127 (1958).

L. Genzel and R. Weber, Spektroskopie im fernen Ultrarot durch Interferenz-Modulation, *Z. angew. Phys.* **10**, 195 (1958).

L. Genzel, H. Happ, and R. Weber, Dispersionsmessumgen An NaCl, KCl, and KBr Zwischen 0.3 and 3 mm Wellenlange, *Z. Phys.* **154**, 13 (1959).

J. E. Gibbs and H. A. Gebbie, Calculation of the Power-Spectrum from an Interferogram Sampled at Points None of Which Coincides Exactly with Zero Path-Difference, *Infrared Phys.* **5**, 187 (1965).

A. Girard and P. Jacquinot, *in* "Advanced Optical Techniques." (A. C. S., Van Hell, ed.). North-Holland Publ., Amsterdam, 1967.

S. Goldman, "Information Theory." Prentice-Hall, Englewood Cliffs, New Jersey, 1953.

I. J. Good, The Interaction Algorithm and Practical Fourier Series, *J. Roy. Statist. Soc. Ser. B.* **20**, 361 (1958).

J. W. Goodman, "Introduction to Fourier Optics." McGraw-Hill, New York, 1968.

C. C. Grimes, P. L. Richards, and S. Shapiro, Josephson-Effect Far Infrared Detector, *J. Appl. Phys.* **39**, 3905 (1968).

E. A. Guilliemin, "The Mathematics of Circuit Analysis." Wiley, New York, 1949.

A. Hadni, Far Infrared Transpose Spectroscopy, *Appl. Opt.* **4**, 1374 (1965).

A Hadni, "Essentials of Modern Physics Applied to the Study of the Infrared." Pergamon Press, Oxford, 1967.

R. T. Hall, D. Vrabec, and J. Dowling, High Resolution Far-Infrared Lamellar-Grating Spectrometer, *J. Opt. Soc. Amer.* **54**, 1390 (1964).

R. T. Hall and J. M. Dowling, Pure Rotational Spectrum of Nitric Oxide, *J. Chem. Phys.* **45**, 1899 (1966).

R. T. Hall and J. M. Dowling, Pure Rotational Spectrum of Water Vapor, *J. Chem. Phys.* **47**, 2454 (1967).

R. T. Hall, D. Vrabec, and J. M. Dowling, A High-Resolution Far-Infrared Double-Beam Lamellar Grating Interferometer, *Appl. Opt.* **5**, 1147 (1966).

R. A. Hanel, B. Schlachman, F. D. Clark, C. H. Prokesh, J. B. Taylor, W. M. Wilson, and L. Chaney, The Nimbus III Michelson Interferometer, *Aspen Int. Conf. on Fourier Spectrocs., 1970*, p. 231. (G. A. Vanasse, A. T. Stair, Jr., and D. J. Baker, eds.). AFCRL-71-0019, 5 Jan. 1971, Spec. Rep. No. 114.

N. P. Hansen and J. Strong, Performance of a Simple Spherical Lamellar Grating Beamsplitter by Wavefront Division, *Aspen Int. Conf. on Fourier Spectrosc., 1970*, p. 215. (G. A. Vanasse, A. T. Stair, Jr., and D. J. Baker, eds.). AFCRL-71-0019, 5 Jan. 1971, Spec. Rep. No. 114.

H. Happ and L. Genzel, Interferenz-Modulation Mit Monochromatischen Millimeter-Wellen, *Infrared Phys.* **1**, 39 (1961).

N. J. Harrick, "Internal Reflection Spectroscopy." Wiley (Interscience), New York, 1967.

M. Hass and B. W. Hevis, Infrared Lattice Reflection Spectra of III–IV Compound Semiconductors, *J. Phys. Chem. Solids* **23**, 1099 (1962).

R. Haswell, A. Martin, and G. Sharp, Correction of Phase Errors in Interferograms, *Appl. Opt.* **3**, 1195 (1964).

J. R. Hickey and O. B. Daniels, Modified Optical System for the Golay Detector, *Appl. Opt.* **8**, 732 (1969).

Hillel Bar-Lev, A Dual-Beam Infrared Interferometer-Spectrometer, *Infrared Phys.* **7**, 93 (1967).

B. F. Hochheimer and C. F. Bradley, A Fourier Transform Spectrometer for the 10 to 10,000 $cm^{-1}$ Spectral Region, *Appl. Opt.* **8**, 557 (1969).

P. G. Hoel, "Introduction to Mathematical Statistics," 2nd ed., pp. 166–167. Wiley, New York, 1954.

J. E. Hoffman and G. A. Vanasse, Real-Time Spectral Synthesis in Fourier Spectroscopy, *Appl. Opt.* **5**, 1167 (1966).

J. E. Hoffman, Jr., Real-Time Fourier Spectroscopy, *Appl. Opt.* **8**, 323 (1969).

R. D. Hudson, "Infrared System Engineering," p. 270. Wiley, New York, 1969.

D. M. Hunten, Fourier Spectroscopy of Planets, *Science* **162**, 313 (1968).

W. J. Hurley, Interferometric Spectroscopy in the Far Infrared, *J. Chem. Educ.* **43**, 236 (1966).

J. D. Jackson, "Classical Electrodynamics." Wiley, New York, 1962.

N. Jacobi, Resolution and Noise in Fourier Spectroscopy, *J. Opt. Soc. Amer.* **58**, 495 (1968).

P. Jacquinot, Caracteres Communs aus Nouvelles Methodes d'Spectroscopie Interferentielle Facteur Merite, *J. Phys. Radium* **19**, 223 (1958).

P. Jacquinot, Appendix E, Apodization, *in* "Concepts of Classical Optics," (J. Strong, ed.). Freeman, San Francisco, California, 1958.

P. Jacquinot, New Developments in Interference Spectroscopy, *Progr. Phys.* **23**, 267 (1960).

P. Jacquinot, Interferometry and Grating Spectroscopy: an Introductory Survey, *Appl. Opt.* **8**, 497 (1969).

P. Jacquinot and C. Dufour, Optical Conditions in the Use of Photo-Electric Cells in Spectrographs and Interferometers, *J. Rech. CNRS* **6**, 91 (1948).

P. Jacquinot and B. Roizen-Dossier, Apodization, *in* "Progress in Optics" (E. Wolf, ed.), Vol. III, p. 33 ff. North-Holland Publ. Amsterdam, 1964.

D. James and J. Ring, The Efficiency of Beamsplitters in The Far Infra-Red Michelson Interferometer, *J. Phys.* (*Suppl.*) **28**, C2-150 (1967).

F. A. Jenkins and H. E. White, "Fundamentals of Optics," 3rd ed. McGraw-Hill, New York, 1957.

R. Kaplan, Magnetooptical Studies of Solids Using Fourier Transform Spectroscopy, *Appl. Opt.* **6**, 685 (1957).

G. A. Korn and T. M. Korn, "Handbook of Mathematical Physics," 2nd ed., p. 879. McGraw-Hill, New York, 1968.

F. K. Kneubuhl, J. F. Moser, and H. Steffen, High Resolution Grating Spectrometer for the Far Infrared, *J. Opt. Soc. Amer.* **56**, 760 (1966).

F. K. Kneubuhl, J. F. Moser, and H. Steffen, Note on a Letter by Dowling and Hall, *J. Opt. Soc. Amer.* **57**, 271 (1967).

P. W. Kruse, L. D. McGlauchlin, and R. B. McQuistan, "Elements of Infrared Technology, Generation, Transmission, and Detection." Wiley, London, 1962.

M. Lax and E. Brustein, Infrared Lattice Absorption in Ionic and Homopolar Crystals, *Phys. Rev.* **97**, 39 (1955).

J. Lecomte, Spectroscopic dans l'Infrarouge, "Handbuch der Physik," Vol. 24, p. 283, Springer-Verlag, Heidelberg, 1957.

D. C. Lewis and W. D. Westwood, The Spectra of Some Iron Oxide Films Between 30,000 $cm^{-1}$ and 4000 $cm^{-1}$, *Can. J. Phys.* **42**, 2367 (1964).

F. Levy, R. C. Milward, S. Bras, and R. le Toullec, "Real Time" Far Infrared Fourier Spectroscopy Using a Small Digital Computer, *Aspen Int. Conf. on Fourier Spectrosc., 1970*, p. 331. (G. A. Vanasse, A. T. Stair, Jr., and D. J. Baker, eds.). AFCRL-71-0019, 5 Jan. 1971, Spec. Rep. 114.

B. Liu and T. Kaneko, Error Analysis of Digital Filters Realized with Floating-Point Arithmetic, *Proc. IEEE* **57**, 1735 (1969).

E. Loewenstein, Interferometric Spectra of Ammonia and Carbon Monoxide in the Far Infrared, *J. Opt. Soc. Amer.* **50**, 1163 (1960).

E. Loewenstein, Optical Properties of Sapphire in the Far-Infrared, *J. Opt. Soc. Amer.* **51**, 108 (1961).

E. Loewenstein, On the Correction of Phase Errors in Interferograms, *Appl. Opt.* **2**, 491 (1963).

E. V. Loewenstein, The History and Current Status of Fourier Transform Spectroscopy, *Appl. Opt.* **5**, 845 (1966).

E. V. Loewenstein, Fourier Spectroscopy: An Introduction, *Aspen Int. Conf. on Fourier Spectrosc., 1970*, p. 3. (G. A. Vanasse, A. T. Stair, Jr., and D. J. Baker, eds.). AFCRL-71-0019, 5 Jan. 1971, Spec. Rep. No. 114.

E. V. Loewenstein and A. Engelsrath, Polarization Properties of Far Infra-Red Beamsplitters, *J. Phys.* (*Suppl.*) **28**, C2-153 (1967).

R. C. Lord and T. K. McCubbin, Infrared Spectroscopy from 5 to 200 Microns with a Small Grating Spectrometer, *J. Opt. Soc. Amer.* **47**, 689 (1952).

F. Low, Low Temperature Germanium Bolometer, *J. Opt. Soc. Amer.* **51**, 1300 (1961).

M. F. D. Low, Fourier Transform Spectroscopy, *J. Chem. Ed.* **47**, A163, A255, A349, A415 (1970).

R. Madden and J. Strong, Cylindrical Interference Filter of Variable Thickness, *J. Opt. Soc. Amer.* **44**, 352 (1954).

D. Malz, Interference Phenomena in Plane Parallel Plates in Fourier Spectroscopy and Their Effect on Spectroscopic Investigations on Semiconductors in the Middle and Far Infrared, *Exp. Tech. Phys.* (*Germany*) **13**, 257 (1965).

L. Mandel, Fluctuations of Light Beams, *in* "Progress in Optics" (E. Wolf, ed.). North-Holland Publ., Amsterdam, 1963.

A. E. Martin, "Infra Red Instrumentation and Techniques." Elsevier, New York, 1966.

D. H. Martin, ed.,"Spectroscopic Techniques for Far Infrared, Submillimeter and Millimeter Waves." North-Holland Publ. Amsterdam, 1967.

D. H. Martin and E. Puplett, Polarized Interferometric Spectroscopy for the Millimeter and Submillimetre Spectrum, *Infrared Phys.* **10**, 57 (1966).

T. K. McCubbin and W. M. Sinton, Recent Investigations in the Far Infrared, *J. Opt. Soc. Amer.* **40**, 538 (1950).

D. E. McCarthy, The Reflection and Transmission of Infrared Materials: 1, Spectra from 2 to 50 Microns, *Appl. Opt.* **2**, 591 (1963).

D. E. McCarthy, Transmission of Irtran Materials from 50 $\mu$m to 300 $\mu$m, *Appl. Opt.* **5**, 472 (1966).

D. E. McCarthy, Black Polyethylene as a Far-Infrared Filter, *J. Opt. Soc. Amer.* **57**, 699 (1967).

D. E. McCarthy, The Reflection and Transmission of Infrared Materials. V: Spectra from 2$\mu$ to 50$\mu$, *Appl. Opt.* **7**, 1997 (1968).

L. Mertz, Correction of Phase Errors in Interferograms, *Appl. Opt.* **2**, 1331 (1963).

L. Mertz, "Transformations in Optics." Wiley, New York, 1965.

L. Mertz, Auxilliary Computation of Fourier Spectrometry, *Infrared Phys.* **7**, 17 (1967).

L. Mertz, Rapid Scanning Fourier Transform Spectroscopy, *J. Phys. Colloq. C2 suppl. 3–4* **28**, C2-87 (1967).

L. Mertz, Fourier Spectroscopy, Past, Present, and Future, *Aspen Int. Conf. on Fourier Spectrosc., 1970*, p. 142. (G. A. Vanasse, A. T. Stair, Jr., and D. J. Baker, eds.). AFCRL-71-0019, 5 Jan. 1971, Spec. Rep. No. 114.

L. Mertz, N. O. Young, and J. Armitage, Mock Interferometry, *in* "Optical Instruments and Techniques"(K. J. Habell, ed.). p. 51. Chapman and Hall, London, 1962.

A. A. Michelson, *Astrophys. J.* **7**, 131 (1898).

A. A. Michelson, *Phil. Mag. Ser. 5* **31**, 256 (1891).

A. Michelson, On the Application of Interference Methods to Spectroscopic Measurements, *Phil. Mag.* **34**, 280 (1892).

A. Michelson, "Studies in Optics." Univ. of Chicago Press, Chicago, Illinois, 1927.

A. A. Michelson, excerpt from "Light Waves and Their Uses." Univ. of Chicago Press, Chicago, Illinois, 1902; *also in Appl. Opt.* **6**, 692 (1967).

R. C. Milward, A Complete Fourier Spectrophotometer System for the Far-Infra-Red, "Molecular Spectroscopy," p. 81. Inst. Petroleum, Adlard and Son, Ltd., Dorking, England, 1968.

R. C. Milward, A Small Lamellar Grating Interferometer for the Very Far-Infrared, *Infrared Phys.* **9**, 59 (1969).

A. Mitsuishi, Y. Yamada, and H. Yoshinaga, Reflection Measurements on Reststrahlen Crystals in the Far-Infrared Region, *J. Opt. Soc. Amer.* **52**, 14 (1962).

A. Mitsuishi, Y. Otsuka, S. Fujita, and H. Yoshinaga, Metal Mesh Filters in the Far Infrared Region, *Jap. J. Appl. Phys.* **2**, 574 (1963).

K. D. Moller and R. V. McKnight, Far-Infrared Transmission Filter Gratings, *J. Opt. Soc. Amer.* **53**, 760 (1963).

K. D. Moller and W. G. Rothschild, "Far-Infrared Spectroscopy." Wiley-Interscience, New York, 1971.

T. S. Moss, "Optical Properties of Semiconductors." Butterworths, London and Washington, D.C., 1959.

R. E. Murphy and H. Sakai, Application of the Fourier Spectroscopy Technique to the Study of Relaxation Phenomena, *Aspen Int. Conf. on Fourier Spectrosc., 1970*, p. 301. (G. A. Vanasse, A. T. Stair, Jr., and D. J. Baker, eds.). AFCRL-71-0019, 5 Jan. 1971, Spec. Rep. No. 114.

M. V. R. K. Murty, Modification of Michelson Interferometer Using Only One Cube-Corner Prism, *J. Opt. Soc. Amer.* **50**, 83 (1960).

E. D. Palik, A Far Infrared Bibliography, Naval Res. Lab., Washington, D.C. (1967).

W. F. Parks and R. W. Alexander, Interferometers with Spherical Mirrors (to be published).

E. R. Peck, A New Principle in Interferometer Design, *J. Opt. Soc. Amer.* **38**, 66 (1948).

E. R. Peck, Theory of the Corner-Cube Interferometer, *J. Opt. Soc. Amer.*, **38**, 1015 (1948).

E. R. Peck, Uncompensated Corner-Reflector Interferometer, *J. Opt. Soc. Amer.* **47**, 250 (1957).

C. H. Perry, R. Geick, and E. F. Young, Solid State Studies by Means of Fourier Transform Spectroscopy, *Appl. Opt.* **5**, 1171 (1966).

E. Plyler, D. Yates, and H. Gebbie, Radiant Energy from Sources in the Far Infrared, *J. Opt. Soc. Amer.* **52**, 859 (1962).

J. L. Pritchard, H. Sakai, W. H. Steel, and G. A. Vanasse, "Colloque sur les Methods Nouvelles de Spectroscopie Instrumentate," Paris, CNRS (1966).

E. H. Putley, Indium Antimonide Submillimeter Photoconductive Detectors, *Appl. Opt.* **4**, 649 (1965).

C. M. Randall, Computational Techniques in Fourier Transform Spectroscopy, Air Force Rep. No. SSD-TR-67-75, Aerospace Rep. No. TR-1001 (9260-1001 (9260-01)-8, Aerospace Corp. (March 1967).

C. M. Randall, Fast Fourier Transform for Unequal Number of Input and Output Points, *Appl. Opt.* **6**, 1432 (1967).

C. M. Randall and R. D. Rawcliffe, Refractive Indices of Germanium, Silicon, and Fused Quartz in the Far Infrared, *Appl. Opt.* **6**, 1889 (1967).

K. Rasmusseu, Conformations and Vibration Spectra of TRIS-(DIAMINE) Chelate Complexes, Chem. Dept., The Technical Univ. of Denmark, 2800 Lyngby (1970).

Lord Rayleigh (J. W. Strutt), *Phil. Mag.* **24**, 864 (1912).

K. F. Renk and L. Genzel, Interference Filters and Fabry-Perot Interferometers for the Far Infrared, *Appl. Opt.* **1**, 643 (1962).

P. L. Richards, Antiferromagnetic Resonance in $CoF_2$, $NiF_3$, and $MnCO_3$, *J. Appl. Phys.* **35**, 850 (1963).

P. L. Richards, Far-Infrared Magnetic Resonance in $CoF_2$, $NiF_2$, $KNiF_3$, and YbIG, *J. Appl. Phys.* **34**, 1237 (1964).

P. L. Richards, High-Resolution Fourier Transform Spectroscopy in the Far-Infrared, *J. Opt. Soc. Amer.* **54**, 1474 (1964)

P. L. Richards, High Resolution Far Infrared Interferometry with Application to Solid State Physics, *Jap. J. Appl. Phys. Suppl. 1* **4**, 417 (1965).

P. L. Richards, Computing Reliable Power Spectra, *IEEE Spectrum* **4**, 83 (1967).

P. L. Richards, Far Infrared Detectors, *in* "Far Infrared Properties of Solids," (S. S. Mitra and S. Nudelman, eds.) p. 103. Plenum, New York, 1970.

A. K. Rigler, Note on the Fast Fourier Transform, *J. Opt. Soc. Amer.* **58**, 274 (1968).

W. H. Robinette and R. B. Sanderson, Dipole Moment of HBr From Far Infrared Dispersion Measurements, *Appl. Opt.* **8**, 711 (1969).

G. Roland, Spectroscopie par Transformation de Fourier, Thesis, Liege, Belgium (1965).

H. Rubens and R. W. Wood, *Phil. Mag.* **21**, 249 (1911).

H. N. Rundle, Construction of a Michelson Interferometer for Fourier Spectroscopy, *J. Res. Nat. Bur. Std.* **69C**, **5** (1965).

E. E. Russell and E. E. Bell, Measurement of the Far Infrared Optical Properties of Solids with a Michelson Interferometer Used in the Asymmetric Mode: Part II, The Vacuum Interferometer, *Infrared Phys.* **6**, 75 (1966).

E. E. Russell and E. E. Bell, Optical Constants of Sapphire in the Far Infrared, *J. Opt. Soc. Amer.* **57**, 543 (1967).

E. E. Russell and E. E. Bell, Measurements of the Optical Constants of Crystal Quartz in the Far Infrared with the Asymmetric Fourier-Transform Method, *J. Opt. Soc. Amer.* **57**, 341 (1967).

H. Sakai, Consideration of the Signal-To-Noise Ratio in Fourier Spectroscopy, *Aspen Int. Conf. on Fourier Spectrosc., 1970*, p. 19. (G. A. Vanasse, A. T. Stair, Jr., and D. J. Baker, eds.). AFCRL-71-0019, 5 Jan. 1971, Spec. Rep. No. 114.

H. Sakai and G. A. Vanasse, Direct Determination of the Transfer Function of an Infrared Spectrometer, *J. Opt. Soc. Amer.* **56**, 357 (1966).

H. Sakai and G. A. Vanasse, Hilbert Transform in Fourier Spectroscopy, *J. Opt. Soc. Amer.* **56**, 131 (1966).

H. Sakai and C. A. Vanasse, Accuracy of Zero-Path-Difference-Determination in Fourier Spectroscopy for Narrow-Band Spectra, *J. Opt. Soc. Amer.* **57**, 844 (1967).

H. Sakai, G. A. Vanasse, and M. L. Forman, Spectral Recovery in Fourier Spectroscopy, *J. Opt. Soc. Amer.* **58**, 84 (1968).

B. D. Saksena, D. R. Pahwa, M. M. Pradhan, and K. Lal, Reflection and Transmission Characteristics of Wire Gratings in the Far Infrared, *Infrared Phys.* **9**, 43 (1969).

R. B. Sanderson, Measurement of Rotational Line Strengths in HCl by Asymmetric Fourier Transform Techniques, *Appl. Opt.* **6**, 1527 (1967).

R. B. Sanderson and H. E. Scott, A High Resolution Far Infrared Interferometer, *Appl. Opt.* **10**, 1097 (1971).

A. Schawlow, "Lasers and Light." Freeman, San Francisco, California, 1969.

R. A. Schlindler, A Small, High Speed Interferometer for Aircraft, Balloon, and Spacecraft Applications, *Appl. Opt.* **9**, 301 (1970).

M. J. Selby, Mock Interferometer—I, Theory, *Infrared Phys.*, **6**, 21 (1966).

J. E. Shaw, Spectroradiometry over Broad Spectral Regions by Fourier Spectroscopy, *J. Opt. Soc. Amer.* **57**, 1136 (1967).

W. M. Sinton, Recent Infrared Spectra of Mars and Venus, *J. Quantum Spectrosc. Radiat. Transfer* **3**, 551 (1963).

D. R. Smith, R. L. Morgan, and E. V. Loewenstein, Comparison of the Radiance of Far-Infrared Sources, *J. Opt. Soc. Amer.* **58**, 433 (1968).

M. Smith, Optimization of Bias in Thermistor Radiator Detectors. II Practical Networks, *Appl. Opt.* **8**, 1213 (1969).

W. H. Steel, Interferometers without Collimation for Fourier Spectroscopy, *J. Opt. Soc. Amer.* **54**, 151 (1964).

W. H. Steel, "Interferometry." Cambridge Univ. Press, London and New York, 1967.

W. H. Steel, Interferometers for Fourier Spectroscopy, *Aspen Int. Conf. on Fourier Spectrosc., 1970*, p. 43. (G. A. Vanasse, A. T. Stair, Jr., and D. J. Baker, eds.). AFCRL-71-0019, 5 Jan. 1971, Spec. Rep. No. 114.

W. H. Steel and M. L. Forman, Example of Errors Occurring in Fourier Spectroscopy Due to Hilbert-Transform Effect, *J. Opt. Soc. Amer.* **55**, 982 (1966).

R. S. Sternberg and J. F. James, A New Type of Michelson Interference Spectrometer, *J. Sci. Instrum.* **41**, 225 (1964).

J. E. Stewart, Field of View and Resolution of a Polarization Fourier Spectrometer, *J. Opt. Soc. Amer.* **58**, 434 (1968).

J. M. Stone, "Radiation and Optics." McGraw-Hill, New York, 1963.

J. D. Strong, Investigations in the Far Infrared, *Phys. Rev.* **38**, 1818 (1931).

J. Strong, Interferometric Modulator, *J. Opt. Soc. Amer.* **44**, 352 (1954).

J. D. Strong, Interferometry in the Far-Infrared, *J. Opt. Soc. Amer.* **47**, 354 (1957).

J. D. Strong, "Concepts of Classical Optics." Freeman, San Francisco, California, 1958.

J. D. Strong and G. Vanasse, Modulation Interferentielle et Calculateur Analogique pour un Spectrometre Intereferentiel, *J. Phys. Radium* **19**, 192 (1958).

J. D. Strong and G. Vanasse, Interferometric Spectroscopy in the Far-Infrared, *J. Opt. Soc. Amer.* **49**, 844 (1959).

J. D. Strong and G. Vanasse, Lamellar Grating Far Infrared Interferometer, *J. Opt. Soc. Amer.* **50**, 113 (1960).

M. T. Surh, The Effect of Time Jitter in Sampling of an Interferogram, *Appl. Opt.* **5**, 880 (1966).

A. G. Tescher, Beamsplitter Optimization for Fourier Spectroscopy, *Aspen Int. Conf. on Fourier Spectrosc., 1970*, p. 225. (G. A. Vanasse, A. T. Stair, Jr., and D. J. Baker, eds.). AFCRL-71-0019, 5 Jan. 1971, Spec. Rep. No. 114.

L. W. Thorpe, D. J. Neale, and G. C. Hayward, New Instrumentation for Far Infrared Fourier Spectrometers, *Aspen Int. Conf. on Fourier Spectrosc., 1970*, p. 187. (G. A. Vanasse, A. T. Stair, Jr., and D. J. Baker, eds.). AFCRL-71-0019, 5 Jan. 1971, Spec. Rep. No. 114.

J. S. Toll, Causality and Dispersion Relation: Logical Foundations, *Phys. Rev.* **104**, 1760 (1956).

R. Ulrich, Far Infrared Properties of Metallic Mesh and Its Complementary Structure, *Infrared Phys.* **7**, 37 (1967).

R. Ulrich, Effective Low-Pass Filters for Far Infrared Frequencies, *Infrared Phys.* **7**, 65 (1967).

R. Ulrich, Interference Filters for the Far Infrared, *Appl. Opt.* **7**, 1987 (1968).

R. Ulrich, Preparation of Grids for Far Infrared Filters, *Appl. Opt.* **8**, 319 (1969).

G. Vanasse, Fourier Series Approximation to Spectral Functions Obtained by Interferometric Techniques, *J. Opt. Soc. Amer.* **52**, 572 (1962).

G. Vanasse and E. Loewenstein, Pure Rotation Spectra of HCN and $NH_3$. *J. Opt. Soc. Amer.* **49**, 512 (1959).

G. A. Vanasse and H. Sakai, Fourier Spectroscopy, *in* "Progress in Optics" (E. Wolf, ed.). Vol. 6, p. 261. North-Holland Publ., Amsterdam, 1967.

G. Vanasse, J. Strong, and E. Loewenstein, Far Infrared Spectra of $H_2O$ and $H_2S$ Taken with an Interferometric Spectrograph, *J. Opt. Soc. Amer.* **49**, 309 (1959).

G. A. Vanasse, A. T. Stair, Jr., and D. J. Baker, eds., *Aspen Int. Conf. on Fourier Spectrosc., 1970.* AFCRL-71-0019, 5 Jan. 1971, Spec. Rep. No. 114.

D. G. Vickers, E. I. Robson, and J. E. Beckman, Freestanding Polarizers for Submillimeter Interferometry, *Appl. Opt.* **10**, 682 (1971).

E. A. Vinogradov, E. M. Dianov, and N. A. Irisova, Fabry-Perot Interferometer for the Short Millimeter and Submillimeter Bands with Metallic Grids Having Periods Smaller Than the Wavelength, *JETP Lett.* **2**, 205 (1965).

E. A. Vinogradov, E. M. Dianov, and N. A. Irisova, "Michelson Interferometer for Measuring the Refractive Index of Dielectric Materials in the 2 mm Wavelength Range," *Radiotekhn. Elektron.* **10**, 1804 (1965); *Radio Engn. Electron. Phys.* **10**, 1547 (1965).

P. Vogel and L. Genzel, A Method for Performing the Analog Fourier Transform for Interference Modulation Spectroscopy, *Appl. Opt.* **3**, 367 (1964).

P. Vogel and L. Genzel, Transmission and Reflection of Metallic Mesh in the Far Infrared, *Infrared Phys.* **4**, 257 (1964).

R. G. Wheeler and J. C. Hill, Spectroscopy in the 5 to 400 Wavenumber Region with the Grubbs-Parsons Interferometric Spectrometer, *J. Amer. Opt. Soc.* **56**, 657 (1966).

C. S. Williams, Mirror Misalignment in Fourier Spectroscopy Using a Michelson Interferometer with Circular Aperture, *Appl. Opt.* **5**, 1084 (1966).

R. A. Williams and W. S. C. Chang, Resolution and Noise in Fourier-Transform Spectroscopy, *J. Opt. Soc. Amer.* **56**, 167 (1966).

T. Williams, Direct Determination of the Line Shapes of Rotational Spectra from Interferometric Measurements, *J. Opt. Soc. Amer.* **50**, 1153 (1960).

D. E. Williamson, Cone Channel Condenser Optics, *J. Opt. Soc. Amer.* **42**, 712 (1952).

R. W. Wood, "Physical Optics." Macmillan, New York, 1936.

C. Wu and G. Andermann, Improved Kramers-Kranig Dispersion Analysis of Infrared Reflectance Data for Lithium Fluoride, *J. Opt. Soc. Amer.* **58**, 519 (1968).

Y. Yamada, A. Mitsuishi, and H. Yoshinaga, Transmission Filters in the Far-Infrared Region, *J. Opt. Soc. Amer.* **52**, 17 (1962).

H. Yoshinaga, Recent Techniques in Far Infrared Spectroscopy, *Jap. J. Appl. Phys., Suppl. I*, **4**, 420 (1965).

H. Yoshinaga, S. Fujita, S. Minami, Y. Suemoto, M. Inoue, K. Chiba, K. Nakano, S. Yoshida, and H. Sugimori, A Far Infrared Interferometric Spectrometer with a Special Electronic Computer, *Appl. Opt.* **5**, 1159 (1966).

# *Subject Bibliography*

CATEGORIES

Apodization and Mathematical Filtering
Applications
Asymmetric Interferometers
Basic Analysis
Beamsplitters
Books, Reviews, and Long Reports
Computer Techniques
Detectors
Devices and Techniques
Errors
Filter Materials
Historical Papers and Sketches
Interferometers versus Spectrometers
Lamellar Grating Interferometers
Real Time Fourier Spectroscopy
Resolution and Signal to Noise
Sources and Source Associated Effects

## APODIZATION AND MATHEMATICAL FILTERING

J. Connes, Computing Problems in Fourier Spectroscopy, *Aspen Int. Conf. on Fourier Spectrosc., 1970*, p. 83. (G. A. Vanasse, A. T. Stair, Jr., and D. J. Baker, eds.). AFCRL-71-0019, 5 Jan. 1971, Spec. Rep. No. 114.

J. Connes and V. Nozal, Mathematical Filtering in Spectroscopy by the Fourier Transformation, *J. Phys. Radium* **22**, 359 (1961).

A. Filler, Apodization and Interpolation in Fourier Transform Spectroscopy, *J. Opt. Soc. Amer.* **54**, 762 (1964).

H. Happ and L. Genzel, Interferenz-Modulation Mit Monochromatischen Millimeter-Wellen, *Infrared Phys.* **1**, 39 (1961).

P. Jacquinot, Appendix E, Apodization, "Concepts of Classical Optics," (J. Strong, ed.). Freeman, San Francisco, California, 1958.

P. Jacquinot and B. Roizen-Dossier, Apodization, *in* "Progress in Optics" (E. Wolf, ed.) Vol. III, p. 33 ff. North-Holland Publ. Company, Amsterdam, 1964.

E. V. Loewenstein, Fourier Spectroscopy: An Introduction, *Aspen Int. Conf. on Fourier Spectrosc., 1970*, p. 3. (G. A. Vanasse, A. T. Stair, Jr., and D. J Baker, eds.). AFCRL-71-0019, 5 Jan. 1971, Spec. Rep. No. 114.

A. E. Martin, "Infra Red Instrumentation and Techniques." Elsevier, New York, 1966.

K. Rasmussen, Conformations and Vibration Spectra of TRIS-(DIAMINE) Chelate Complexes, Chem. Dept., The Technical Univ. of Denmark, 2800 Lyngby (1970).

Lord Rayleigh (J. W. Strutt), *Phil. Mag.* **24**, 864 (1912).

## APPLICATIONS

E. Archbold and H. A. Gebbie, The Use of a Michelson Interferometer to Determine Luminescence Spectra of Optical Maser Materials, *Proc. Phys. Soc.* **80**, 793 (1962).

E. E. Bell, Measurement of Spectral Transmittance and Reflectance with a Far Infrared Michelson Interferometer, *Jap. J. Appl. Phys. Suppl. 1* **4**, 412 (1965).

E. E. Bell, The Use of Asymmetric Interferograms in Transmittance Measurements, *J. Phys. Colloq. C-2* **28**, C2-18 (1967).

D. R. Bosomworth and H. D. Gush, Far Infrared Spectroscopy of Compressed Gases Using a Michelson Interferometer, *Jap. J. Appl. Phys. Suppl. 1*, **4**, 588 (1965).

H. Bowers, Atmospheric Turbulence and Interferometric Spectroscopy, *Appl. Opt.* **3**, 627 (1964).

J. E. Chamberlain, On a Relation Between Absorption Strength and Refractive Index, *Infrared Phys.* **5**, 175 (1965).

J. E. Chamberlain, On the Calculation of Integrated Absorption Strengths From Refraction Data, *J. Quantum Spectrosc. Radiat. Transfer* **7**, 151 (1967).

J. E. Chamberlain and H. A. Gebbie, Sub-Millimetre Dispersion and Rotational Line Strengths of the Hydrogen Halides, *Nature* **208**, 480 (1965).

J. E. Chamberlain and H. A. Gebbie, Dispersion Measurements on Polytetrafluorethylene in the Far Infrared, *Appl. Opt.* **5**, 393 (1966).

J. E. Chamberlain, A. E. Costley, and H. A. Gebbie, Submillimeter Dispersion of Liquid Tetrabromoethane, *Spectrochim. Acta* **23A**, 2255 (1967).

J. E. Chamberlain, H. A. Gebbie, G. W. F. Pardoc, and M. Davies, Millimetre-Wave Fourier Transform Spectroscopy for Studies of the Liquid State, *Chem. Phys. Lett.* **1**, 523 (1968).

J. E. Chamberlain, J. E. Gibbs, and H. A. Gebbie, The Determination of Refractive Index Spectra by Fourier Spectroscopy, *Infrared Phys.* **9**, 185 (1969).

L. W. Chaney, S. R. Drayson, and C. Young, Fourier Transform Spectrometer—Radiative Measurements and Temperature Inversion, *Appl. Opt.* **6**, 347 (1967).

G. W. Chantry, H. M. Evans, J. Chamberlain, and H. A. Gebbie, Absorption and Dispersive Studies in the Range 10–10,000 $\mu$m Using a Modular Michelson Interferometer, *Infrared Phys.* **9**, 85 (1969).

J. Connes and P. Connes, Near-Infrared Planetary Spectra by Fourier Spectroscopy. I. Instruments and Results, *J. Opt. Soc. Amer.* **56**, 896 (1966).

J. Dowling, Direct Interpretation of Far-Infrared Interferograms with Application to Diatomic and Linear Molecules, *J. Opt. Soc. Amer.* **54**, 663 (1964).

J. M. Dowling and R. T. Hall, Far Infrared Interferometry: Upper Limit for Line Widths for the Pure Rotational Band of Carbon Monoxide, *J. Mol. Spectrosc.* **19**, 108 (1966).

H. Gebbie, Spectres D'Absorption Atmospherique dans l'Infrared Lointain par Interferometric a Duex Ondes, *J. Phys. Radium* **19**, 197 (1958).

H. A. Gebbie, Submillimeter Wave Spectroscopy Using a Michelson Interferometer, *in* "Advances in Quantum Electronics." Columbia Univ. Press, New York, 1961.

H. Gebbie, N. Stone, and D. Williams, A Interferometric Study of the Far-Infrared Spectrum of Compressed Nitrogen, *Mol. Phys.* **6**, 215 (1963).

R. T. Hall and J. M. Dowling, Pure Rotational Spectrum of Nitric Oxide, *J. Chem. Phys.* **45**, 1899 (1966).

R. T. Hall and J. M. Dowling, Pure Rotational Spectrum of Water Vapor, *J. Chem. Phys.* **47**, 2454 (1967).

R. A. Hanel, B. Schlachman, F. D. Clark, C. H. Prokesh, J. B. Taylor, W. M. Wilson, and L. Chaney, The Nimbus III Michelson Interferometer, *Aspen Int. Conf. on Fourier Spectrosc., 1970*, p. 231. (G. A. Vanasse, A. T. Stair, Jr., and D. J. Baker, eds.). AFCRL-71-0019, 5 Jan. 1971, Spec. Rep. No. 114.

H. Happ and L. Genzel, Interferenz-Modulation Mit Monochromatischen Millimeter-Wellen, *Infrared Phys.* **1**, 39 (1961).

D. M. Hunten, Fourier Spectroscopy of Planets, *Science* **162**, 313 (1968).

W. J. Hurley, Interferometric Spectroscopy in the Far Infrared, *J. Chem. Educ.* **43**, 236 (1966).

R. Kaplan, Magnetooptical Studies of Solids Using Fourier Transform Spectroscopy, *Appl. Opt.* **6**, 685 (1957).

E. Loewenstein, Interferometric Spectra of Ammonia and Carbon Monoxide in the Far Infrared, *J. Opt. Soc. Amer.* **50**, 1163 (1960).

E. Loewenstein, Optical Properties of Sapphire in the Far-Infrared, *J. Opt. Soc. Amer.* **51**, 108 (1961).

M. F. D. Low, Fourier Transform Spectroscopy, *J. Chem. Educ.* **47**, A163, A255, A349, A415 (1970).

D. Malz, Interference Phenomena in Plane Parallel Plates in Fourier Spectroscopy and Their Effect on Spectroscopic Investigations on Semiconductors in the Middle and Far Infrared, *Exp. Tech. Phys.* (*Germany*) **13**, 257 (1965).

K. D. Moller and W. G. Rothschild, "Far-Infrared Spectroscopy." Wiley (Interscience) New York, 1971.

R. E. Murphy and H. Sakai, Application of the Fourier Spectroscopy Technique to the Study of Relaxation Phenomena, *Aspen Int. Conf. on Fourier Spectrosc., 1970*, p. 301. (G. A. Vanasse, A. T. Stair, Jr., and D. J. Baker, eds.). AFCRL-71-0019, 5 Jan. 1971, Spec. Rep. No. 114.

C. H. Perry, R. Geick, and E. F. Young, Solid State Studies by Means of Fourier Transform Spectroscopy, *Appl. Opt.* **5**, 1171 (1966).

E. Plyler, D. Yates, and H. Gebbie, Radiant Energy from Sources in the Far Infrared, *J. Opt. Soc. Amer.* **52**, 859 (1962).

C. M. Randall and R. D. Rawcliffe, Refractive Indices of Germanium, Silicon, and Fused Quartz in the Far Infrared, *Appl. Opt.* **6**, 1889 (1967).

P. L. Richards, Antiferromagnetic Resonance in $CoF_2$, $NiF_3$, and $MnCO_3$, *J. Appl. Phys.* **35**, 850 (1963).

P. L. Richards, Far-Infrared Magnetic Resonance in $CoF_2$, $NiF_2$, $KNiF_3$, and YbIG, *J. Appl. Phys.* **34**, 1237 (1964).

P. L. Richards, High Resolution Fourier Transform Spectroscopy in the Far-Infrared, *J. Opt. Soc. Amer.* **54**, 1474 (1964).

P. L. Richards, High Resolution Far Infrared Interferometry with Application to Solid State Physics, *Jap. J. Appl. Phys. Suppl. 1*, **4**, 417 (1965).

W. H. Robinette and R. B. Sanderson, Dipole Moment of HBr From Far Infrared Dispersion Measurements, *Appl. Opt.* **8**, 711 (1969).

E. E. Russell and E. E. Bell, Measurements of the Optical Constants of Crystal Quartz in the Far Infrared with the Asymmetric Fourier-Transform Method, *J. Opt. Soc. Amer.* **57**, 341 (1967).

E. E. Russell and E. E. Bell, Optical Constants of Sapphire in the Far Infrared, *J. Opt. Soc. Amer.* **57**, 543 (1967).

R. B. Sanderson, Measurement of Rotational Line Strengths in HCl by Asymmetric Fourier Transform Techniques, *Appl. Opt.* **6**, 1527 (1967).

J. E. Shaw, Spectroradiometry over Broad Spectral Regions by Fourier Spectroscopy, *J. Opt. Soc. Amer.* **57**, 1136 (1967).

W. M. Sinton, Recent Infrared Spectra of Mars and Venus, *J. Quantum Spectrosc. Radiat. Transfer* **3**, 551 (1963).

G. Vanasse, Fourier Series Approximation to Spectral Functions Obtained by Interferometric Techniques, *J. Opt. Soc. Amer.* **52**, 572 (1962).

G. Vanasse and E. Loewenstein, Pure Rotation Spectra of HCN and $NH_3$, *J. Opt. Soc. Amer.* **49**, 512 (1959).

G. Vanasse, J. Strong, and E. Loewenstein, Far Infrared Spectra of $H_2O$ and $H_2S$ Taken with an Interferometric Spectrograph, *J. Opt. Soc. Amer.* **49**, 309 (1959).

G. A. Vanasse, A. T. Stair, Jr., and D. J. Baker, eds., *Aspen Int. Conf. on Fourier Spectrosc., 1970.* AFCRL-71-0019, 5 Jan. 1971, Spec. Rep. No. 114.

E. A. Vinogradov, E. N. Dianov, and N. A. Irisova, Michelson Interferometer for Measuring the Refractive Index of Dielectric Materials in the 2 mm Wavelength Range, *Radiotekh. Elektron.* **10**, 1804 (1965); *Radio Eng. Electron. Phys.* **10**, 1547 (1965).

T. Williams, Direct Determination of the Line Shapes of Rotational Spectra from Interferometric Measurements, *J. Opt. Soc. Amer.* **50**, 1153 (1960).

## ASYMMETRIC INTERFEROMETERS

E. E. Bell, Measurement of Spectral Transmittance and Reflectance with a Far Infrared Michelson Interferometer, *Jap. J. Appl. Phys. Suppl. 1* **4**, 412 (1965).

E. E. Bell, Measurement of the Far Infrared Optical Properties of Solids with a Michelson Interferometer Used in the Asymmetric Mode: Part I, Mathematical Formulation, *Infrared Phys.* **6**, 57 (1966).

E. E. Bell, The Use of Asymmetric Interferograms in Transmittance Measurements, *J. Phys. Colloq. C-2* **28**, C2-18 (1967).

E. E. Bell, "Optical Constants and Their Measurement" (L. Genzel, ed.) Bd. XXV/2a, Springer-Verlag, Heidelberg, 1967.

E. E. Bell, Amplitude Spectroscopy, *Aspen Int. Conf. on Fourier Spectrosc., 1970*, p. 71. (G. A. Vanasse, A. T. Stair, Jr., and D. J. Baker, eds.). AFCRL-71-0019, 5 Jan. 1971, Spec. Rep. No. 114.

J. E. Chamberlain, On a Relation Between Absorption Strength and Refractive Index, *Infrared Phys.* **5**, 175 (1965).

J. E. Chamberlain, On a Convergence Correction for Dispersive Fourier Spectrometry, *Appl. Opt.* **6**, 980 (1967).

J. E. Chamberlain, On the Calculation of Integrated Absorption Strengths From Refraction Data, *J. Quantum Spectrosc. Radiative Transfer* **7**, 151 (1967).

J. E. Chamberlain and H. A. Gebbie, Sub-Millimetre Dispersion and Rotational Line Strengths of the Hydrogen Halides, *Nature* **208**, 480 (1965).

J. E. Chamberlain and H. A. Gebbie, Dispersion Measurements on Polytetrafluorethylene in the Far Infrared, *Appl. Opt.* **5**, 393 (1966).

J. E. Chamberlain, J. E. Gibbs, and H. A. Gebbie, Refractometry in the Far Infrared Using a Two Beam Interferometer, *Nature* **198**, 874 (1963).

J. E. Chamberlain, G. W. Chantry, F. D. Findlay, H. A. Gebbie, J. E. Gibbs, N. W. B. Stone, and A. J. Wright, The Spectra Transmission at Infra-Red Wavelengths of Michelson Interferometers with Dielectric Film Beam Splitters, *Infrared Phys.* **6**, 195 (1966).

J. E. Chamberlain, A. E. Costley, and H. A. Gebbie, Submillimeter Dispersion of Liquid Tetrabromoethane, *Spectrochim. Acta* **23A**, 2255 (1967).

J. E. Chamberlain, H. A. Gebbie, G. W. F. Pardoc, and M. Davies, Millimetre-Wave Fourier Transform Spectroscopy for Studies of the Liquid State, *Chem. Phys. Lett.* **1**, 523 (1968).

J. E. Chamberlain, J. E. Gibbs, and H. A. Gebbie, The Determination of Refractive Index Spectra by Fourier Spectroscopy, *Infrared Phys.* **9**, 185 (1969).

H. A. Gebbie and N. W. B. Stone, A Michelson Interferometer for Far-Infrared Spectroscopy of Gases, *Infrared Phys.* **4**, 85 (1964).

H. A. Gebbie, and R. Q. Twiss, Two-Beam Interferometric Spectroscopy, *Rep. Progr. Phys.* **29**, 729 (1966).

E. E. Russell and E. E. Bell, Measurement of the Far Infrared Optical Properties of Solids with a Michelson Interferometer Used in the Asymmetric Mode: Part II, The Vacuum Interferometer, *Infrared Phys.* **6**, 75 (1966).

E. E. Russell and E. E. Bell, Measurements of the Optical Constants of Crystal Quartz in the Far Infrared with the Asymmetric Fourier-Transform Method, *J. Opt. Soc. Amer.* **57**, 341 (1967).

E. E. Russell and E. E. Bell, Optical Constants of Sapphire in the Far Infrared, *J. Opt. Soc. Amer.* **57**, 543 (1967).

R. B. Sanderson and H. E. Scott, A High Resolution Far Infrared Interferometer, *Appl. Opt.* **10**, 1097 (1971).

J. S. Toll, Causality and Dispersion Relation: Logical Foundations, *Phys. Rev.* **104**, 1760 (1956).

R. Ulrich, Far Infrared Properties of Metallic Mesh and Its Complementary Structure, *Infrared Phys.* **7**, 37 (1967).

R. Ulrich, Effective Low-Pass Filters for Far Infrared Frequencies, *Infrared Phys.* **7**, 65 (1967).

R. Ulrich, Interference Filters for the Far Infrared, *Appl. Opt.* **7**, 1987 (1968).

## BASIC ANALYSIS

J. Connes, Domaine d'Utilisation de la Methode par Transformee Fourier, *J. Phys. Radium* **19**, 197 (1958).

J. Connes, Recherches sur la Spectroscopie par Transformation de Fourier, *Rev. Opt.* **40**, 45–78, 116–140, 171–190, 231–265 (1961).

J. Connes and V. Nozal, Mathematical Filtering in Spectroscopy by the Fourier Transformation, *J. Phys. Radium* **22**, 359 (1961).

P. Connes, Increase of the Product of Luminosity × Resolution with Interferometers by Employing Path Differences Independent of the Incidence, *Rev. Opt.* **35**, 37 (1956).

P. Connes, How Light is Analyzed, *in* "Lasers and Light" (A. Shawlow, ed.), p. 35. Freeman, San Francisco, California, 1969.

P. Fellgett, A Propos de la Theorie du Spectrometre Interferentiel Multiplex, *J. Phys. Radium* **19**, 187 (1958).

P. Fellgett, Spectrometry Interferentiel Multiplex Pour Measures Infrared Sur Les Etoiles, *J. Phys. Radium* **19**, 237 (1958).

H. A. Gebbie, Submillimeter Wave Spectroscopy Using a Michelson Interferometer, "Advances in Quantum Electronics." Columbia Univ. Press, New York, 1961.

L. Genzel and R. Weber, Zur Theorie der Interference-Modulation fur Zweistrahl-Interferenzen, *Z. angew. Phys.* **10**, 127 (1958).

L. Genzel Aperiodic and Periodic Interference Modulation for Spectrographic Purposes, *J. Mol. Spectrosc.* **4**, 241 (1960).

L. Genzel, Fourier-Spectroskopie, Plenarvortrage, p. 128, 33 Physiker-Tagung 1968, Karlsruhe, W. Germany, B. G. Teubner, Stuttgart, 1969.

L. Genzel and R. Weber, Spektroskopie im fernen Ultrarot durch Interferenz-Modulation, *Z. angew. Phys.* **10**, 195 (1958).

L. Genzel, H. Happ, and R. Weber, Dispersionsmessumgen An NaCl, KCl and KBr Zwischen 0.3 and 3 mm Wellenlange, *Z. Phys.* **154**, 13 (1959).

A. Hadni, Far Infrared Transpose Spectroscopy, *Appl. Opt.* **4**, 1374 (1965).

P. Jacquinot, Caracteres Communs aus Nouvelles Methodes d'Spectroscopie Interferentielle Facteur Merite, *J. Phys. Radium* **19**, 223 (1958).

P. Jacquinot, New Developments in Interference Spectroscopy, *Progr. Phys.* **23**, 267 (1960).

J. Lecomte, Spectroscopie dans l'Infrarouge, "Handbuch der Physik," Vol. 24, p. 283, Springer-Verlag, Berlin, 1957.

E. V. Loewenstein, The History and Current Status of Fourier Transform Spectroscopy, *Appl. Opt.* **5**, 845 (1966).

E. V. Loewenstein, Fourier Spectroscopy: An Introduction, *Aspen Int. Conf. on Fourier Spectrosc., 1970*, p. 3. (G. A. Vanasse, A. T. Stair, Jr., and D. J. Baker, eds.). AFCRL-71-0019, 5 Jan. 1971, Spec. Rep. No. 114.

A. E. Martin, "Infra Red Instrumentations and Techniques." Elsevier, New York, 1966.

L. Mertz, "Transformations in Optics." Wiley, New York, 1965.

J. L. Pritchard, H. Sakai, W. H. Steel, and G. A. Vanasse, "Colloque sur les Methodes Nouvelles de Spectroscopie Instrumentale." Paris, CNRS, 1966.

P. L. Richards, High-Resolution Fourier Transform Spectroscopy in the Far Infrared, *J. Opt. Soc. Amer.* **54**, 1474 (1964).

H. Sakai and G. A. Vanasse, Direct Determination of the Transfer Function of an Infrared Spectrometer, *J. Opt. Soc. Amer.* **56**, 357 (1966).

H. Sakai and G. A. Vanasse, Hilbert Transform in Fourier Spectroscopy, *J. Opt. Soc. Amer.* **56**, 131 (1966).

J. D. Strong, "Concepts of Classical Optics." Freeman, San Francisco, California, 1958.

J. D. Strong and G. Vanasse, Interferometric Spectroscopy in the Far-Infrared, *J. Opt. Soc. Amer.* **49**, 844 (1959).

G. A. Vanasse and H. Sakai, Fourier Spectroscopy, *in* "Progress in Optics" (E. Wolf, ed.), Vol. 6, p. 261. North-Holland Publ., Amsterdam, 1967.

G. A. Vanasse, A. T. Stair, Jr., and D. J. Baker, eds., *Aspen Int. Conf. on Fourier Spectrosc., 1970.* AFCRL-71-0019, 5 Jan. 1971, Spec. Rep. No. 114.

## BEAMSPLITTERS

H. A. Beggie and N. W. B. Stone, A Michelson Interferometer for Far Infrared Spectroscopy of Gases, *Infrared Phys.* **4**, 85 (1964).

J. E. Chamberlain, G. W. Chantry, F. D. Findlay, H. A. Gebbie, J. E. Gibbs, N. W. B. Stone, and A. J. Wright, The Spectra Transmission at Infra-Red Wavelengths of Michelson Interferometers with Dielectric Film Beam Splitters, *Infrared Phys.* **6**, 195 (1966).

D. James and J. Ring, The Efficiency of Beamsplitters in the Far Infra-Red Michelson Interferometer, *J. Phys. Suppl.* **28**, C2-150 (1967).

M. Lax and E. Brustein, Infrared Lattice Absorption in Ionic and Homopolar Crystals, *Phys. Rev.* **97**, 39 (1955).

D. C. Lewis and W. D. Westwood, The Spectra of Some Iron Oxide Films Between 30,000 $cm^{-1}$ and 4000 $cm^{-1}$, *Can. J. Phys.* **42**, 2367 (1964).

E. V. Loewenstein and A. Engelsrath, Polarization Properties of Far Infra-Red Beamsplitters, *J. Phys. Suppl.* **28**, C2-153 (1967).

D. H. Martin and E. Puplett, Polarized Interferometric Spectroscopy for the Millimetre and Submillimetre Spectrum, *Infrared Phys.* **10**, 57 (1966).

P. L. Richards, High-Resolution Fourier Transform Spectroscopy in the Far Infrared, *J. Opt. Soc. Amer.* **54**, 1474 (1964).

G. Roland, Spectroscopie par transformation de Fourier, Thesis, Liege, Belgium (1965).

E. E. Russell and E. E. Bell, Measurement of the Far Infrared Optical Properties of Solids with a Michelson Interferometer Used in the Asymmetric Mode: Part II, The Vacuum Interferometer, *Infrared Phys.* **6**, 75 (1966).

A. G. Tescher, Beamsplitter Optimization for Fourier Spectroscopy, *Aspen Int. Conf. on Fourier Spectrosc., 1970*, p. 225 (G. A. Vanasse, A. T. Stair, Jr., and D. J. Baker, eds.). AFCRL-71-0019, 5 Jan. 1971, Spec. Rep. No. 114.

W. H. Steel, Interferometry. Cambridge Univ. Press, New York, 1967.

W. H. Steel, Interferometers for Fourier Spectroscopy, *Aspen Int. Conf. on Fourier Spectrosc., 1970*, p. 43. (G. A. Vanasse, A. T. Stair, Jr., and D. J. Baker, eds.). AFCRL-71-0019, 5 Jan. 1971 Spec. Rep. No. 114.

D. G. Vickers, E. I. Robson, and J. E. Beckman, Freestanding Polarizers for Submillimeter Interferometry, *Appl. Opt.* **10**, 682 (1971).

P. Vogel and L. Genzel, Transmission and Reflection of Metallic Mesh in the Far Infrared, *Infrared Phys.* **4**, 257 (1964).

## BOOKS, REVIEWS, AND LONG REPORTS

D. Bloor, Bibliography of Far Infrared Spectroscopy, *Infrared Phys.* **10**, 1 (1970).

M. Born and E. Wolf, "Principles of Optics," 2nd ed., p. 61. Pergamon Press, Macmillan Co., New York, 1964.

R. Bracewell, "The Fourier Transform and Its Applications." McGraw-Hill, New York, 1965.

J. E. Chamberlain, J. E. Gibbs, and H. A. Gebbie, The Determination of Refractive Index Spectra by Fourier Spectroscopy, *Infrared Phys.* **9**, 185 (1969).

E. E. Bell, Measurement of the Far Infrared Optical Properties of Solids with a Michelson Interferometer Used in the Asymmetric Mode: Part I, Mathematical Formulation, *Infrared Phys.* **6**, 57 (1966).

E. E. Bell, "Optical Constants and Their Measurement" (L. Genzel, ed.), Bd XXV/2a. Springer-Verlag, Heidelberg, 1967.

E. E. Bell, Amplitude Spectroscopy, *Aspen Int. Conf. on Fourier Spectrosc., 1970*, p. 71. (G. A. Vanasse, A. T. Stair, Jr., and D. J. Baker, eds.). AFCRL-71-0019, 5 Jan. 1971, Spec. Rep. No. 114.

J. Connes, Computing Problems in Fourier Spectroscopy, *Aspen Int. Conf. on Fourier Spectrosc., 1970*, p. 83. (G. A. Vanasse, A. T. Stair, Jr., and D. J. Baker, eds.). AFCRL-71-0019, 5 Jan. 1971, Spec. Rep. No. 114.

P. Connes, High Resolution and High Information Fourier Spectroscopy, *Aspen Int. Conf. on Fourier Spectrosc., 1970*, p. 121. (G. A. Vanasse, A. T. Stair, Jr., and D. J. Baker, eds.). AFCRL-71-0019, 5 Jan. 1971, Spec. Rep. No. 114.

J. Connes, P. Connes, and J. P. Maillard, "Atlas des Spectres dans le Proche Infrarouge de Venus, Mars, Jupiter, et Saturne." Centre National de la Recherche Scientique, Paris, 1969.

J. M. Dowling, Investigations in the Far Infrared with a Lamellar Grating Interferometer, Air Force Rep. No. SSD-TR-67-30; Aerospace Rep. No. TR-1001 (9260-01)-7; March, 1967; J. M. Dowling, Space Phys. Lab., Lab. Div., Lab. Operations, Aerospace Corp., Los Angeles, California.

J. M. Dowling, Double Beaming in Fourier Spectroscopy, *Aspen Int. Conf. on Fourier Spectrosc., 1970*, p. 55. (G. A. Vanasse, A. T. Stair, Jr., and D. J. Baker, eds.). AFCRL-71-0019, 5 Jan. 1971, Spec. Rep. No. 114.

Fast Computational Techniques Issue of *IEEE Trans. Audio Electroacoust.* **AU-15(2)** (1967).

M. Francon (translated by I. Wilmanns), "Optical Interferometry." Academic Press, New York, 1966.

L. Genzel, Fourier-Spectroskopie, Plenarvortrage, p. 128, 33 Physiker-Tagung 1968, Karlsruhe, W. Germany, B. G. Teubner, Stuttgart, 1969.

A. Girard and P. Jacquinot, *in* "Advanced Optical Techniques" (A. C. S. Van Hell, ed.). North-Holland Publ., Amsterdam, 1967.

S. Goldman, "Information Theory." Prentice Hall, Englewood Cliffs, New Jersey, 1953.

J. W. Goodman, "Introduction to Fourier Optics." McGraw-Hill, New York, 1968.

E. A. Guilliemin, "The Mathematics of Circuit Analysis." Wiley, New York, 1949.

A. Hadni, "Essentials of Modern Physics Applied to the Study of the Infrared." Pergamon Press, Oxford, 1967.

N. J. Harrick, "Internal Reflection Spectroscopy." Wiley (Interscience), New York, 1967.

P. G. Hoel, "Introduction to Mathematical Statistics," 2nd ed., pp. 166–167. Wiley, New York, 1954.

R. D. Hudson, "Infrared System Engineering," p. 270. Wiley, New York, 1969.

J. D. Jackson, "Classical Electrodynamics." Wiley, New York, 1962.

F. A. Jenkins and H. E. White, "Fundaments of Optics," 3rd ed. McGraw-Hill, New York, 1957.

G. A. Korn and T. M. Korn, "Handbook of Mathematical Physics," 2nd ed., p. 879. McGraw-Hill, New York, 1968.

P. W. Kruse, L. D. McGlauchlin, and R. B. McQuistan, "Elements of Infrared Technology, Generation, Transmission, and Detection." Wiley, London, 1962.

E. V. Loewenstein, Fourier Spectroscopy: An Introduction, *Aspen Int. Conf. on Fourier Spectrosc., 1970,* p. 3 (G. A. Vanasse, A. T. Stair, Jr., and D. J. Baker, eds.). AFCRL-71-0019, 5 Jan. 1971, Spec. Rep. No. 114.

L. Mandel, Fluctuations of Light Beams, *in* "Progress in Optics" (E. Wolf, ed.). North-Holland Publ., Amsterdam, 1963.

A. E. Martin, "Infra Red Instrumentation and Techniques." Elsevier, New York, 1966.

D. H. Martin, ed. "Spectroscopic Techniques for Far Infrared, Submillimeter and Millimeter Waves." North-Holland Publ. Amsterdam, 1967.

L. Mertz, "Transformations in Optics." Wiley, New York, 1965.

A. Michelson, "Studies in Optics." Univ. of Chicago Press, Chicago, Illinois, 1927.

K. D. Moller and W. G. Rothschild, "Far-Infrared Spectroscopy." Wiley (Interscience), New York, 1971.

T. S. Moss, "Optical Properties of Semiconductors." Butterworths, London and Washington, D.C. , 1959.

E. D. Palik, A Far Infrared Bibliography, Naval Res. Lab., Washington, D.C. (1967).

E. E. Russell and E. E. Bell, Measurement of the Far Infrared Optical Properties of Solids with a Michelson Interferometer Used in the Asymmetric Mode: Part II, The Vacuum Interferometer, *Infrared Phys.* **6**, 75 (1966).

A. Schawlow, "Lasers and Light." Freeman, San Francisco, California, 1969.

W. H. Steel, "Interferometry." Cambridge Univ. Press, New York, 1967.

W. H. Steel, Interferometers for Fourier Spectroscopy, *Aspen Int. Conf. on Fourier Spectrosc., 1970*, p. 43. (G. A. Vanasse, A. T. Stair, Jr., and D. J. Baker, eds.). AFCRL-71-0019, 5 Jan. 1971 Spec. Rep. No. 114.

John M. Stone, "Radiation and Optics." McGraw-Hill, New York, 1963.

G. A. Vanasse and H. Sakai, Fourier Spectroscopy, *in* "Progress in Optics" (E. Wolf, ed.). Vol. 6, p. 261. North-Holland Publ., Amsterdam, 1967.

G. A. Vanasse, A. T. Stair, Jr., and D. J. Baker, eds., *Aspen Int. Conf. on Fourier Spectrosc., 1970*. AFCRL-71-0019, 5 Jan. 1971, Spec. Rep. No. 114.

R. W. Wood, "Physical Optics." Macmillan, New York, 1936.

## COMPUTER TECHNIQUES

G. D. Bergland, The Fast Fourier Transform Recursive Equations for Arbitrary Length Records, *Math. Comput.* **21**, 236 (1967).

G. D. Bergland, A Fast Fourier Transform Algorithm Using Base 8 Iterations, *Math. Comput.* **22**, 275 (1968).

E. O. Brigham and R. E. Morrow, The Fast Fourier Spectrum, *IEEE Trans. Audio Electroacoust.* **AU-15(2)**, *63* (1967).

H. L. Buijs, Fast Fourier Transformation of Large Arrays of Data, *Appl. Opt.* **8**, 211 (1969).

W. T. Cochran, James W. Cooley, David L. Favin, Howard D. Helms, Reg. A. Kaenel, William W. Lang, George C. Maling, Jr., David E. Nelson, Charles M. Rader, and Peter D. Welch, What is the Fast Fourier Transform?, *Proc. IEEE* **55**, 1664 (1967).

J. Connes, Domaine d'Utilisation de la Methode par Transformee Fourier, *J. Phys. Radium*, **19**, 197 (1958).

J. Connes, Recherches sur la Spectroscopie par Transformation de Fourier, *Rev. Opt.* **40**, 45–78, 116–140, 171–190, and 231–265 (1961).

J. Connes, Computing Problems in Fourier Spectroscopy, *Aspen Int. Conf. on Fourier Spectrosc., 1970*, p. 83. (G. A. Vanasse, A. T. Stair, Jr., and D. J. Baker, eds.). AFCRL-71-0019, 5 Jan. 1971, Spec. Rep. No. 114.

P. Connes and G. Michel, Real Time Computer for Fourier Spectroscopy, *Aspen Int. Conf. on Fourier Spectrosc., 1970*, p. 313. (G. A. Vanasse A. T., Stair, Jr., and D. J. Baker, eds.). AFCRL-71-0019, 5 Jan. 1971, Spec. Rep. No. 114.

J. W. Cooley and J. W. Tukey, An Algorithm for the Machine Calculation of Complex Fourier Series, *Math. Comput.* **19,** 297 (1965).

Fast Computational Techniques Issue of *IEEE Trans. Audio Electroacoust.* **AU-15(2)** (1967).

M. L. Forman, Fast Fourier-Transform Technique and Its Application to Fourier Spectroscopy, *J. Opt. Soc. Amer.* **56**, 978 (1966).

I. J. Good, The Interaction Algorithm and Practical Fourier Series, *J. Roy. Statist. Soc. Ser. B.* **20**, 361 (1958).

F. Levy, R. C. Milward, S. Bras, and R. le Toullec, "Real Time" Far Infrared Fourier Spectroscopy Using a Small Digital Computer, *Aspen Int. Conf. on Fourier Spectrosc., 1970*, p. 331. (G. A. Vanasse, A. T. Stair, Jr., and D. J. Baker, eds.). AFCRL-71-0019, 5 Jan. 1971, Spec. Rep. 114.

B. Liu and T. Kaneko, Error Analysis of Digital Filters Realized with Floating-Point Arithmetic, *Proc. IEEE* **57**, 1735 (1969).

L. Mertz, Auxilliary Computation of Fourier Spectrometry, *Infrared Phys.* **7**, 17 (1967).

C. M. Randall, Computational Techniques in Fourier Transform Spectroscopy, Air Force Rep. No. SSD-TR-67-75, Aerospace Rep. No. TR-1001 (9260-1001 (9260-01)-8, Aerospace Corp., (March 1967).

C. M. Randall, Fast Fourier Transform for Unequal Number of Input and Output Points, *Appl. Opt.* **6**, 1432 (1967).

P. L. Richards, Computing Reliable Power Spectra, *IEEE Spectrum* **4**, 83 (1967).

A. K. Rigler, Note on the Fast Fourier Transform, *J. Opt. Soc. Amer.* **58**, 274 (1968).

H. Sakai and G. A. Vanasse, Hilbert Transform in Fourier Spectroscopy, *J. Opt. Soc. Amer.* **56**, 131 (1966).

J. D. Strong and G. Vanasse, Modulation Interferentielle et Calculateur Analogique pour un Spectrometre Intereferentiel, *J. Phys. Radium* **19**, 192 (1958).

G. Vanasse, Fourier Series Approximation to Spectral Functions Obtained by Interferometric Techniques, *J. Opt. Soc. Amer.* **52**, 472 (1962).

H. Yoshinaga, S. Fujita, S. Minami, Y. Suemoto, M. Inoue, K. Chiba, K. Nakano, S. Yoshida, and H. Sugimori, A Far Infrared Interferometric Spectrometer with a Special Electronic Computer, *Appl. Opt.* **5**, 1159 (1966).

## DETECTORS

R. Bachman, H. D. Kirsch, and T. H. Geballe, Low Temperature Silicon Thermometer and Bolometer, *Rev. Sci. Instrum.* **41**, 547 (1970).

W. Boyle and K. Rodgers, Performance Characteristics of New Low-Temperature Bolometer, *J. Opt. Soc. Amer.* **52**, 859 (1962).

C. C. Grimes, P. L. Richards, and S. Shapiro, Josephson-Effect Far Infrared Detector, *J. Appl. Phys.* **39**, 3905 (1968).

R. D. Hudson, "Infrared System Engineering," p. 270. Wiley, New York, 1969.

P. W. Kruse, L. D. McGlauchlin, and R. B. McQuistan, "Elements of Infrared Technology, Generation, Transmission, and Detection." Wiley, London, 1962.

F. Low, Low Temperature Germanium Bolometer, *J. Opt. Soc. Amer.* **51**, 1300 (1961).

E. H. Putley, Indium Antimonide Submillimeter Photoconductive Detectors, *Appl. Opt.* **4**, 649 (1965).

P. L. Richards, Far Infrared Detectors, "Far Infrared Properties of Solids," p. 103 (S. S. Mitra and S. Nudelman, eds.). Plenum, New York, 1970.

F. Schwarz and R. R. Poole, Performance Characteristics of a Small TGS Detector Operated in the Pyroelectric Mode, *Appl. Opt.* **9**, 1940 (1970).

M. Smith, Optimization of Bias in Thermistor Radiator Detectors. II Practical Networks, *Appl. Opt.* **8**, 1213 (1969).

## DEVICES AND TECHNIQUES

J. M. Ballontyne, Double Beam Operation of a Fourier Spectrometer, *Appl. Opt.* **6**, 587 (1967).

H. A. Beggie and N. W. B. Stone, A Michelson Interferometer for Far Infrared Spectroscopy of Gases, *Infrared Phys.* **4**, 85 (1964).

R. Beer and D. Marjaniemi, Wavefronts and Construction Tolerances for a Cat's-Eye Retroreflector, *Appl. Opt.* **7**, 1191 (1966).

R. J. Bell and T. E. Gilmer, Jr., A New Radiation Chopper Principle, *Appl. Opt.* **4**, 45 (1965).

M. Bottema and H. J. Bolle, An Interferometer with Spherical Mirrors for Fourier Spectroscopy, *Aspen Int. Conf. on Fourier Spectrosc., 1970*, p. 210. (G. A. Vanasse, A. T. Stair, Jr., and D. J. Baker, eds.). AFCRL-71-0019, 5 Jan. 1971, Spec. Rep. No. 114.

R. Cano and M. Mattioli, Far Infrared Spectroscopy with a Michelson Interferometer and an InSb Detector, *Infrared Phys.* **7**, 25 (1967).

P. Connes, ICO Conference, "Optical Instruments and Techniques." Oriel Press, Newcastle upon Tyne (to be published).

M. Cuisenier, Etude et realisation d'un interferometre de Michelson permetlant l'utilisation de la spectroscopie par transformation de Fourier a haute resolution sur un intervalle spectral etendu, These de 3[e] cycle, Paris.

M. Cuisenier and J. Pinard, Apectrometre de Fourier a 'Oeils de Chat' et a Balayage Rapide, *J. Phys. Colloq. C-2* **28**, C-297 (1967).

J. M. Dowling and R. T. Hall, The Techniques of High Resolution, Double-Beam, Two-Beam Interferometry with a Far Infrared Lamellar Grating Interferometer, *J. Phys. Colloq. C2, suppl.* are *M° 3–4* **28**, C2-156 (1967).

S. I. Drasky and R. J. Bell, Radiation Chopper for the Submillimeter Wavelength Region, *Infrared Phys.* **5**, 137 (1965).

H. A. Gebbie and N. W. B. Stone, A Michelson Interferometer for Far-Infrared Spectroscopy of Gases, *Infrared Phys.* **4**, 85 (1964).

R. T. Hall, D. Vrabec, and J. Dowling, High Resolution Far-Infrared Lamellar-Grating Spectrometer, *J. Opt. Soc. Amer.* **54**, 1390 (1964).

R. T. Hall, D. Vrabec, and J. M. Dowling, A High-Resolution Far-Infrared Double-Beam Lamellar Grating Interferometer, *Appl. Opt.* **5**, 1147 (1966).

R. A. Hanel, B. Schlachman, F. D. Clark, C. H. Prokesh, J. B. Taylor, W. M. Wilson, and L. Chaney, The Nimbus III Michelson Interferometer, *Aspen Int. Conf. on Fourier Spectrosc., 1970*, p. 231. (G. A. Vanasse, A. T. Stair, Jr., and D. J. Baker, eds.). AFCRL-71-0019, 5 Jan. 1971, Spec. Rep. No. 114.

N. P. Hansen and J. Strong, Performance of a Simple Spherical Lamellar Grating Beamsplitter by Wavefront Division, *Aspen Int. Conf. on Fourier Spectrosc., 1970*, p. 215. (G. A. Vanasse, A. T. Stair, Jr., and D. J. Baker, eds.). AFCRL-71-0019, 5 Jan. 1971, Spec. Rep. No. 114.

H. Happ and L. Genzel, Interferenz-Modulation Mit Monochromatischen Millimeter-Wellen, *Infrared Phys.* **1**, 39 (1961).

Hillel Bar-Lev, A Dual-Beam Infrared Interferometer-Spectrometer, *Infrared Phys.* **7**, 93 (1967).

B. F. Hochheimer and C. F. Bradley, A Fourier Transform Spectrometer for the 10 to 10,000 $cm^{-1}$ Spectral Region, *Appl. Opt.* **8**, 557 (1969).

M. F. D. Low, Fourier Transform Spectroscopy, *J. Chem. Educ.* **47**, A163, A255, A349, A415 (1970).

D. H. Martin and E. Puplett, Polarized Interferometric Spectroscopy for the Millimetre and Submillimetre Spectrum, *Infrared Phys.* **10**, 57 (1966).

L. Mertz, Rapid Scanning Fourier Transform Spectroscopy, *J. Phys. Colloq. C2, suppl. 3–4* **28**, C2-87 (1967).

L. Mertz, N. O. Young, and J. Armitage, Mock Interferometry, *in* "Optical Instruments and Techniques" (K. J. Habell, ed.). p. 51. Chapman and Hall, London, 1962.

R. C. Milward, A Complete Fourier Spectrophotometer System for the Far-Infra-Red, "Molecular Spectroscopy," p. 81. Inst. Petroleum, Adlard and Son, Dorking, England, 1968.

R. C. Milward, A Small Lamellar Grating Interferometer for the Very Far-Infrared, *Infrared Phys.* **9**, 59 (1969).

R. E. Murphy and H. Sakai, Application of the Fourier Spectroscopy Technique to the Study of Relaxation Phenomena, *Aspen Int. Conf. on Fourier Spectrosc., 1970*, p. 310. (G. A. Vanasse, A. T. Stair, Jr., and D. J. Baker, eds.). AFCRL-71-0019, 5 Jan. 1971, Spec. Rep. No. 114.

M. V. R. K. Murty, Modification of Michelson Interferometer Using Only One Cube-Corner Prism, *J. Opt. Soc. Amer.* **50**, 83 (1960).

W. F. Parks and R. W. Alexander, Interferometers with Spherical Mirrors (to be published).

E. R. Peck, A New Principle in Interferometer Design, *J. Opt. Soc. Amer.* **38**, 66 (1948).

E. R. Peck, Theory of the Corner-Cube Interferometer, *J. Opt. Soc. Amer.* **38**, 1015 (1948).

E. R. Peck, Uncompensated Corner-Reflector Interferometer, *J. Opt. Soc. Amer.* **47**, 250 (1957).

R. D. Rawcliffe, Refractive Indices of Germanium, Silicon, and Fused Quartz in the Far Infrared, *Appl. Opt.* **6**, 1889 (1967).

P. L. Richards, High-Resolution Fourier Transform Spectroscopy in the Far-Infrared, *J. Opt. Soc. Amer.* **54**, 1474 (1964).

H. N. Rundle, Construction of a Michelson Interferometer for Fourier Spectroscopy, *J. Res. Nat. Bur. Std.* **69C**, 5 (1965).

B. D. Saksena, D. R. Pahwa, M. M. Pradhan, and K. Lal, Reflection and Transmission Characteristics of Wire Gratings in the Far Infrared, *Infrared Phys.* **9**, 43 (1969).

R. B. Sanderson and H. E. Scott, A High Resolution Far Infrared Interferometer, *Appl. Opt.* **10**, 1097 (1971).

R. A. Schindler, A Small, High Speed Interferometer for Aircraft, Balloon, and Spacecraft Applications, *Appl. Opt.* **9**, 301 (1970).

M. J. Selby, Mock Interferometer—I, Theory, *Infrared Phys.* **6**, 21 (1966).

W. H. Steel, Interferometers without Collimation for Fourier Spectroscopy, *J. Opt. Soc. Amer.* **54**, 151 (1964).

W. H. Steel, "Interferometry." Cambridge Univ. Press, New York, 1967.

W. H. Steel, Interferometers for Fourier Spectroscopy, *Aspen Int. Conf. on Fourier Spectrosc., 1970*, p. 43. (G. A. Vanasse, A. T. Stair, Jr., and D. J. Baker, eds.). AFCRL-71-0019, 5 Jan. 1971, Spec. Rep. No. 114.

R. S. Sternberg and J. F. James, A New Type of Michelson Interference Spectrometer, *J. Sci. Instrum.* **41**, 225 (1964).

D. G. Vickers, E. I. Robson, and J. E. Beckman, Freestanding Polarizers for Submillimeter Interferometry, *Appl. Opt.* **10**, 682 (1971).

E. A. Vinogradov, E. M. Dianov, and N. A. Irisova, Fabry-Perot Interferometer for the Short Millimeter and Submillimeter Bands with Metallic Grids Having Periods Smaller Than the Wavelength, *JETP Lett.* **2**, 205 (1965).

E. A. Vinogradov, E. M. Dianov, and N. A. Irisova, Michelson Interferometer for Measuring the Refractive Index of Dielectric Materials in the 2 mm Wavelength Range, *Radiotekhn. Elektron.* **10**, 1804 (1965); *Radio Eng. Electron. Phys.* **10**, 1547 (1965).

R. G. Wheeler and J. C. Hill, Spectroscopy in the 5 to 400 Wavenumber Region with the Grubbs-Parsons Interferometric Spectrometer, *J. Opt. Soc. Amer.* **56**, 657 (1966).

D. E. Williamson, Cone Channel Condenser Optics, *J. Opt. Soc. Amer.* **42**, 712 (1952).

H. Yoshinaga, Supplement I, Recent Techniques in Far Infrared Spectroscopy, *Jap. J. Appl. Phys.* **4**, 420 (1965).

## ERRORS

J. E. Chamberlain, On a Convergence Correction for Dispersive Fourier Spectrometry, *Appl. Opt.* **6**, 980 (1967).

J. M. Dowling, Investigations in the Far Infrared with a Lamellar Grating Interferometer, Air Force Rep. No. SSD-TR-67-30; Aerospace Rep. No. TR-1001 (9260-01)-7 (March, 1967); J. M. Dowling, Space Phys. Lab., Lab. Div., Lab. Operations, Aerospace Corp., Los Angeles, California.

M. L. Forman *et al.*, Nonlinear Phase Correction of Interferograms Obtained in Fourier Spectroscopy, Utah State Univ., Logan, Utah (1 July 1965), Defense Documentation Center, Defense Supply Agency, AD 624 086.

M. L. Forman, W. H. Steel, and G. A. Vanasse, Correction of Asymmetric Interferograms Obtained in Fourier Spectroscopy, *J. Opt. Soc. Amer.* **56**, 59 (1966).

J. E. Gibbs and H. A. Gebbie, Calculation of the Power-Spectrum from an Interferogram Sampled at Points None of Which Coincides Exactly with Zero Path-Difference, *Infrared Phys.* **5**, 187 (1965).

R. Haswell, A. Martin, and G. Sharp, Correction of Phase Errors in Interferograms, *Appl. Opt.* **3**, 1195 (1964).

E. Loewenstein, On the Correction of Phase Errors in Interferograms, *Appl. Opt.* **2**, 491 (1963).

L. Mertz, Correction of Phase Errors in Interferograms, *Appl. Opt.* **2**, 1331 (1963).

P. L. Richards, Computing Reliable Power Spectra, *IEEE Spectrum* **4**, 83 (1967).

H. Sakai and C. A. Vanasse, Accuracy of Zero-Path-Difference-Determination in Fourier Spectroscopy for Narrow-Band Spectra, *J. Opt. Soc. Amer.* **57**, 844 (1967).

H. Sakai, G. A. Vanasse, and M. L. Forman, Spectral Recovery in Fourier Spectroscopy, *J. Opt. Soc. Amer.* **58**, 84 (1968).

W. H. Steel and M. L. Forman, Example of Errors Occurring in Fourier Spectroscopy Due to Hilbert-Transform Effect, *J. Opt. Soc. Amer.* **55**, 982 (1966).

C. S. Williams, Mirror Misalignment in Fourier Spectroscopy Using a Michelson Interferometer with Circular Aperature, *Appl. Opt.* **5**, 1084 (1966).

## FILTER MATERIALS

R. J. Bell and G. M. Goldman, Optical Constants of Black Polyethylene, *J. Opt. Soc. Amer.* **57**, 1552 (1967).

R. J. Bell and H. V. Romero, A Study of an Array of Square Openings, *Appl. Opt.* **9**, 2341 (1970).

R. J. Bell, H. V. Romero, and J. M. Blea, Theory and Experiments for Multi-element Grid Filters in a Dielectric, *Appl. Opt.* **9**, 2350 (1970).

D. W. Berreman, Low Pass Filters for Far Infrared Spectroscopy, *Rev. Sci. Instrum.* **37**, 513 (1966).

J. M. Blea, W. F. Parks, P. A. R. Ade, and R. J. Bell, Optical Properties of Black Polyethylene from 3 to 4000 $cm^{-1}$, *J. Opt. Soc. Amer.* **60**, 603 (1970).

S. I. Drasky and R. J. Bell, Radiation Chopper for the Submillimeter Wavelength Region, *Infrared Phys.* **5**, 137 (1965).

H. Y. Fan, M. L. Shepherd, and W. Spitzer, "Photoconductivity Conference." Wiley, New York, 1956.

R. Geick, Der Brechungsindex Von Kristallinem Und Geschmolzenem Quarz Im Spektralbereich Um 100 $\mu$m, *Z. Phys.* **161**, 116 (1961).

L. Genzel, Application of Frustrated Total Reflection as a Filter in the Far Infrared, *Infrared Phys.* **4**, 253 (1964).

L. Genzel, H. Happ, and R. Weber, Dispersionsmessumgen An NaCl, KCl, and KBr Zwischen 0.3 and 3 mm Wellenlange, *Z. Phys.* **154**, 13 (1959).

A. Hadni, "Essentials of Modern Physics Applied to the Study of the Infrared." Pergamon Press, Oxford, 1967.

M. Hass and B. W. Hevis, Infrared Lattice Reflection Spectra of III–IV Compound Semiconductors, *J. Phys. Chem. Solids* **23**, 1099 (1962).

M. Lax and E. Brustein, Infrared Lattice Absorption in Ionic and Homopolar Crystals, *Phys. Rev.* **97**, 39 (1955).

R. C. Lord and T. K. McCubbin, Infrared Spectroscopy from 5 to 200 Microns with a Small Grating Spectrometer, *J. Opt. Soc. Amer.* **47**, 689 (1952).

D. E. McCarthy, The Reflection and Transmission of Infrared Materials: 1, Spectra from 2 to 50 Microns, *Appl. Opt.* **2**, 591 (1963).

D. E. McCarthy, Transmission of Irtran Materials from 50 $\mu$m to 300 $\mu$m, *Appl. Opt.* **5**, 472 (1966).

D. E. McCarthy, Black Polyethylene as a Far-Infrared Filter, *J. Opt. Soc. Amer.* **57**, 699 (1967).

D. E. McCarthy, The Reflection and Transmission of Infrared Materials. V: Spectra from $2\mu$ to $50\mu$, *Appl. Opt.* **7**, 1997 (1968).

T. K. McCubbin and W. M. Sinton, Recent Investigations in the Far Infrared, *J. Opt. Soc. Amer.* **40**, 538 (1950).

A. Mitsuishi, Y. Yamada, and H. Yoshinaga, Reflection Measurements on Reststrahlen Crystals in the Far-Infrared Region, *J. Opt. Soc. Amer.* **52**, 14 (1962).

A. Mitsuishi, Y. Otsuka, S. Fujita, and H. Yoshinaga, Metal Mesh Filters in the Far Infrared Region, *Jap. J. Appl. Phys.* **2**, 574 (1963).

K. D. Moller and R. V. McKnight, Far-Infrared-Transmission Filter Gratings, *J. Opt. Soc. Amer.* **53**, 760 (1963).

K. D. Moller and W. G. Rothschild, "Far-Infrared Spectroscopy." Wiley-Interscience, New York, 1971.

T. S. Moss, "Optical Properties of Semiconductors." Butterworths, London and Washington, D.C., 1959.

R. D. Rawcliffe, Refractive Indices of Germanium, Silicon, and Fused Quartz in the Far Infrared, *Appl. Opt.* **6**, 1889 (1967).

K. F. Renk and L. Genzel, Interference Filters and Fabry-Perot Interferometers for the Far Infrared, *Appl. Opt.* **1**, 643 (1962).

J. D. Strong, Investigations in the Far Infrared, *Phys. Rev.* **38**, 1818 (1931).

R. Ulrich, Far Infrared Properties of Metallic Mesh and Its Complementary Structure, *Infrared Phys.* **7**, 37 (1967).

R. Ulrich, Effective Low-Pass Filters for Far Infrared Frequencies, *Infrared Phys.* **7**, 65 (1967).

R. Ulrich, Interference Filters for the Far Infrared, *Appl. Opt.* **7**, 1987 (1968).

R. Ulrich, Preparation of Grids for Far Infrared Filters, *Appl. Opt.* **8**, 319 (1969).

E. A. Vinogradov, E. M. Dianov, and N. A. Irisova, Fabry-Perot Interferometer for the Short Millimeter and Submillimeter Bands with Metallic Grids Having Periods Smaller Than the Wavelength, *JETP Lett.* **2**, 205 (1965).

C. Wu and G. Andermann, Improved Kramers-Kranig Dispersion Analysis of Infrared Reflectance Data for Lithium Fluoride, *J. Opt. Soc. Amer.* **58**, 519 (1968).

Y. Yamada, A. Mitsuishi, and H. Yoshinaga, Transmission Filters in the Far-Infrared Region, *J. Opt. Soc. Amer.* **52**, 17 (1962).

H. Yoshinaga, Recent Techniques in Far Infrared Spectroscopy, *Jap. J. Appl. Phys. Suppl. 1* **4**, 420 (1965).

## HISTORICAL PAPERS AND SKETCHES

J. E. Chamberlain, J. E. Gibbs, and H. A. Gebbie, Refractometry in the Far Infrared Using a Two Beam Interferometer, *Nature* **198**, 874 (1963).

J. W. Cooley and J. W. Tukey, An Algorithm for the Machine Calculation of Complex Fourier Series, *Math. Computat.* **19**, 297 (1965).

P. Fellgett, Thesis, Cambridge Univ. (1951).

M. L. Forman, Fast Fourier-Transform Technique and Its Application to Fourier Spectroscopy, *J. Opt. Soc. Amer.* **56**, 978 (1966).

H. Gebbie and G. Vanasse, Interferometric Spectroscopy in the Far-Infrared, *Nature* **178**, 432 (1956).

L. Genzel and R. Weber, Zur Theorie der Interferenz-Modulation für Zweistrahl-Interferenzen, *Zeit. Physik* **10**, 127 (1957).

I. J. Good, The Interaction Algorithm and Practical Fourier Series, *J. Roy. Statist. Soc. Ser. B.* **20**, 361 (1958).

P. Jacquinot and C. Dufour, Optical Conditions in the Use of Photo-Electric Cells in Spectrographs and Interferometers, *J. Rech. CNRS* **6**, 91 (1948).

R. Madden and J. Strong, Cylindrical Interference Filter of Variable Thickness, *J. Opt. Soc. Amer.* **44**, 352 (1954).

L. Mertz, Rapid Scanning Fourier Transform Spectroscopy, *J. Phys. Colloq. C2, suppl. 3–4* **28**, C2-87 (1967).

L. Mertz, Fourier Spectroscopy, Past, Present, and Future, *Aspen Int. Conf. on Fourier Spectrosc., 1970*, p. 142. (G. A. Vanasse, A. T. Stair, Jr., and D. J. Baker, eds.). AFCRL-71-0019, 5 Jan. 1971, Spec. Rep. No. 114.

A. A. Michelson, *Phil. Mag. Ser. 5* **31**, 256 (1891).

A. Michelson, On the Application of Interference Methods to Spectroscopic Measurements, *Phil. Mag.* **34**, 280 (1892).

A. A. Michelson, *Astrophys. J.* **7**, 131 (1898).

A. A. Michelson, excerpt from "Light Waves and Their Uses." Univ. of Chicago Press. Chicago, Illinois, 1902; *also in Appl. Opt.* **6**, 692 (1967).

Lord Rayleigh (J. W. Strutt), *Phil. Mag.* **24**, 864 (1912).

H. Rubens and R. W. Wood, *Phil. Mag.* **21**, 249 (1911).

J. Strong, Interferometric Modulator, *J. Opt. Soc. Amer.* **44**, 352 (1954).

J. D. Strong, Interferometry in the Far-Infrared, *J. Opt. Soc. Amer.* **47**, 354 (1957).

J. D. Strong, "Concepts of Classical Optics." Freeman, San Francisco, California, 1958.

J. D. Strong and G. Vanasse, Modulation Interferentielle et Calculateur Analogique pour un Spectrometre Intereferentiel, *J. Phys. Radium* **19**, 192 (1958).

J. D. Strong and G. Vanasse, Interferometric Spectroscopy in the Far-Infrared, *J. Opt. Soc. Amer.* **49**, 844 (1959).

J. D. Strong and G. Vanasse, Lamellar Grating Far Infrared Interferometer, *J. Opt. Soc. Amer.* **50**, 113 (1960).

## INTERFEROMETERS VERSUS SPECTROMETERS

P. Connes, Spectrometre Interferentiel a Selection par 1; Amplitude de Modulation, *J. Phys. Radium* **19**, 215 (1958).

J. M. Dowling and R. T. Hall, Comments on High Resolution Grating Spectrometer for the Far Infrared, *J. Opt. Soc. Amer.* **57**, 269 (1967).

H. A. Gebbie, Fourier Transforms versus Grating Spectroscopy, *Appl. Opt.* **8**, 501 (1969).

P. Jacquinot, Interferometry and Grating Spectroscopy: an Introductory Survey, *Appl. Opt.* **8**, 497 (1969).

P. Jacquinot and C. Dufour, Optical Conditions in the Use of Photo-Electric Cells in Spectrographs and Interferometers, *J. Rech. CNRS* **6**, 91 (1948).

F. K. Kneubuhl, J. F. Moser, and H. Steffen, High Resolution Grating Spectrometer for the Far Infrared, *J. Opt. Soc. Amer.* **56**, 760 (1966).

F. K. Kneubuhl, J. F. Moser, and H. Steffen, Note on a Letter by Dowling and Hall, *J. Opt. Soc. Amer.* **57**, 271 (1967).

P. L. Richards, High-Resolution Fourier Transform Spectroscopy in the Far-Infrared, *J. Opt. Soc. Amer.* **54**, 1474 (1964).

LAMELLAR GRATING INTERFEROMETERS

P. Connes, How Light is Analyzed, *in* "Lasers and Light" (A. Schawlow, ed.). Freeman, San Francisco, California, 1969.

J. Dowling, Direct Interpretation of Far-Infrared Interferograms with Application to Diatomic and Linear Molecules, *J. Opt. Soc. Amer.* **54**, 663 (1964).

J. M. Dowling, Investigations in the Far Infrared with a Lamellar Grating Interferometer, Air Force Rep. No. SSD-TR-67-30; Aerospace Rep. No. TR-1001 (9260-01)-7, (March, 1967); J. M. Dowling, Space Phys. Lab., Lab. Div., Lab. Operations, Aerospace Corp., Los Angeles, California.

J. M. Dowling, Signal and Noise in Two-Beam Interferometry, *Appl. Opt.* **6**, 1580 (1967).

J. M. Dowling, Double Beaming in Fourier Spectroscopy, *Aspen Int. Conf. on Fourier Spectrosc., 1970*, p. 55. (G. A. Vanasse, A. T. Stair, Jr., and D. J. Baker, eds.). AFCRL-71-0019, 5 Jan. 1971, Spec. Rep. No. 114.

J. M. Dowling and R. T. Hall, Far Infrared Interferometry: Upper Limit for Line Widths for the Pure Rotational Band of Carbon Monoxide, *J. Mol. Spectrosc.* **19**, 108 (1966).

J. M. Dowling and R. T. Hall, The Techniques of High Resolution, Double-Beam, Two-Beam Interferometry with a Far Infrared Lamellar Grating Interferometer, *J. Phys. Colloq. C2, suppl. are M° 3–4* **28**, C2-156 (1967).

L. Genzel and R. Weber, Zur Theorie der Interferenz-Modulation für Zweistrahl-Interferenzen, *Zeit. Physik* **10**, 127 (1957).

L. Genzel and R. Weber, Spektroskopie im fernen Ultrarot durch Interferenz-Modulation, *Zeit. Physik* **10**, 195 (1958).

L. Genzel, Aperiodic and Periodic Interference Modulation for Spectographic Purposes, *J. Mol. Spec.* **4**, 241 (1960).

R. T. Hall, D. Vrabec, and J. Dowling, High Resolution Far-Infrared Lamellar-Grating Spectrometer, *J. Opt. Soc. Amer.* **54**, 1390 (1964).

R. T. Hall, D. Vrabec, and J. M. Dowling, A High-Resolution Far-Infrared Double-Beam Lamellar Grating Interferometer, *Appl. Opt.* **5**, 1147 (1966).

N. P. Hansen and J. Strong, Performance of a Simple Spherical Lamellar Grating Beamsplitter by Wavefront Division, *Aspen Int. Conf. on Fourier Spectrosc., 1970*, p. 215. (G. A. Vanasse, A. T. Stair, Jr., and D. J. Baker, eds.). AFCRL-71-0019, 5 Jan. 1971, Spec. Rep. No. 114.

W. F. Parks and R. W. Alexander, Interferometers with Spherical Mirrors (to be published).

P. L. Richards, High-Resolution Fourier Transform Spectroscopy in the Far-Infrared, *J. Opt. Soc. Amer.* **54**, 1474 (1964).

John D. Strong, "Concepts of Classical Optics." Appendix F, coauthored with G. A. Vanasse. Freeman, San Francisco, California, 1958.

J. D. Strong and G. Vanasse, Lamellar Grating Far Infrared Interferometer, *J. Opt. Soc. Amer.* **50**, 113 (1960).

## REAL TIME FOURIER SPECTROSCOPY

P. Connes, High Resolution and High Information Fourier Spectroscopy, *Aspen Int. Conf. on Fourier Spectrosc., 1970*, p. 121. (G. A. Vanasse, A. T. Stair, Jr., and D. J. Baker, eds.). AFCRL-71-0019, 5 Jan. 1971, Spec. Rep. No. 114.

R. Curbelo and C. Foskett, An Automated Interferometer Spectrometer: Real Time Data Acquisition and On-Line Computation to 0.5 cm, *Aspen Int. Conf. on Fourier Spectrosc., 1970*, p. 221. (G. A. Vanasse, A. T. Stair, Jr., and P. J. Baker, eds.). AFCRL-71-0019, 5 Jan. 1971, Spec. Rep. No. 114.

J. E. Hoffman and G. A. Vanasse, Real-Time Spectral Synthesis in Fourier Spectroscopy, *Appl. Opt.* **5**, 1167 (1966).

J. E. Hoffman, Jr., Real-Time Fourier Spectroscopy, *Appl. Opt.* **8**, 323 (1969).

F. Levy, R. C. Milward, S. Bras, and R. le Toullec, "Real Time" Far Infrared Fourier Spectroscopy Using a Small Digital Computer, *Aspen Int. Conf. on Fourier Spectrosc., 1970*, p. 331. (G. A. Vanasse, A. T. Stair, Jr., and D. J. Baker, eds.). AFCRL-71-0019, 5 Jan. 1971, Spec. Rep. 114.

L. W. Thorpe, D. J. Neale, and G. C. Hayward, New Instrumentation for Far Infrared Fourier Spectrometers, *Aspen Int. Conf. on Fourier Spectrosc., 1970*, p. 187. (G. A. Vanasse, A. T. Stair, Jr., and D. J. Baker, eds.). AFCRL-71-0019, 5 Jan. 1971, Spec. Rep. No. 114.

H. Yoshinaga, S. Fujita, S. Minami, Y. Suemoto, M. Inoue, K. Chiba, K. Nakano, S. Yoshida, and H. Sugimori, A Far Infrared Interferometric Spectrometer with a Special Electronic Computer, *Appl. Opt.* **5**, 1159 (1966).

## RESOLUTION AND SIGNAL TO NOISE

J. M. Dowling, Investigations in the Far-Infrared with a Lamellar Grating Interferometer, Air Force Rep. No. SSD-TR-67-30. Aerospace Corp. Rep. No. TR-1001 (9260-01)-7 (March 1967).

J. M. Dowling, Signal and Noise in Two-Beam Interferometry, *Appl Opt.* **6**, 1580 (1967).

S. Goldman, "Information Theory." Prentice Hall, Englewood Cliffs, New Jersey, 1953.

N. Jacobi, Resolution and Noise in Fourier Spectroscopy, *J. Opt. Soc. Amer.* **58**, 495 (1968).

A. E. Martin, "Infra-Red Instrumentation and Techniques." Elsevier, New York, 1966.

H. Sakai, Consideration of the Signal-To-Noise Ratio in Fourier Spectroscopy, *Aspen Int. Conf. on Fourier Spectrosc., 1970*, p. 19. (G. A. Vanasse, A. T. Stair, Jr., and D. J. Baker, eds.). AFCRL-71-0019, 5 Jan. 1971, Spec. Rep. No. 114.

J. E. Stewart, Field of View and Resolution of a Polarization Fourier Spectrometer, *J. Opt. Soc. Amer.* **58**, 434 (1968).

M. T. Surh, The Effect of Time Jitter in Sampling of an Interferogram, *Appl. Opt.* **5**, 880 (1966).

R. A. Williams and W. S. C. Chang, Resolution and Noise in Fourier-Transform Spectroscopy, *J. Opt. Soc. Amer.* **56**, 167 (1966).

T. Williams, Direct Determination of the Line Shapes of Rotational Spectra from Interferometric Measurements, *J. Opt. Soc. Amer.* **50**, 1153 (1960).

## SOURCES AND SOURCE ASSOCIATED EFFECTS

R. Cano and M. Mattioli, Far Infrared Spectroscopy with a Michelson Interferometer and an InSb Detector, *Infrared Phys.* **7**, 25 (1967).

D. W. Faries, K. A. Gehring, P. L. Richards, and Y. R. Shen, Tunable Far-Infrared Radiation Generated from the Difference Frequence Between Two Ruby Lasers, *Phys. Rev.* **180**, 363 (1969).

L. Genzel, Far Infrared Sources, *NATO Conf. Far Infrared Properties Solids, Delft, Netherlands* (August 1968).

A. Girard and P. Jacquinot, *in* "Advanced Optical Techniques" (A. C. S. Van Hell, ed.). North-Holland Publ., Amsterdam, 1967.

J. R. Hickey and O. B. Daniels, Modified Optical System for the Golay Detector, *Appl. Opt.* **8**, 732 (1969).

E. Plyler, D. Yates, and H. Gebbie, Radiant Energy from Sources in the Far Infrared, *J. Opt. Soc. Amer.* **52**, 859 (1962).

D. R. Smith, R. L. Morgan, and E. V. Loewenstein, Comparison of the Radiance of Far-Infrared Sources, *J. Opt. Soc. Amer.* **58**, 433 (1968).

J. E. Stewart, Field of View and Resolution of a Polarization Fourier Spectrometer, *J. Opt. Soc. Amer.* **58**, 434 (1968).

C. S. Williams, Mirror Misalignment in Fourier Spectroscopy Using a Michelson Interferometer with Circular Aperture, *Appl. Opt.* **5**, 1084 (1966).

# Author Index

# Subject Index

## A

## C

## M



## N



## O

## R



## S

## T



## U